The Deep Mixing Method

The Deep Mixing Method

Masaki Kitazume
Tokyo Institute of Technology, Tokyo, Japan

Masaaki Terashi
Consultant, Tokyo, Japan

CRC Press
Taylor & Francis Group
Boca Raton London New York Leiden

CRC Press is an imprint of the
Taylor & Francis Group, an **informa** business

A BALKEMA BOOK

Cover illustrations:
Photo (left): Land machine, Courtesy of Cement Deep Mixing Method Association
Photo (right): the CDM vessel, September 2012, Masaki Kitazume

First issued in paperback 2017

CRC Press/Balkema is an imprint of the Taylor & Francis Group, an informa business

© 2013 Taylor & Francis Group, London, UK

Typeset by MPS Limited, Chennai, India

CIP data applied for

Published by: CRC Press/Balkema
          P.O. Box 11320, 2301 EH, Leiden, The Netherlands
          e-mail: Pub.NL@taylorandfrancis.com
          www.crcpress.com – www.taylorandfrancis.com

ISBN 13: 978-1-138-07579-5 (pbk)
ISBN 13: 978-1-138-00005-6 (hbk)

Table of contents

Preface   xvii
List of technical terms and symbols   xix

**1   Overview of ground improvement – evolution of deep mixing
      and scope of the book**   1

1   Introduction   1
2   Classification of ground improvement technologies   2
    2.1   Replacement   3
    2.2   Densification   3
    2.3   Consolidation/dewatering   4
    2.4   Grouting   5
    2.5   Admixture stabilization   6
    2.6   Thermal stabilization (heating and freezing)   7
    2.7   Reinforcement   7
    2.8   Combined uses of various techniques   7
    2.9   Limitation of traditional ground improvement
          techniques   8
3   Development of deep mixing in Japan – historical review   8
    3.1   Development of the deep mixing method   8
    3.2   Development of high pressure injection deep mixing
          method   12
4   Diversified admixture stabilization techniques without
    compaction   13
    4.1   Classification of admixture stabilization techniques   13
    4.2   In-situ mixing   15
          4.2.1   Surface treatment   15
          4.2.2   Shallow mixing   15
          4.2.3   Deep mixing method   17
    4.3   Ex-situ mixing   19
          4.3.1   Premixing method   19
          4.3.2   Lightweight Geo-material   20
          4.3.3   Dewatered stabilized soil   22
          4.3.4   Pneumatic flow mixing method   23
5   Scope of the text   24
References   26

2 Factors affecting strength increase   **29**

   1   Introduction   29
   2   Influence of various factors on strength of lime stabilized soil   30
       2.1   Mechanism of lime stabilization   30
       2.2   Characteristics of lime as a binder   31
             2.2.1   Influence of quality of quicklime   32
       2.3   Characteristics and conditions of soil   34
             2.3.1   Influence of soil type   34
             2.3.2   Influence of grain size distribution   35
             2.3.3   Influence of humic acid   36
             2.3.4   Influence of potential Hydrogen (pH)   36
             2.3.5   Influence of water content   37
       2.4   Mixing conditions   38
             2.4.1   Influence of amount of binder   38
             2.4.2   Influence of mixing time   39
       2.5   Curing conditions   39
             2.5.1   Influence of curing period   39
   3   Influence of various factors on strength of cement stabilized soil   40
       3.1   Mechanism of cement stabilization   40
             3.1.1   Characteristics of binder   41
             3.1.2   Influence of chemical composition of binder   42
             3.1.3   Influence of type of binder   44
             3.1.4   Influence of type of water   45
       3.2   Characteristics and conditions of soil   47
             3.2.1   Influence of soil type   47
             3.2.2   Influence of grain size distribution   49
             3.2.3   Influence of humic acid   50
             3.2.4   Influence of ignition loss   51
             3.2.5   Influence of pH   51
             3.2.6   Influence of water content   54
       3.3   Mixing conditions   56
             3.3.1   Influence of amount of binder   56
             3.3.2   Influence of mixing time   56
             3.3.3   Influence of time and duration of mixing and
                     holding process   56
       3.4   Curing conditions   59
             3.4.1   Influence of curing period   59
             3.4.2   Influence of curing temperature   61
             3.4.3   Influence of maturity   63
             3.4.4   Influence of overburden pressure   67
   4   Prediction of strength   68
   References   69

3 Engineering properties of stabilized soils   **73**

   1   Introduction   73
   2   Physical properties   73
       2.1   Change of water content   73

	2.2	Change of unit weight	76
	2.3	Change of consistency of soil-binder mixture before hardening	78
3		Mechanical properties (strength characteristics)	79
	3.1	Stress–strain curve	79
	3.2	Strain at failure	82
	3.3	Modulus of elasticity (Yong's modulus)	83
	3.4	Residual strength	83
	3.5	Poisson's ratio	84
	3.6	Angle of internal friction	86
	3.7	Undrained shear strength	87
	3.8	Dynamic property	87
	3.9	Creep strength	88
	3.10	Cyclic strength	90
	3.11	Tensile and bending strengths	94
	3.12	Long term strength	96
		3.12.1 Strength increase	97
		3.12.2 Strength decrease	100
		3.12.2.1 Strength distribution	100
		3.12.2.2 Calcium distribution in specimens	102
		3.12.2.3 Depth of deterioration	104
4		Mechanical properties (consolidation characteristics)	105
	4.1	Void ratio – consolidation pressure curve	105
	4.2	Consolidation yield pressure	106
	4.3	Coefficient of consolidation and coefficient of volume compressibility	107
	4.4	Coefficient of permeability	110
		4.4.1 Permeability of stabilized clay	110
		4.4.2 Influence of grain size distribution on the coefficient of permeability of stabilized soil	112
5		Environmental properties	113
	5.1	Elution of contaminant	113
	5.2	Elution of Hexavalent chromium (chromium VI) from stabilized soil	115
	5.3	Resolution of alkali from stabilized soil	119
6		Engineering properties of cement stabilized soil manufactured in situ	122
	6.1	Mixing degree of in-situ stabilized soils	122
	6.2	Water content distribution	122
	6.3	Unit weight distribution	123
	6.4	Variability of field strength	124
	6.5	Difference in strength of field produced stabilized soil and laboratory prepared stabilized soil	126
	6.6	Size effect on unconfined compressive strength	128
	6.7	Strength and calcium distributions at overlapped portion	131
		6.7.1 Test conditions	131
		6.7.2 Calcium distribution	132
		6.7.3 Strength distribution	132
		6.7.4 Effect of time interval	133

7   Summary   134
     7.1   Physical properties   134
            7.1.1   Change of water content and density   134
            7.1.2   Change of consistency of soil-binder mixture before
                     hardening   135
     7.2   Mechanical properties (strength characteristics)   135
            7.2.1   Stress–strain behavior   135
            7.2.2   Poisson's ratio   135
            7.2.3   Angle of internal friction   135
            7.2.4   Undrained shear strength   135
            7.2.5   Dynamic property   136
            7.2.6   Creep and cyclic strengths   136
            7.2.7   Tensile and bending strengths   136
            7.2.8   Long term strength   136
     7.3   Mechanical properties (consolidation characteristics)   137
            7.3.1   Void ratio – consolidation pressure curve   137
            7.3.2   Coefficient of consolidation and coefficient of volume
                     compressibility   137
            7.3.3   Coefficient of permeability   137
     7.4   Environmental properties   137
            7.4.1   Elution of contaminant   137
            7.4.2   Resolution of alkali from a stabilized soil   138
     7.5   Engineering properties of cement stabilized soil
            manufactured in situ   138
            7.5.1   Water content and unit weight by stabilization   138
            7.5.2   Variability of field strength   138
            7.5.3   Difference in the strength of field produced stabilized
                     soil and laboratory prepared stabilized soil   138
            7.5.4   Size effect on unconfined compressive strength   138
            7.5.5   Strength distributions at overlapped portion   138
     References   139

4   Applications   143

     1   Introduction   143
     2   Patterns of applications   143
          2.1   Size and geometry of the stabilized soil element   143
          2.2   Column installation patterns by the mechanical
                 deep mixing method   144
                 2.2.1   Group column type improvement   145
                 2.2.2   Wall type improvement   147
                 2.2.3   Grid type improvement   147
                 2.2.4   Block type improvement   148
          2.3   Column installation pattern by high pressure injection   150
     3   Improvement purposes and applications   150
          3.1   Mechanical deep mixing method   150
          3.2   High pressure injection   153

4   Applications in Japan   154
    4.1   Statistics of applications   154
          4.1.1   Mechanical deep mixing   154
          4.1.2   Statistics of high pressure injection   157
    4.2   Selected case histories   157
          4.2.1   Group column type – individual columns – for
                  settlement reduction   158
                  4.2.1.1   Introduction and ground condition   158
                  4.2.1.2   Ground improvement   158
          4.2.2   Group column type – tangent block – for embankment
                  stability   159
                  4.2.2.1   Introduction and ground condition   159
                  4.2.2.2   Ground improvement   160
          4.2.3   Grid type improvement for liquefaction prevention   162
                  4.2.3.1   Introduction and ground condition   162
                  4.2.3.2   Ground improvement   163
          4.2.4   Block type improvement to increase bearing capacity of
                  a bridge foundation   165
                  4.2.4.1   Introduction and ground condition   165
                  4.2.4.2   Ground improvement   165
          4.2.5   Block type improvement for liquefaction mitigation   167
                  4.2.5.1   Introduction and ground condition   167
                  4.2.5.2   Ground improvement   168
          4.2.6   Grid type improvement for liquefaction prevention   168
                  4.2.6.1   Introduction and ground condition   168
                  4.2.6.2   Ground improvement   169
          4.2.7   Block type improvement for the stability of a
                  revetment   171
                  4.2.7.1   Introduction and ground condition   171
                  4.2.7.2   Ground improvement   172
          4.2.8   Jet grouting application to shield tunnel   174
                  4.2.8.1   Introduction and ground condition   174
                  4.2.8.2   Ground improvement   175
5   Performance of improved ground in the 2011 Tohoku earthquake   176
    5.1   Introduction   176
    5.2   Improved ground by the wet method of deep mixing   176
          5.2.1   Outline of survey   176
          5.2.2   Performance of improved ground   177
                  5.2.2.1   River embankment in Saitama Prefecture   177
          5.2.3   River embankment in Ibaraki Prefecture   177
          5.2.4   Road embankment in Chiba Prefecture   177
    5.3   Improved ground by the dry method of deep mixing   180
          5.3.1   Outline of survey   180
          5.3.2   Performance of improved ground   181
                  5.3.2.1   River embankment in Chiba Prefecture   181
                  5.3.2.2   Road embankment in Chiba Prefecture   182
                  5.3.2.3   Box culvert in Chiba Prefecture   182

5.4 Improved ground by Grouting method 182
  5.4.1 Outline of survey 182
  5.4.2 Performance of improved ground 183
    5.4.2.1 River embankment at Tokyo 183
    5.4.2.2 Approach road to immerse tunnel in Kanagawa Prefecture 184
5.5 Summary 184
References 184

5 Execution – equipment, procedures and control **187**

1 Introduction 187
  1.1 Deep mixing methods by mechanical mixing process 187
  1.2 Deep mixing methods by high pressure injection mixing process 188
2 Classification of deep mixing techniques in Japan 189
3 Dry method of deep mixing for on-land works 189
  3.1 Dry jet mixing method 189
    3.1.1 Equipment 189
      3.1.1.1 System and specifications 189
      3.1.1.2 Mixing tool 192
      3.1.1.3 Binder plant 194
      3.1.1.4 Control unit 195
    3.1.2 Construction procedure 196
      3.1.2.1 Preparation of site 196
      3.1.2.2 Field trial test 196
      3.1.2.3 Construction work 196
      3.1.2.4 Quality control during production 199
    3.1.3 Quality assurance 200
4 Wet method of deep mixing for on-land works 200
  4.1 Ordinary cement deep mixing method 201
    4.1.1 Equipment 201
      4.1.1.1 System and specifications 201
      4.1.1.2 Mixing tool 201
      4.1.1.3 Binder plant 205
      4.1.1.4 Control unit 205
    4.1.2 Construction procedure 206
      4.1.2.1 Preparation of site 206
      4.1.2.2 Field trial test 206
      4.1.2.3 Construction work 207
      4.1.2.4 Quality control during production 209
      4.1.2.5 Quality assurance 210
  4.2 CDM-LODIC method 210
    4.2.1 Equipment 210
      4.2.1.1 System and specifications 210
      4.2.1.2 Mixing tool 212
      4.2.1.3 Binder plant 213
      4.2.1.4 Control unit 213

		4.2.2	Construction procedure	213
			4.2.2.1 Preparation of site	213
			4.2.2.2 Field trial test	213
			4.2.2.3 Construction work	213
		4.2.3	Quality control during production	215
		4.2.4	Quality assurance	215
		4.2.5	Effect of method – horizontal displacement during execution	215
	4.3	CDM-Lemni 2/3 method		216
		4.3.1	Equipment	216
			4.3.1.1 System and specifications	216
			4.3.1.2 Mixing tool	218
			4.3.1.3 Binder plant	220
			4.3.1.4 Control unit	220
		4.3.2	Construction procedure	220
			4.3.2.1 Preparation of site	220
			4.3.2.2 Field trial test	220
			4.3.2.3 Construction work	220
		4.3.3	Quality control during execution	220
			4.3.3.1 Quality assurance	221
			4.3.3.2 Effect of method	221
5	Wet method of deep mixing for in-water works			222
	5.1	Cement deep mixing method		222
		5.1.1	Equipment	222
			5.1.1.1 System and specifications	222
			5.1.1.2 Mixing tool	225
			5.1.1.3 Plant and pumping unit	226
			5.1.1.4 Control room	227
		5.1.2	Construction procedure	227
			5.1.2.1 Site exploration and examination of execution circumstances	227
			5.1.2.2 Positioning	228
			5.1.2.3 Field trial test	228
			5.1.2.4 Construction work	228
		5.1.3	Quality control during production	230
			5.1.3.1 Quality assurance	231
6	Additional issues to be considered in the mechanical mixing method			231
	6.1	Soil improvement method for locally hard ground		231
	6.2	Noise and vibration during operation		232
	6.3	Lateral displacement and heave of ground by deep mixing work		232
		6.3.1	On-land work	232
		6.3.2	In-water work	232
7	High pressure injection method			235
	7.1	Single fluid technique (CCP method)		236
		7.1.1	Equipment	236
		7.1.2	Construction procedure	237

		7.1.2.1 Preparation of site	237
		7.1.2.2 Construction work	237
		7.1.2.3 Quality control during production	238
		7.1.2.4 Quality assurance	238
	7.2	Double fluid technique (JSG method)	239
		7.2.1 Equipment	239
		7.2.2 Construction procedure	241
		7.2.2.1 Preparation of site	241
		7.2.2.2 Construction work	241
		7.2.2.3 Quality control during production	243
		7.2.2.4 Quality assurance	243
	7.3	Double fluid technique (Superjet method)	244
		7.3.1 Equipment	244
		7.3.2 Construction procedure	244
		7.3.2.1 Preparation of site	244
		7.3.2.2 Construction work	245
		7.3.2.3 Quality control during production	246
		7.3.2.4 Quality assurance	246
	7.4	Triple fluid technique (CJG method)	247
		7.4.1 Equipment	247
		7.4.2 Construction procedure	248
		7.4.2.1 Preparation of site	248
		7.4.2.2 Construction work	249
		7.4.2.3 Quality control during production	249
		7.4.2.4 Quality assurance	250
	7.5	Triple fluid technique (X-jet method)	251
		7.5.1 Equipment	251
		7.5.2 Construction procedure	252
		7.5.2.1 Preparation of site	252
		7.5.2.2 Construction work	252
		7.5.2.3 Quality control during production	253
		7.5.2.4 Quality assurance	253
8	Combined technique		254
	8.1	JACSMAN method	255
		8.1.1 Equipment	255
		8.1.1.1 System and specifications	255
		8.1.1.2 Mixing shafts and mixing blades	255
		8.1.1.3 Plant and pumping unit	256
		8.1.1.4 Control unit	257
		8.1.2 Construction procedure	258
		8.1.2.1 Preparation of site	258
		8.1.2.2 Field trial test	258
		8.1.2.3 Construction work	259
		8.1.2.4 Quality control during production	259
		8.1.2.5 Quality assurance	260
		8.1.2.6 Effect of method	260
References			261

6   Design of improved ground by the deep mixing method   **263**

1   Introduction   263
2   Engineering behavior of deep mixed ground   264
     2.1   Various column installation patterns and their applications   264
     2.2   Engineering behavior of block (grid and wall) produced by
            overlap operation   266
            2.2.1   Engineering behavior of improved ground leading to
                     external instability   266
            2.2.2   Engineering behavior of improved ground leading to
                     internal instability   268
            2.2.3   Change of failure mode   269
                     2.2.3.1   Influence of strength ratio q_{ub}/q_{us} on vertical
                                bearing capacity   270
                     2.2.3.2   Influence of load inclination   272
                     2.2.3.3   Influence of overlap joint on mode of failure   274
                     2.2.3.4   Influence of overlap joint on external stability   274
                     2.2.3.5   Influence of overlap joint on internal stability   277
                     2.2.3.6   Summary of failure modes for block type
                                improvement   278
     2.3   Engineering behavior of a group of individual columns   280
            2.3.1   Stability of a group of individual columns   280
                     2.3.1.1   Bearing capacity of a group of individual
                                columns   282
                     2.3.1.2   Embankment stability on a group of
                                individual columns   284
                     2.3.1.3   Numerical simulation of stability of
                                embankment   288
     2.4   Summary of failure modes for a group of stabilized soil columns   291
3   Work flow of deep mixing and design   292
     3.1   Work flow of deep mixing and geotechnical design   292
            3.1.1   Work flow of deep mixing   292
            3.1.2   Strategy – selection of column installation pattern 294
4   Design procedure for embankment support, group column type
     improved ground   295
     4.1   Introduction   295
     4.2   Basic concept   296
     4.3   Design procedure   296
            4.3.1   Design flow   296
            4.3.2   Trial values for dimensions of improved ground   297
            4.3.3   Examination of sliding failure   299
            4.3.4   Slip circle analysis   300
            4.3.5   Examination of horizontal displacement   302
            4.3.6   Examination of bearing capacity   302
            4.3.7   Examination of settlement   303
                     4.3.7.1   Amount of settlement for fixed type improved
                                ground   303

	4.3.8	Amount of settlement for floating type improved ground	305
		4.3.8.1 Rate of settlement	307
	4.3.9	Important issues on design procedure	307
		4.3.9.1 Strength of stabilized soil column, improvement area ratio and width of improved ground	307
		4.3.9.2 Limitation of design procedure based on slip circle analysis	308
5	Design procedure for block type and wall type improved grounds		309
	5.1	Introduction	309
	5.2	Basic concept	310
	5.3	Design procedure	311
		5.3.1 Design flow	311
		5.3.2 Examination of the external stability of a superstructure	312
		5.3.3 Trial values for the strength of stabilized soil and geometric conditions of improved ground	314
		5.3.4 Examination of the external stability of improved ground	314
		5.3.4.1 Sliding and overturning failures	315
		5.3.4.2 Bearing capacity	318
		5.3.5 Examination of the internal stability of improved ground	320
		5.3.5.1 Subgrade reaction at the front edge of improved ground	321
		5.3.5.2 Average shear stress along a vertical plane	322
		5.3.5.3 Allowable strengths of stabilized soil	323
		5.3.5.4 Extrusion failure	325
		5.3.6 Slip circle analysis	327
		5.3.7 Examination of immediate and long term settlements	328
		5.3.8 Determination of strength and specifications of stabilized soil	329
	5.4	Sample calculation	329
	5.5	Important issues on design procedure	330
6	Design procedure for block type and wall type improved grounds, reliability design		330
	6.1	Introduction	330
	6.2	Basic concept	331
	6.3	Design procedure	331
		6.3.1 Design flow	331
		6.3.2 Examination of external stability of a superstructure	333
		6.3.2.1 Sliding failure	333
		6.3.2.2 Overturning failure	335
		6.3.3 Setting of geometric conditions of improved ground	336
		6.3.4 Evaluation of seismic coefficient for verification	336
		6.3.4.1 For level 1 performance verification	336
		6.3.4.2 For level 2 performance verification	337
		6.3.5 Examination of the external stability of improved ground	337
		6.3.5.1 Sliding failure	338
		6.3.5.2 Overturning failure	341
		6.3.5.3 Bearing capacity	343
		6.3.6 Examination of internal stability of improved ground	344

		6.3.6.1	Subgrade reactions at front edge of improved ground	345
		6.3.6.2	Average shear stress along a vertical shear plane	345
		6.3.6.3	Allowable strengths of stabilized soil	347
		6.3.6.4	Extrusion failure	348
	6.3.7	Slip circle analysis		349
	6.3.8	Examination of immediate and long term settlements		350
	6.3.9	Determination of strength and specifications of stabilized soil		350

7 Design procedure of grid type improved ground for liquefaction prevention — 350

7.1	Introduction		350
7.2	Basic concept		351
7.3	Design procedure		351
	7.3.1	Design flow	351
	7.3.2	Design seismic coefficient	352
	7.3.3	Determination of width of grid	353
	7.3.4	Assumption of specifications of improved ground	353
	7.3.5	Examination of the external stability of improved ground	353

		7.3.5.1	Sliding and overturning failures	353
		7.3.5.2	Bearing capacity	358
	7.3.6	Examination of the internal stability of improved ground		360
		7.3.6.1	Subgrade reaction at the front edge of improved ground	360
		7.3.6.2	Average shear stress along a horizontal shear plane	360
		7.3.6.3	Average shear stress along the horizontal plane of the rear most grid wall	361
		7.3.6.4	Average shear stress along a vertical shear plane	362
	7.3.7	Slip circle analysis		363
	7.3.8	Important issues on design procedure		364
		7.3.8.1	Effect of grid wall spacing on liquefaction prevention	364
References				365

7 QC/QA for improved ground – Current practice and future research needs — **369**

1	Introduction		369
2	Flow of a deep mixing project and QC/QA		369
3	QC/QA for stabilized soil – current practice		371
	3.1	Relation of laboratory strength, field strength and design strength	371
	3.2	Flow of quality control and quality assurance	373
		3.2.1 Laboratory mix test	374

		3.2.2	Field trial test	374
		3.2.3	Quality control during production	375
		3.2.4	Quality verification	376
	3.3	Technical issues on the QC/QA of stabilized soil		378
		3.3.1	Technical issues with the laboratory mix test	378
			3.3.1.1 Effect of rest time	381
			3.3.1.2 Effect of molding	381
			3.3.1.3 Effect of curing temperature	382
		3.3.2	Impact of diversified execution system on QC/QA	383
		3.3.3	Verification techniques	385
4	QC/QA of improved ground – research needs			388
	4.1	Embankment support by group of individual columns		388
		4.1.1	QC/QA associated with current design practice	388
		4.1.2	QC/QA for sophisticated design procedure considering the actual failure modes of group column type improved ground	389
		4.1.3	Practitioners' approach	390
	4.2	Block type and wall type improvements for heavy structures		391
5	Summary			391
References				392

Appendix A   Japanese laboratory mix test procedure   **395**

1	Introduction		395
2	Testing equipment		395
	2.1	Equipment for making specimen	395
		2.1.1 Mold	395
		2.1.2 Mixer	396
		2.1.3 Binder mixing tool	396
	2.2	Soil and binder	397
		2.2.1 Soil	397
		2.2.2 Binder	398
3	Making and curing of specimens		398
	3.1	Mixing materials	398
	3.2	Making specimen	399
	3.3	Curing	400
	3.4	Specimen removal	400
4	Report		405
5	Use of specimens		405
References			405

Subject index   407

Preface

The deep mixing method is a deep in-situ admixture stabilization technique using lime, cement or lime-based and cement-based special binders. Compared to the other ground improvement techniques deep mixing has advantages such as the large strength increase within a month period, little adverse impact on environment and high applicability to any kind of soil if binder type and amount are properly selected. The application covers on-land and in-water constructions ranging from strengthening the foundation ground of buildings, embankment supports, earth retaining structures, retrofit and renovation of urban infrastructures, liquefaction hazards mitigation, man-made island constructions and seepage control. Due to the versatility, the total volume of stabilized soil by the mechanical deep mixing method from 1975 to 2010 reached 72.3 million m^3 for the wet method of deep mixing and 32.1 million m^3 for the dry method of deep mixing in the Japanese market.

Improved ground by the method is a composite system comprising stiff stabilized soil and un-stabilized soft soil, which necessitates geotechnical engineers to fully understand the interaction of stabilized and unstabilized soil and the engineering characteristics of in-situ stabilized soil. Based on the knowledge, the geotechnical engineer determines the geometry (plan layout, verticality and depth) of stabilized soil elements, by assuming/establishing the engineering properties of stabilized soil, so that the improved ground may satisfy the performance criteria of the superstructure. The success of the project, however, cannot be achieved by the well determined geotechnical design alone. The success is guaranteed only when the quality and geometric layout envisaged in the design is realized with an acceptable level of accuracy.

The strength of the stabilized soil is influenced by many factors including original soil properties and stratification, type and amount of binder, curing conditions and mixing process. The accuracy of the geometric layout heavily depends upon the capability of mixing equipment, mixing process and contractor's skill. Therefore the process design, production with careful quality control and quality assurance are the key to the deep mixing project. Quality assurance starts with the soil characterization of the original soil and includes various activities prior to, during and after the production. QC/QA methods and procedures and acceptance criteria should be determined before the actual production and their meanings should be understood precisely by all the parties involved in a deep mixing project.

Until the end of the 1980s, deep mixing has been developed and practiced only in Japan and Nordic countries with a few exceptions. In the 1990s deep mixing gained popularity also in Southeast Asia, the United States of America and central Europe.

To enhance the international exchange of information on the technology, the first international specialty conference on deep mixing was co-organized by the Japanese Geotechnical Society and the ISSMGE TC-17 in 1996 in Tokyo. This landmark conference was followed by a series of specialty conferences/symposia in 1999 Stockholm, 2000 Helsinki, 2002 Tokyo, 2003 New Orleans, 2005 Stockholm and 2009 Okinawa. The authors contributed to these international forums by a number of technical papers and keynote lectures and emphasized the importance of the collaboration of owner, designer and contractor for the success of a deep mixing project.

The current book is intended to provide the state of the art and practice of deep mixing rather than a user friendly manual. The book covers the factors affecting the strength increase by deep mixing, the engineering characteristics of stabilized soil, a variety of applications and associated column installation patterns, current design procedures, execution systems and procedures, and QC/QA methods and procedures based on the experience and research efforts accumulated in the past 40 years in Japan.

The authors wish the book is useful for practicing engineers to understand the current state of the art and also useful for academia to find out the issues to be studied in the future.

August 2012
Masaki Kitazume
Masaaki Terashi

List of technical terms and symbols

DEFINITION OF TECHNICAL TERMS

additive	chemical material to be added to stabilizing agent for improving characteristics of stabilized soil
binder	chemically reactive material that can be used for mixing with in-situ soils to improve engineering characteristics of soils such as lime, cement, lime-based and cement-based special binders. Also referred to as stabilizer or stabilizing agent.
binder content	ratio of weight of dry binder to the volume of soil to be stabilized. (kg/m^3)
binder factor	ratio of weight of dry binder to the dry weight of soil to be stabilized. (%)
binder slurry	slurry-like mixture of binder and water
DM machine	a machine to be used to construct stabilized soil column
external stability	overall stability of the stabilized body
field strength	strength of stabilized soil produced in-situ
fixed type	a type of improvement in which a stabilized soil column reaches a bearing layer
floating type	a type of improvement in which a stabilized soil column ends in a soft soil layer
improved ground	a region with stabilized body and surrounding original soil
internal stability	stability on internal failure of improved ground
laboratory strength	strength of stabilized soil produced in the laboratory
original soil	soil left without stabilization
stabilizing agent	chemically reactive materials (lime, cement, etc.)
stabilized body	a sort of underground structure constructed by the stabilized columns
stabilized soil	soil stabilized by mixing with binder
stabilized soil column	column of stabilized soil constructed by a single operation of a deep mixing machine

LIST OF SYMBOLS

a_s improvement area ratio
aw binder factor (%)

B_i width of improved ground (m)

B_{is} width of a vertical shear plane from toe of improved ground (m)

C/W_t ratio of the weight of the binder to the total weight of water including pore water and mixing water

C_c compression index of soft soil

C_g subsoil condition factor

C_s importance factor

c_u undrained shear strength

c_{ub} undrained shear strength of soil beneath improved ground (kN/m^2)

c_{uc} undrained shear strength of soft soil (kN/m^2)

c_{us} undrained shear strength of stabilized soil (kN/m^2)

c_{vs} coefficient of consolidation of stabilized soil

c_{vu} coefficient of consolidation of unstabilized soil

D_{50} 50% diameter on the grain size diagram

D_a allowable displacement (cm)

D_r reference displacement ($=10\,cm$)

d_s diameter of stabilized soil column (m)

e eccentricity (m)

e void ratio

e_0 initial void ratio of soil beneath improved ground

E_{50} modulus of elasticity,

f average shear stress along a vertical shear plane (kN/m^2)

f_c design compressive strength (kN/m^2)

F_c fine fraction content

f_m coefficient of friction of mound

F_{Ri} total shear force per unit length mobilized on bottom of improved ground (kN/m)

F_{Ru} total shear force per unit length mobilized on bottom of unstabilized soil (kN/m)

f'_{ru} internal friction angle incorporating excess pore water pressure

F_s safety factor

F_{se} safety factor against extrusion failure

f_{sh} design shear strength of stabilized soil (kN/m^2)

Fs_o safety factor against overturning failure

Fs_s safety factor against sliding failure

Fs_{sp} safety factor against slip circle failure

f_t design tensile strength of stabilized soil (kN/m^2)

G_c specific gravity of binder

$G_{Ca(OH)_2}$ specific gravity of $Ca(OH)_2$

G_{eq} equivalent shear modulus

G_{max} maximum shear modulus

G_s specific gravity of soil particle

G_{sec} secant shear modulus

G_w specific gravity of water

h depth from water surface (m)

H length of stabilized soil column (m)

H_c thickness of ground (m)

H_{cb}	thickness of soil beneath improved ground (m)
H_e	height of embankment (m)
h_{eq}	damping ratio
H_f	height of periphery of improved ground mobilizing cohesion (m)
H_i	height of improved ground (m)
HK_{bf}	total seismic inertia force per unit length of backfill (kN/m)
HK_e	total seismic inertia force per unit length of embankment (kN/m)
HK_f	total seismic inertia force per unit length of fill (kN/m)
HK_i	total seismic inertia force per unit length of improved ground (kN/m)
HK_m	total seismic inertia force per unit length of mound (kN/m)
HK_{pr}	total seismic inertia force per unit length of soil prism (kN)
HK_s	total seismic inertia force per unit length of stabilized soil (kN/m)
HK_{sp}	total seismic inertia force per unit length of superstructure (kN/m)
HK_u	total seismic inertia force per unit length of unstabilized soil (kN/m)
Hpr	height of assumed prism (m)
H_s	height of short wall of improved ground (m)
H_w	water depth (m)
I_p	plasticity index
K	coefficient of efficiency of soil removal
k	coefficient of permeability
k	mobilization factor of soil strength
K_A	coefficient of static active earth pressure
K_{EA}	coefficient of dynamic active earth pressure
K'_{EA}	coefficient of dynamic active earth pressure incorporating pore water pressure generation
K_{EP}	coefficient of dynamic passive force per unit length
K'_{EP}	coefficient of dynamic passive earth pressure incorporating pore water pressure generation
k_h	seismic coefficient
k_{h0}	regional seismic coefficient
k_{h0}	seismic coefficient at the surface of ground
k_{h1k}	seismic coefficient for superstructure
k_{h2k}	seismic coefficient for external forces acting on DM improved ground
k'_{h2k}	seismic coefficient for dynamic force acting on superstructure
k_{h3k}	seismic coefficient for dynamic force acting on DM improved ground
K_P	coefficient of static passive earth pressure
l	length of improved wall (m)
L_l	thickness of long wall of improved ground (m)
L_s	thickness of short wall of improved ground (m)
L_T	thickness of grid of improved ground (m)
L_u	unit length of improved ground (m)
M	maturity
m	ratio of generated heat for evaporating water in soil
m_{vc}	coefficient of volume compressibility of unstabilized soil (m^2/kN)
m_{vs}	coefficient of volume compressibility of stabilized soil (m^2/kN)
N	number of rotation of helical screw
n	stress concentration ratio (σ_s/σ_c)

N_c	bearing capacity factor of soil beneath improved ground
N_d	number of rotation of mixing shaft during penetration (N/min)
N_f	number of loadings at failure
N_γ	bearing capacity factor of soil beneath improved ground
N_q	bearing capacity factor of soil beneath improved ground
N_u	number of rotation of mixing shaft during withdrawal (N/min)
P	pitch of helical screw (m)
p	subgrade reaction at bottom of improved ground (kN/m^2)
p_0	initial subgrade reaction at bottom of improved ground (kN/m^2)
P_{Ac}	total static active force per unit length of soft ground (kN/m)
P_{Ae}	total static active force per unit length of embankment (kN/m)
P_{AHbf}	total static active force per unit length of backfill (kN/m)
P_{AHc}	horizontal component of total static active force per unit length of soft ground (kN/m)
P_{AVc}	vertical component of total static active force per unit length of soft ground (kN/m)
P_{DAH}	horizontal component of total dynamic active earth and pore water forces per unit length (kN/m)
P_{DAHbf}	total dynamic active force per unit length of backfill (kN/m)
P_{DAHc}	horizontal component of total dynamic active force per unit length of soft ground (kN/m)
P_{DAV}	vertical component of total dynamic active earth and pore water forces per unit length (kN/m)
P_{DAVc}	vertical component of total dynamic active force per unit length of soft ground (kN/m)
P_{DPH}	horizontal component of total dynamic passive earth and pore water forces per unit length (kN/m)
P_{DPHc}	horizontal component of total dynamic passive force per unit length acting on the prism (kN/m)
P_{DPV}	vertical component of total dynamic passive and pore water forces per unit length (kN/m)
P_{DPVc}	vertical component of total dynamic passive force per unit length (kN/m)
P_{Dw}	total dynamic water force per unit length (kN/m)
P_{Pc}	total static passive force per unit length of soft ground (kN/m)
P_{PHc}	horizontal component of total static passive force per unit length of soft ground (kN/m)
P_{PVc}	vertical component of total static passive force per unit length of soft ground (kN/m)
P_{Rw}	total residual water force per unit length (kN/m)
P_{su}	total surcharge force per unit length (kN/m)
p_y	consolidation yield pressure (the pseudo pre-consolidation pressure)
Q	amount of binder (m^3)
q	volume of jet (m^3/min.)
q_a	allowable bearing capacity (kN/m^2)
q_{ar}	bearing capacity (kN/m^2)
q_c	cone resistance,
q_c	volume of injected binder (m^3/min.)

q_f	bearing capacity of soil beneath improved ground (kN/m^2)
$q_{f(Bi)}$	bearing capacity of strip foundation with width of improved ground, B_i (kN/m^2)
$q_{f(Ll)}$	bearing capacity of strip foundation with thickness of long wall, L_l (kN/m^2)
q_u	unconfined compressive strength,
q_{ua}	allowable unconfined compressive strength (kN/m^2)
q_{uck}	design unconfined compressive strength of stabilized soil (kN/m^2)
q_{uf}	unconfined compressive strength of in-situ stabilized soil (kN/m^2)
q_{ul}	unconfined compressive strength of stabilized soil manufactured in laboratory (kN/m^2)
q_w	volume of high pressured water injected (m^3/min.)
RQD	rock quality designation index
R_u	bearing capacity of soil beneath stabilized soil column (kN/m)
r_u	excess pore water pressure ratio
S	sectional area of helical screw (m^2)
S	settlement (m)
S_c	consolidation settlement of soft ground without improvement (m)
S_l	spacing of long walls of improved ground (m)
T	blade rotation number (N/m)
t	drilling time (min.)
t_1	subgrade reaction at front edge (kN/m^2)
t_2	subgrade reaction at rear edge (kN/m^2)
t_c	curing period (day)
T_c	curing temperature (°C)
T_{c0}	reference temperature (-10°C)
t_m	mixing time of binder-slurry
t_r	rest time on the strength of stabilized soil
V	amount of soil removed (m^3)
V	volume of slime (m^3)
v	withdrawal speed (min./m)
V_1	volume of slime due to column construction (m^3)
V_2	volume of slime due to drilling (m^3)
V_d	penetration speed of mixing shaft (m/min)
V_u	withdrawal speed of mixing shaft (m/min)
W/C	water to cement ratio
W_{bf}	weight per unit length of backfill (kN/m)
W_c	dry weight of cement added to original soil of 1 m^3
W_e	weight per unit length of embankment (kN/m)
W_f	weight per unit length of fill (kN/m)
W_i	weight per unit length of improved ground (kN/m)
w_L	liquid limit (%)
W_m	weight per unit length of mound (kN/m)
w_o	water content of original soil (%)
w_p	plastic limit (%)
w_s	water content of stabilized soil (%)
W_s	weight per unit length of stabilized soil (kN/m)

W_{sp}	weight per unit length of superstructure (kN/m)
W_u	weight per unit length of unstabilized soil (in case of wall type improvement) (kN/m)
γ	partial factor
α	binder content
α	characteristic of helical screw (m^3)
α	shape factor of foundation
α	coefficient of effective width of stabilized soil column
α_c	modified maximum seismic acceleration (cm/s^2)
β	settlement reduction factor
β	shape factor of foundation
β	water binder ratio
β	reliability coefficient of overlapping
δ	friction angle of boundary of improved ground and unstabilized soil (°)
δ_{ru}	friction angle of boundary of improved ground and unstabilized soil incorporating excess pore water pressure
Δe	increment of void ratio of soil beneath improved ground
Δu	excess pore water pressure (kN/m^2)
ε_f	axial strain at failure (%)
ε_{vf}	volumetric strain at failure (%)
ϕ'	internal friction angle
ϕ'_m	internal friction angle of mound
γ	correction factor for strength variability
γ_c	unit weight of soil (kN/m^3)
γ_a	structural analysis factor
γ_d	reduction factor
γ_e	unit weight of embankment (kN/m^3)
γ_i	structural factor
γ_{SA}	pulsating shear strain
γ_w	unit weight of water (kN/m^3)
η	amount of water evaporated due to heat by unit weight of CaO (0.478 g/g)
η	ratio of required water for cement hydration
λ	ratio of q_{uf}/q_{ul}
μ	Poisson's ratio
μ_k	coefficient of friction of soil beneath improved ground
θ	resultant angle of seismic coefficient (°)
ρ_b	density of soil beneath improved ground (g/cm^3)
ρ_c	density of soft soil (g/cm^3)
ρ_s	density of stabilized soil (g/cm^3)
ρ_w	density of water (g/cm^3)
σ	standard deviation (kN/m^2)
σ	vertical stress (kN/m^2)
σ'_c	effective confining pressure (kN/m^2)
σ_c	vertical stress acting on soft ground between stabilized soil columns (kN/m^2)
σ_{ca}	allowable compressive strength of stabilized soil (kN/m^2)
ΣM	total number of mixing blades

σ_s vertical stress acting on stabilized soil columns (kN/m^2)
σ_{ta} allowable tensile strength of stabilized soil (kN/m^2)
σ_{tb} tensile strength measured by bending test
σ_{td} tensile strength measured by simple tension test
σ_{ts} tensile strength measured by split test
σ_v' effective overburden pressure (kN/m^2)
τ average shear stress along vertical shear plane (kN/m^2)
τ average strength of improved ground (kN/m^2)
τ_c shear strength of soft ground (kN/m^2)
τ_{ca} allowable shear strength of stabilized soil (kN/m^2)
τ_e shear strength of embankment (kN/m^2)

Chapter 1

Overview of ground improvement – evolution of deep mixing and scope of the book

1 INTRODUCTION

It is difficult to locate a new infrastructure on a good ground due to the over-population in urban areas throughout the world. Renovation or retrofit of old infrastructures should often be carried out in close proximity of existing structures. Good quality material for constructions is becoming a precious resource to be left for the next generation. Due to these reasons and environmental restrictions on public works, ground improvement is becoming a necessary part of infrastructure development projects both in the developed and developing countries. This situation is especially pronounced in Japan, where many construction projects must locate on soft alluvial clay grounds, artificial lands reclaimed with soft dredged clays, highly organic soils and so on. These ground conditions would pose serious problems of large ground settlement and/or instability of structures. Apart from clayey or highly organic soils, loose sand deposits under the water table would cause a serious problem of liquefaction under seismic condition. When these problems are anticipated to violate the performance and function of the structure, the foundation ground is called a 'soft ground' and needs to be improved. The required performance and function of the ground are, however, different for different structures. It is not appropriate to define a 'soft ground' by its geotechnical characteristics alone, but by incorporating the size, type, function and importance of structure, and construction period. Only if the type of structure is specified it is possible to define 'soft ground'. Table 1.1 provides a rough idea of 'soft ground' for several types of structures in terms of water content, unconfined compressive strength, SPT N-value, ground thickness and bearing capacity (Japanese Society of Soil Mechanics and Foundation Engineering, 1988).

Figures 1.1(a) and 1.1(b) show the typical physical characteristics of soft clayey soils often encountered at on-land and off-shore constructions in Japan respectively (Watanabe, 1974; Ogawa and Matsumoto, 1978). In the figures, the relationships between the plasticity index (I_p) and liquid limit (w_L) are plotted. The figures clearly show that the I_p increases almost linearly with increasing w_L, and many soft soils have quite large I_p and w_L values. The w_L values of most clayey soils vary in a quite large range of about 50 to 200%. It has been generally known that clayey soils with high liquid limit, w_L cause both stability and deformation problems during and after construction.

Table 1.1 Definition of soft soil for several structures (Japanese Society of Soil Mechanics and Foundation Engineering, 1988).

| | Highway | | | Railway | | Building bearing capacity (kN/m²) | Fill dam SPT N-value |
	Water content (%)	UCS, q_u (kN/m²)	SPT N-value	SPT N-value	Thickness (m)		
organic soil	>100	<50	<4	0	>2	<100	<20
clayey soil	>50	<50	<4	2	>5	<100	–
sandy soil	>30	≒0	<10	4	>10	–	–

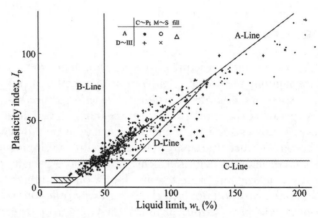

(a) in Japanese railway construction (Watanabe, 1974).

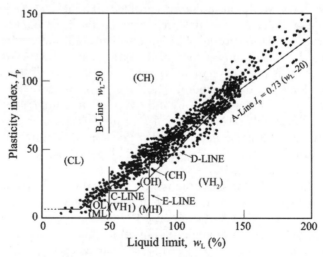

(b) in Japanese coastal area (Ogawa and Matsumoto, 1978).

Figure 1.1 Plasticity of soft clay ground in Japan.

2 CLASSIFICATION OF GROUND IMPROVEMENT TECHNOLOGIES

If a structure to be constructed would be unstable under given conditions of external loads and of original ground, or if the expected deformation during and/or after construction would exceed the allowable value from the viewpoint of the expected function of the structure, necessary countermeasures must be undertaken. The following four approaches can be applied: (a) changing the type of structure and/or type of its foundation, (b) replacing soft soil by better quality soil, (c) improving the properties of soft soil, and (d) introducing reinforcing material into soft soil. 'Ground improvement' covers (b), (c) and (d) above, and defined as any countermeasures given to soft soil in order to attain the successful performance of structure if otherwise unattainable. Ground improvement techniques are classified, based on their working principles, into replacement, densification, consolidation/dewatering, grouting, admixture stabilization, thermal stabilization, reinforcement, and miscellaneous. These techniques have been introduced to or originally developed in Japan during the past decades. Table 1.2 shows the historical evolution in Japan. Brief descriptions of ground improvement techniques are provided in the followings.

2.1 Replacement

Replacement is the most simple in concept and reliable technique if employed properly. Soft soil, mostly soft clay or highly organic soil under or near the expected structure is removed and replaced by a good quality foreign material up to the extent required to achieve stability and/or to avoid unfavorable settlement of the structure. Natural sand and gravel were preferred as foreign materials initially. It is because even loosely placed sand and gravel exhibit good performance in comparison with soft clay and organic soil as far as the static problems are concerned. However, loose sand if saturated might cause serious problems in a seismic region, which was exemplified in past large earthquakes. Due to the shortage of good quality granular materials and due to concern on the dynamic problems, engineered soils are becoming popular in Japan recently. A typical example is the cement stabilized soils which will be described later in the paragraph of admixture stabilization.

2.2 Densification

Densification of loose granular soil, heterogeneous soil, municipal waste, or potentially liquefiable soil is quite a common practice. The purpose of densification is to increase strength, to reduce settlement of loose granular soil and to prevent liquefaction. Often, improvement of uniformity of originally heterogeneous soil becomes the purpose of densification. A group of short wooden piles driven into a loose sand layer beneath the building may be the forerunner of this category. Vibro-rod, vibro-flotation, sand compaction pile method (Kitazume, 2005), compaction grouting, and heavy tamping are well established techniques of recent days. Development of efficient vibrating equipment enhanced the techniques and is an indispensable technique to tame the potentially liquefiable loose sandy deposit. These techniques, however, produce large noise and vibration and may give unfavorable influence on nearby structures and residents. Thus the applicability of these techniques to urban renovation or retrofit is restricted.

Table 1.2 Evolution of ground improvement methods in Japan.

Improvement principle	Engineering method	Work examples	Period practical application introduced								
			1930s	1940s	1950s	1960s	1970s	1980s	1990s	2000s	2010s
Replacement	Excavation replacement	Dredging replacement method									
	Forced replacement	Sand compaction pile method				1966					
Densification — Dewatering/compaction	Compaction by displacement and vibration	Sand compaction pile method			1956						
		Gravel compaction pile method				1965					
Densification — Compaction	Vibration compaction	Vibro-flotation method			1955						
	Impact compaction	Dynamic consolidation method					1973				
Consolidation/ dewatering	Preloading	Preloading method	1928								
	Preloading with vertical drains	Sand drain method			1952						
		Packed sand drain method				1967					
		Board drain method				1963					
	Dewatering	Deep well method		1944							
		Well point method			1953						
		Vacuum consolidation method					1971				
	Chemical dewatering	Quick lime pile method				1963					
Grouting	Grouting	Chemical grouting method				1961					
		High pressure injection method				1965					
Solidification (Admixture stabilization)	In-situ mixing	Shallow mixing method					1972				
		Deep mixing method					1974				
	Plant mixing	Pre-mix mixing method							1990		
		Light weight soil method							1996		
	Pipe mixing	Pneumatic flow mixing method							1999		
Thermal stabilization	Heating										
	Freezing					1962					
Reinforcement							1972				

2.3   Consolidation/dewatering

When a foundation ground is cohesive soil with low strength and low permeability, structures constructed on the ground will experience a stability problem and/or long term unfavorable settlement. These soils however increase the strength and improve their compressibility with time under sustained loading. An applied external load causes an increase of total stress in the ground. The increment of the total stress is initially sustained by the excess pore water pressure if the soil is saturated. Then the excess pore water pressure dissipates with time and results in the reduction of soil volume and increase of effective stress and increase of strength. This is the consolidation. If time allows, the preloading of the ground prior to the construction by embankment fill whose load intensity is equivalent to or exceeding that of the expected structure will solve the problem.

The preloading by embankment fill is one of the oldest techniques to improve this kind of soils. As the soil to be improved has a low strength, it is not always possible to place the required embankment fill in a single stage. Most often, preloading is done by staged constructions to avoid the instability of embankment. As mentioned above, the preloading is given to the ground with a final target of increasing effective stress. The same can be achieved by alternative techniques of decreasing the pore water pressure in the ground. The alternatives are the application of vacuum, dewatering the ground water, electro-osmosis and quick lime piling. When there is a certain limitation on the space for the required embankment construction or there is limited resource for fill material, these alternatives are most effective. The further merit of these alternatives is that they will not accompany the increased shear stress and hence will not create a stability problem.

With increasing thickness of cohesive soil (with increasing drainage path), consolidation time becomes longer and may become unacceptable. The idea of accelerating the consolidation by reducing the length of drainage path was born in the USA and in Nordic countries in the 1930s. Commonly used artificial drainages are vertical drainage by means of a sand drain or prefabricated drain. As some amount of ground settlement take place at the surrounding area due to the settlement at the improved area, which can cause adverse influence to the surroundings, the applicability of the technique to neighboring construction is restricted. Apart from the application to cohesive soils, vertical drainage is recently employed to dissipate quickly the excess pore water pressure induced by earthquake in order to tame the liquefaction.

2.4 Grouting

By the American Society of Civil Engineers Committee on Grouting (1995), grouting is broadly defined as the placement of a pumpable material which will subsequently set or gel in pre-existing natural or artificial openings (permeation grouting) or openings created by the grouting process (displacement or replacement grouting). The major purpose of the grouting technique is to provide increased strength and/or to retard water seepage of soil or rock formation. Grouting is also used to compensate unfavorable displacement of the existing structure. When grout, a pumpable material is injected into a soil or rock formation, it may permeate into the natural openings such as void spaces of the soils and fissures in the rocks, or create an opening by fracturing the soil mass, or displace the surrounding soil. The final location of the grout and the maximum distance of travel from the injection point are determined by a number of factors; the viscosity of grout, gel time, size of particles in relation to the openings, injection pressure, rate of injection, properties of soil and rock to be grouted, and so forth. As a consequence, the completed grouted zone usually has an irregular shape and inhomogeneity. Hence the R&D efforts have been paid to control the shape and extent of the grout, to control the expected or unexpected displacement caused by the grouting process, and to predict the quality of the grouted formation. These research efforts have led to innovative grouting equipment, development of new grout materials, and application of sophisticated data acquisition and simultaneous computer control of the grouting process.

Jet grouting which was developed in Japan in the 1960s is a technique in between the grouting and the deep mixing (Shibasaki, 1996). The jet grouting is composed of a combined process of cutting soil by high pressure jet and filling the space created by cutting with grout. When most of the soil cut by the jet is discharged to the ground surface, it is thought as a family of the replacement grouting. However during the process the mixing of grout and in situ soil is unavoidable and the strength gain is influenced by the soil type. The jet grouting has been frequently applied to various improvement purposes similar to the deep mixing method, such as stability of ground and liquefaction prevention. At present, jet grouting is classified as a part of the deep mixing method.

2.5 Admixture stabilization

Admixture stabilization is a technique of mixing chemical binder with soil to improve the consistency, strength, deformation characteristics, and permeability of the soil. The improvement becomes possible by the ion exchange at the surface of clay minerals, bonding of soil particles and/or filling of void spaces by chemical reaction products. Although a variety of chemical binders has been developed and used, most frequently used binders nowadays are lime and cement due to their availability and cost. The mechanisms of the lime and cement stabilizations have been studied extensively in the 1960s by highway engineers in relation to the improvement of base and sub-base materials for road construction. The need of rapid construction enhanced the application of the technique to deep in situ soils. The deep mixing method, deep admixture stabilization was developed in Japan in the 1970s. Lime columns developed in Sweden at the same period is the same technology in principle. The deep mixing method utilizes mixing blades or augers to manufacture a stabilized soil column of predetermined size and shape in situ. The strength of stabilized soil is in the order of 100 to 1,000 kN/m^2 in terms of unconfined compressive strength.

To cope with the lack of good quality material for land reclamation or to save the environment at the borrow area, even inappropriate materials must be used for land reclamation. These materials are often come from maintenance dredging of navigation channels or from construction waste soils in the urban areas. The common practice was to improve these fill materials after reclamation to a desired level by the traditional technologies such as densification or drainage. Since the late 1980s to the middle of the 1990s new technologies to improve the fill materials prior to land reclamation work were developed by the port and harbor engineers in Japan. Sometimes, the improvement of such materials prior to land reclamation is a cost saving. One of the new techniques is pre-mixing of a small amount of cement with sandy material to improve liquefaction resistance (Zen et al., 1987). It is reported that the unconfined compressive strength around 100 kN/m^2 is sufficient to prevent liquefaction in most of the cases

(Zen *et al.*, 1987; Coastal Development Institute of Technology, 2003, 2008b). The other is pre-mixing of cement and air foam or EPS (expanded polystyrene) beads with clayey soils to manufacture a good quality fill material with pre-determined density and strength. The stabilized soil is called "Lightweight Geo-material" (Tsuchida *et al.*, 1996; Tsuchida and Egashira, 2004; Coastal Development Institute of Technology, 2008c). Retrofit and rehabilitation of existing structures are increasing demand in the developed countries. To reduce the external load acting on the structure or to the ground, replacement of soil in the vicinity of existing structures by engineered soil is often conducted in Japan. The lightweight geo-materials found the place of application in such a situation.

2.6 Thermal stabilization (heating and freezing)

Thermal stabilization is divided into two groups of heating and freezing. Even at ordinary temperature under the sun shine, properties of fine-grained soils are improved by desiccation. This is often found as a dry crust formed at the surface of reclaimed sludge. When the reclamation process is very slow, the thickness of the desiccated layer becomes several meters (Katagiri *et al.*, 1996). The artificial heating is naturally much more effective and the applications of heating the soil up to 300 to 1,000°C have been reported. Recently heating finds its application in the remediation of contaminated soils. Heating the soil at moderate temperature assists the vapor extraction of volatile organic compounds. Soil vapor extraction performance can be enhanced or improved by injecting heated air or steam into the contaminated soil through injection wells. Heating the soil to extremely high temperature is the in situ vitrification by which electrical current is used to heat and melt the soil in place. The technique is effective for soils contaminated with organic, inorganic and radioactive compounds.

The first reported use of ground freezing was in South Wales in 1862 in conjunction with mine shaft excavation (American Society of Civil Engineers, 1997). The strength of frozen soil is in the order of 1 to 10 MN/m^2, although it depends on a variety of factors such as soil type, water content, rate of freezing, temperature of frozen soil. Frozen soil becomes nearly impermeable materials. The technique is currently used for the temporary increase of strength and temporary shut off of water seepage around open cut, shaft excavation and tunneling.

2.7 Reinforcement

Ground reinforcement consists of creating a composite reinforced soil system by inserting inclusions in predetermined directions to improve the shear strength characteristics and bearing capacity of the existing ground. Ground reinforcement technologies include a constantly increasing diversity of installation techniques and reinforcing materials which, depending upon the target engineering applications, are designed to withstand the required resisting forces (e.g. tension, compression, bending moments or their combinations) over the expected life service of the structure.

2.8 Combined uses of various techniques

Engineers facing real life problems must find a solution for their problems at hand. In difficult situations, combined use of a variety of technologies is a common practice.

In the construction of an embankment on soft compressible soil, the major portion of the ground underneath the embankment is improved by accelerating the consolidation with the aid of vertical drainage. At the same time, the densification method may be employed to improve stability at the embankment toe. In braced excavation, the excavated bottom may be improved by the deep mixing method to shut off the seepage and prevent bottom heave. As the ordinary deep mixing method cannot improve the vicinity of sheet piles, jet grouting may be employed to fill the gaps between the deep mixing stabilized soils and sheet piles.

When ground improvement cannot solve the entire problem, the combination of ground improvements and elaborate foundation systems may be used together. Combination of preloading and lime columns followed by installation of pile foundations for housing and buildings is common practice in Nordic countries. In the huge man-made island of the Kansai International Airport, a soft alluvial clay layer was improved by vertical drainage, fill materials above the clay was improved by a variety of densification techniques (Maeda, 1989; Furudoi, 2005).

2.9 Limitation of traditional ground improvement techniques

All the ground improvement techniques have advantages and disadvantages. The replacement method has the advantage of a short construction period with the aid of modern machinery. However in recent years the disposal of excavated soft soil has become more difficult than ever due to environmental restrictions on civil engineering works and the replacement method cannot be adopted in some cases. In addition, good quality soil for a fill material cannot be obtained at reasonable cost in some cases. The densification techniques can be applicable to not only sandy ground but also clayey ground for various improvement purposes. The techniques, however, produce large noise and vibration and may give unfavorable influence on nearby structures. Thus the applicability to urban renovation or retrofit is limited. The consolidation/dewatering techniques are in most cases more economical than the other techniques. However, they have several demerits, such as long term consolidation process and adverse influence due to consolidation settlement to surrounding structures. The thermal stabilization techniques can be applicable to a wide varieties of soils, but their applications are mainly temporary purposes.

In order to overcome the disadvantages of the conventional ground improvement techniques, and to meet the social demands of rapid infrastructure development during the post-war economic growth, the research and development of the deep mixing method started in the early 1970s in Japan. The admixture stabilization techniques including deep mixing have disadvantages such as relatively high construction cost, but have advantages such as large strength increase, reduction of settlement, low noise and vibration during construction.

3 DEVELOPMENT OF DEEP MIXING IN JAPAN – HISTORICAL REVIEW

3.1 Development of the deep mixing method

Research and development of the deep mixing method in Japan was initiated by the Port and Harbour Research Institute (PHRI) of the Japanese Ministry of Transport.

The concept of lime stabilization of marine clays was first publicized in a technical publication of the PHRI in 1968 (Yanase, 1968). When the feasibility of the method was confirmed by Okumura, Terashi and their colleagues at the PHRI in the early 1970s, the research and development of the deep mixing method was accelerated. The subjects of R/D include 1) investigation of the lime and cement reactivity of Japanese marine clays, 2) development of equipment which permits a constant supply of binder and reasonably uniform mixing at depth, 3) understanding the engineering characteristics of stabilized soil, and 4) establishing a design procedure.

By the extensive laboratory tests on a variety of clays, it was found that most of Japanese marine clays easily gained strength of the order of $100 \, kN/m^2$ to $1 \, MN/m^2$ in terms of unconfined compressive strength (Okumura et al., 1972a, 1972b, 1974; Okumura and Terashi, 1975; Terashi et al., 1977, 1980, 1983a). Terashi and Tanaka at the PHRI continued the study on the engineering properties of lime and cement stabilized soils (Terashi et al., 1979, 1983) and proposed a laboratory mixing test procedure. The procedure was welcomed by Japanese researchers and engineers, and essentially the same procedure was adopted by the Japanese Society of Soil Mechanics and Foundation Engineering (Currently Japanese Geotechnical Society) in 1981 as its Draft Standard JSF: T31-81T (Japanese Society of Soil Mechanics and Foundation Engineering, 1982). The draft was later officially standardized by the Japanese Society of Soil Mechanics and Foundation Engineering in 1990 and experienced minor revisions in 2000 and 2009 (Japanese Geotechnical Society, 2000, 2009). The researches were followed by extensive studies by the research group of Takenaka Co. Ltd. (Kawasaki et al., 1978, 1981a, 1981b; Saitoh, 1988; Saitoh et al., 1985; Niina et al., 1977, 1981).

Terashi, Tanaka and Kitazume extended the study to investigate the behavior of improved ground (Terashi and Tanaka, 1981a, 1981b, 1983; Terashi et al., 1983b, 1985, 1988a, 1988b). During this period in the early 1980s, the Japanese Geotechnical Society established a technical committee to compile the State of the Art of the deep mixing method and its essence was reported in the monthly journal of the Society (Noto et al., 1983; Terashi, 1983). In 1983 the Ministry of Transport established a working group comprising engineers from local port construction bureaus and the PHRI, which spent three years compiling the full details of the design procedure and case histories (Ministry of Transport, 1986). Design procedure for marine works was standardized by the Ministry of Transport in 1989, which was later revised in 1999 and 2007.

For the researches on machinery development, the equipment (Mark I to Mark III) was developed at the PHRI with the collaboration of Toho Chika Koki Co. Ltd. (Okumura et al., 1972a, 1972b, 1974). The first field test was done with the Mark II machine, which was only 2 m high (Figure 1.2). The first trial on the sea was done nearshore at Haneda Airport with the Mark III as shown in Figure 1.3, which was capable of improving the sea bottom sediment up to 10 m from the sea water level. The basic mechanism of the equipment was established by these trials. Finally the Mark IV machine was manufactured by Kobe Steel Co. Ltd. and a marine trial test was done by the PHRI near-shore at Nishinomiya to establish the construction control procedure. These steps in the initial development of the method were continuously publicized through annual meetings of the Japanese Society of Soil Mechanics and Foundation Engineering and later through the PHRI reports (e.g. Okumura et al., 1972a, 1972b, 1974). Okumura and Terashi introduced the technology to the international community in 1975

Figure 1.2 First field test with small on-land machine in the 1970s (by the courtesy of Port and Airport Research Institute).

Figure 1.3 First full-scale test at offshore Haneda in the early 1970s (by the courtesy of Port and Airport Research Institute).

(Okumura and Terashi, 1975). Stimulated by these activities in the development of the new technique, a number of Japanese contractors started their own research and development of this technique in the middle 1970s.

As granular quicklime or powdered hydrated lime was used as a binder in these initial development stages, the method was named the "Deep Lime Mixing (DLM)." The first contractor who put the DLM into practice was Fudo Construction Co. Ltd. The very first application was the use of the Mark IV machine to improve reclaimed soft alluvial clay in Chiba prefecture in June 1974 (Figure 1.4). In the five years before 1978, the DLM was practiced at 21 construction sites including two marine works.

Figure 1.4 Mark IV machine for marine constructions (by the courtesy of Port and Airport Research Institute).

In an effort to improve the uniformity of stabilized soil, cement mortar and cement slurry quickly replaced granular quicklime. The PHRI, Kawasaki Steel Corp. and Fudo Construction Co. Ltd. developed in corporation the deep mixing method with cement mortar as a binder in 1974, which is named the "Clay Mixing Consolidation Method (CMC)." The PHRI also developed the method with cement slurry as a binder in 1975 together with Takenaka Civil Engineering & Construction Co., Ltd. The deep mixing method using binder slurry is now called the wet method of deep mixing. These developments encourage many marine contractors to develop their own method and machine in 1975 to 1977. In 1976, the Second District Port and Harbour Construction Bureau, the Ministry of Transport carried out a large scale experiment on the sea at the Daikoku pier in Yokohama Port, where the properties of the in-situ stabilized soil, the reliability of the overlapped portion, construction ability were confirmed.

A research group at the Public Works Research Institute of the Japanese Ministry of Construction started studies to develop a similar technique from the late 1970s to the early 1980s, inviting staffs of the PHRI to take part as advisory committee members. The technique developed is called the "Dry Jet Mixing (DJM) Method" in which dry powdered cement or lime is used as a binder instead of binder-water slurry (Public Works Research Center, 2004). This is now called the dry method of deep mixing.

Since a variety of equipment was established and standard design procedures became available, the application of the deep mixing method has exploded. Figure 1.5 shows the statistics of the number of deep mixing projects and the accumulative volume of stabilized soil in Japan. The total volume of stabilized soil by the deep mixing method from 1977 to 2010 reached 72.3 million m^3 for the wet method and 32.1 million m^3 for the dry method.

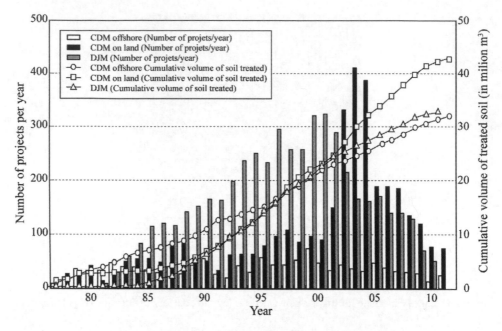

Figure 1.5 Statistics of deep mixing method works in Japan.

Until the end of the 1980s, the deep mixing method has been developed and practiced only in Japan and Nordic countries with a few exceptions. In the 1990s the deep mixing method gained popularity also in the United States of America and central Europe.

The first international specialty conference on deep mixing was co-organized by the Japanese Geotechnical Society and the International Society of Soil Mechanics and Geotechnical Engineering TC-17 in 1996 in Tokyo. The 1996 Tokyo Conference was followed by a series of specialty conferences/symposia in 1999 Stockholm, 2000 Helsinki, 2002 Tokyo, 2003 New Orleans and 2005 Stockholm. Along with these international forums, CEN TC288/WG10 started drafting the European standard of the execution and execution control of deep mixing in 2000. The WG 10 comprising delegates from 9 European countries invited international experts from Japan and USA to take part in their activity and completed an international standard. Recently, the International Symposium on Deep Mixing and Admixture Stabilization, OKINAWA 2009, was held in Okinawa, Japan, which was a continuation of the tradition of the deep mixing community but expanded the scope to cover similar admixture stabilization techniques. Now, the latest information on equipment, material properties, case records, design procedure, quality control (QC) and quality assurance (QA) have been updated and shared by the international deep mixing community by conducting a series of international specialty meetings.

3.2 Development of high pressure injection deep mixing method

The jet grouting technique was developed in circa 1965 in Japan, inspired by the large-scale water jet used in coal mine excavation. In the method, a high-pressure pump is used to convey the binder through an injection pipe to a set of nozzles located just above the drill bit. The high-pressure fluids or binders are injected into the soil at high velocities. They break up the soil structure completely and replace/mix the soil particles in situ to create a homogeneous mass. This ground modification/ground improvement of the soil plays an important role in the fields of foundation stability, particularly in the treatment of foundation ground under new and existing buildings; excavation support; seepage control in tunnel construction; and to solidification of contaminated soils and groundwater.

The jet grouting technique can be used regardless of soil type, permeability, or grain size distribution. It is possible to improve most soils, from soft clays and silts to sands and gravels by the jet grouting technique. Three basic jet grouting techniques currently exist are: single fluid, double fluid and triple fluid methods. In the double fluid method, soil is excavated and filled by an air-coated grout jet. The triple-fluid method excavates soil with an air-coated water jet then mix/replace the soil with a separate binder slurry jet. The latter two methods rapidly spread nationwide in the 1970s. The jet grouting method was then exported to Europe and USA and was accepted worldwide by the 1980s. The historical development and theoretical background have been described in detail by Shibasaki (1996). Currently the double fluid method is most commonly used in Japan.

4 DIVERSIFIED ADMIXTURE STABILIZATION TECHNIQUES WITHOUT COMPACTION

4.1 Classification of admixture stabilization techniques

The mechanism of stabilizations by cement is illustrated in Figure 1.6, which consists of four steps: the hydration of binder, ion exchange reaction, formation of cement hydration products, and formation of pozzolanic reaction products. Lime stabilization is based on similar chemical reactions but without the formation of cement hydration products. The water content of the original soil is decreased by hydration of the binder and subsequent water absorption process. The quicklime pile method, which has often been applied to soft soil with high water content, expects the hydration and absorption

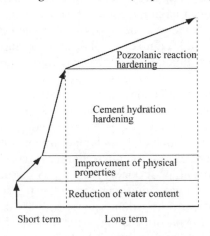

Figure 1.6 Mechanism of cement stabilization.

Table 1.3 Classification of admixture stabilizations.

Place of Mixing		Type of mixing	Method	Application
In-situ	in-situ mixing of surface layer	Mechanical mixing	Surface Treatment	Working platform on extremely soft ground
	in-situ mixing	Mechanical mixing	Shallow Mixing	Stability Settlement reduction Excavation support Seepage shutoff, *etc.*
		Mechanical mixing High pressure injection Hybrid of above two	Deep Mixing	
Ex-situ	Mixing during transportation	Mixing on belt conveyor	Pre-mixing	Improve liquefaction resistance of poor material
		Pneumatic flow mixing (pipe mixing)	Pipe mixing	Reduce compressibility of high water content dredged clay
	Plant mixing	Mechanical mixing	Pre-mixing	Improve liquefaction resistance of poor material
		Mechanical mixing	Lightweight geo-material	Density control of fill material
		Mechanical mixing and High pressure dewatering	Dewatered stabilized soil	Alternate for sand and gravel

process of quicklime piles installed into the ground to reduce the water content of the ground. The ion exchange reaction modifies the physical property of the original soil and results in the decrease of plasticity of the soil. This effect is utilized in the improvement of base or sub-base material with a small amount of lime or cement for road constructions, where the change of consistency of the soil makes compaction easier and more effective. The formation of cement hydration products and pozzolanic reaction products provides the strength increase to the soil binder mixture. The deep mixing method is mostly based on the latter two reactions to increase the strength of stabilized soil, where a relatively large amount of binder is mixed with the original soil.

The success of the deep mixing method for soil stabilization has encouraged the construction industry to develop various types of admixture stabilization techniques in Japan. The currently available admixture stabilization techniques without compaction are classified into in-situ mixing and ex-situ mixing as shown in Table 1.3. In the in-situ mixing, the soil is stabilized in situ with a binder by means of mechanical mixing and/or high pressure injection mixing. In-situ mixing techniques may be sub-divided into surface treatment, shallow mixing and deep mixing, depending upon the depth and purpose of improvement. The ex-situ mixing can be further subdivided into mixing during transportation and batch plant mixing depending upon where soil and binder are mixed. The ex-situ mixing techniques were developed to enhance the beneficial use of dredged soil, poor quality soils and construction surplus soils. These methods are intended to provide additional characteristics such as better liquefaction resistance, smaller density, smaller volume compressibility or extremely high strength to the original soils. In the ex-situ mixing, poor quality soil is once excavated and mixed with binder in a plant or during transportation to the construction site.

★Ⅰ型★

Figure 1.7 Floating type stabilization machine (http://www.chemico.co.jp/02improvement/category01/
pdf/mr001.pdf).

4.2 In-situ mixing

4.2.1 Surface treatment

Construction on extremely soft ground such as reclaimed land with dredged clay or
marsh requires the improvement of the surface layer to create a working platform.
There are several surface treatment techniques including surface drainage, the sheet
and sand mat spreading technique, and surface treatment by cement stabilization.
When the surface soil layer is quite soft with very high water content, a light weight
mixing machine mounted on a pontoon is usually used for stabilization. Figure 1.7
shows a pontoon and a machine which consists of vertical rotary shafts with mixing
blades at their bottom. The binder slurry is manufactured at a slurry plant placed on
the dike nearby and supplied to the mixing machine. In this particular case, the mixing
machine can move laterally on the pontoon to create a stabilized soil slab of about
40 m and 3 to 5 m thick. Then the pontoon is towed by wires from the dike and moved
forward stepwise to expand the slab. The strength and thickness of the stabilized soil
slab are usually designed by the Winkler type slab concept in order to provide sufficient
bearing capacity for the construction machinery working on the slab. The amount of
cement is usually 50 to 150 kg/m^3 to achieve the strength of stabilized soil of about
100 to 200 kN/m^2.

4.2.2 Shallow mixing

The purposes of improvement and applications of shallow mixing and hence the
expected function of stabilized soil do not differ from those of deep mixing, which will
be discussed in the next section. However, when the depth of improvement is smaller
than around 3 m, it is not efficient to use an ordinary deep mixing machine. For the

(a) Bucket mixing type.

(b) Rotary blender type.

(c) Trencher type (by the courtesy of Power Blender Association).

Figure 1.8 Shallow mixing techniques.

improvement of shallow depth, a simpler mixing tool such as bucket mixing, blade mixing and trencher mixing are preferred (Figure 1.8). These simpler machines originally developed for shallow mixing are sometimes used up to a depth around 10 to 13 m.

In the bucket mixing (Figure 1.8(a)), a hydraulic excavator consisting of a boom, a bucket with mixing blades and a cab on a rotating platform is used. As shown in Figure 1.8(a), several mixing blades are installed in the bucket, which rotate vertically to mix soil and binder thoroughly. The binder is usually spread on the ground surface at first and then mixed with the soil by the machine. The soil binder mixture can pass through the slatted plates on the rear of bucket, so that the machine can mix soil and binder thoroughly during excavation work.

In the blade mixing (Figure 1.8(b)), a beam equipped with mixing blades is attached to the arm of a backhoe instead of the bucket. As shown in Figure 1.8(b), two set of mixing blades are installed on the both sides of the beam. The beam is penetrated into the ground up to a depth of about 10 to 13 m while rotating the mixing blades vertically. During the penetration, binder slurry is injected from the outlets close to the mixing blades and is mixed with the soil. A stabilized soil with rectangular parallelepiped shape is constructed by the procedure. Any shape of improved ground can be constructed by successive installations.

In the trencher mixing (Figure 1.8(c)), a sort of chainsaw is used for mixing the soil and binder. The chainsaw cuts and disturbs the soil, so that the soils along the whole depth are mixed uniformly. The binder slurry is injected from the bottom end of the chainsaw together with compressed air to mix the soil and binder. The chainsaw can move vertically and horizontally, which can construct continuous stabilized soil wall and slab. The machine can stabilize the ground up to a depth of about 13 m for various purposes such as improving stability, constructing an impermeable wall and preventing liquefaction.

Common to the shallow mixing is that the operator controls the vertical and horizontal movement of the mixing tool. Hence both the degree of mixing and uniform binder delivery depends upon the skill of the operator to a larger extent in comparison with the deep mixing.

4.2.3 Deep mixing method

In the deep mixing method, soft soil is stabilized in situ with binder without compaction. The deep mixing method (DMM) has usually been applied to improvement of soft clays and organic soils for various purposes such as stability, settlement reduction, excavation support and seepage control (Coastal Development Institute of Technology, 2002, 2008a; Public Works Research Center, 2004). The deep mixing method originally developed in Japan and Nordic countries has now gained popularity in the worldwide market. During the past four decades, a variety of deep mixing processes have been proposed by contractors as their proprietary techniques. The mixing processes are classified in Table 1.4, which follows the classification system first adopted by Bruce *et al.* (2000) but is expanded to include the additional systems available in 2010. The first column from the left shows the method of introducing the binder either by Wet (binder-water slurry) or Dry (dry powder). The second column shows the driving mechanism of mixing tools. The third column shows the type of mixing tool and its location. For the high pressure injection, the second and third columns are combined. The fourth column shows the name of techniques followed by the country or region which was originally developed. The fifth column shows the roots of techniques either originally developed for deep mixing or modified from a diaphragm wall or trench cutter.

The techniques in which dry binder is blown pneumatically into the ground are called the dry method of deep mixing. The dry method employs mechanical mixing which consists of vertical rotary shaft(s) with mixing blades at the end of each shaft. In the penetration and/or withdrawal stage, binder is injected into the ground. The mixing blades rotate in the horizontal plane and mix the soil and the binder. In one operation, a column of stabilized soil is constructed in the ground. The two major techniques for the dry method are the Japanese DJM and the Nordic dry method. The

Table 1.4 Classification of deep mixing based on mixing process.

Binder Type	Type of shaft	Position of mixing	Representative system	Origin
Dry	Vertical rotary shaft	Blades at bottom end of shaft	DJM (Japan), Nordic dry method (Sweden)	Deep mixing
Wet A	Vertical rotary shaft	Blades at bottom end of shaft	CDM (Standard, MEGA, Land 4, LODIC, Column21, Lemni2/3) (Japan), SCC (Japan), Double mixing (Japan), SSM (USA), Keller (Central Europe), MECTOOL (USA)	Deep mixing
Wet B	Vertical rotary shaft assisted by Jet	Blades and high pressure injection at bottom of shaft	JACSMAN (Japan), SWING (Japan), WHJ (Japan), GeoJet (USA), HydraMech (USA), TURBOJET (Italy)	Deep mixing
Wet C	High pressure injection at bottom of shaft		Jet grouting – single fluid, double fluid, triple fluid (Japan), X-jet (Japan)	Deep mixing
Wet D	Vertical rotary shaft	Auger along shaft	SMW (Japan), Bauer Triple Auger (Germany), COLMIX (France), DSM (USA), MULTIMIX (Italy)	Diaphragm wall or Trench cutter
	Horizontal rotary shaft	Vertical mixing by Cutter mixer	CSM (Germany, France)	
	Chainsaw, Trencher	Continuous vertical mixing	Power Blender (Japan, shallow to mid-depth, down to 10 m), FMI (Germany, shallow to mid-depth), TRD (Japan, down to 35 m)	

standard DJM machine is a dual shaft machine and both the penetration/withdrawal speed and rotation speed are fairly slower than the Nordic single shaft machine. The DJM is used extensively in Japan and the Nordic dry method is used mostly in Nordic countries but also used in the other parts of the world in lesser extent. It seems that both Japanese and Nordic dry methods have not experienced substantial change during the last two to three decades.

The techniques in which binder-water slurry is pumped into the ground are generically called the wet method of deep mixing. The wet method, as shown in the table, has a variety and new techniques are continuously appearing in the market (Cement Deep Mixing Method Association, 1999).

The techniques in Wet A in Table 1.4 were originally developed for deep mixing and share the same fundamental mechanism with the dry method mentioned above. The equipment has a single to eight vertical rotary shafts equipped with cutting edge, blades or paddles at the lower part of each shaft. Further modifications/improvements of the basic techniques are purpose oriented. The CDM-LODIC added a continuous auger at the upper portion of the shafts to remove a certain portion of original soft soil during penetration and withdrawal phases in order to reduce the displacement of nearby existing structures. The CDM-MEGA, CDM-Land 4 and CDM-Lemni 2/3 are aimed to improve productivity either by expanding the diameter of mixing blades or by increasing the number of shafts. The CDM-Column 21 and CDM-Double-mixing are employing sophisticated mixing tools to improve the uniformity of the soil-binder mixture.

The techniques in Wet B in Table 1.4 are a hybrid of mechanical mixing and high pressure injection mixing. In these techniques a central portion of deep-mixed column is produced by the same process as those of Wet A and the diameter of which is governed by the size of horizontally rotating blades. In addition to the mechanical mixing, the equipment in this group has the nozzle(s) at the outer end of rotating blade(s) from which the high pressure cement slurry is injected outward to create a ring-shaped treated soil and expand the overall diameter of the deep-mixed column. All the methods except JACSMAN employ horizontal jet and hence the outer diameter of ring-shaped soil depends on soil condition and the applied pressure. The JACSMAN employs a pair of nozzles at two different levels: an upper nozzle inclines downward and a lower one inclines upward in order to make two jets collide at a prescribed point to maintain the constant outer diameter of ring-shaped stabilized soil (JACSMAN Association, 2011). The hybrid method is effective when the overlapping of adjacent deep-mixed soil columns is important or when the contact of stabilized soil to the existing structure is required.

The techniques grouped in Wet C in Table 1.4 are high pressure injection mixing methods called jet grouting. The high-pressure binder slurry with/without the aid of other high pressure fluids is injected into a soil at high velocities from the nozzles located at the bottom of the drill shaft (Japan Jet Grouting Association, 2011). They break up the soil structure completely and replace/mix the soil particles in situ to create a homogeneous mass. When the fluids or binders are injected horizontally, the diameter of completed stabilized soil is difficult to control and that depends on the injection energy and the original soil conditions. X-jet technique injects the binder from the two nozzles at different levels and two jets are designed to collide each other at a prescribed radius in order to create a stabilized column with uniform diameter. As the size of the equipment is extremely smaller than the mechanical deep mixing equipment, the technique is quite useful in a situation with space and head room restrictions.

The techniques grouped in Wet D in Table 1.4 seem to stem from the techniques for diaphragm wall construction or for trench cuttings and are recently modified to meet the deep mixing requirements. Mixing is carried out by various processes such as continuous or discontinuous augers along the shaft, cutter blades rotating around the horizontal shaft, or continuous transportation and mixing of soil-binder mixture by chain-saw type mixing tools.

Figure 1.9 shows the mechanical mixing system by vertical rotary shafts equipped with mixing blades at the bottom of each shaft. Figure 1.9(a) shows the Japanese dry method, DJM, Figure 1.9(b) shows on-land equipment for the Japanese wet method, CDM and Figure 1.9(c) shows the CDM equipment mounted on a special barge for marine construction. Figures 1.10 and 1.11 show the high pressure injection and the hybrid of mechanical and high pressure injection, respectively.

4.3 Ex-situ mixing

4.3.1 Premixing method

The Premixing Method is an admixture stabilization method where a small amount of binder and chemical additives are mixed with sandy material to obtain liquefaction-free fill material for land reclamation (Zen et al., 1987; Coastal Development Institute of Technology, 2003, 2008b). The basic principle of the method is to prevent liquefaction

(a) Dry mixing method for on-land works.

(b) Wet mixing method for on-land works.

(c) Wet mixing method for marine works.

Figure 1.9 Mechanical mixing by vertical rotary shafts and mixing blades.

of ground by a cementation effect between the soil particles and the binder. In the case where soil has some amount of cohesion, c' by the cementation effect, the shear strength does not decrease to zero and liquefaction does not take place even when the pore water pressure is generated up to the overburden pressure. Recently, the pre-mixed material has been also applied for reducing the earth pressure acting on the earth retaining structure, backfilling behind sheet pile walls and concrete structures.

The mixing of sand, binder and separation inhibitor is carried out either in a plant or by belt conveyor during transportation. When the water content of the fill material is relatively high, a double-shaft mixer or other mechanical mixer is used. When the fill material is dry, it is economical and efficient to carry out mixing by dumper chutes between a series of belt conveyors (Figure 1.12(a)). Stabilized soil is transported and placed at the designated area to construct reclaimed ground (Figure 1.12(b)).

4.3.2 Lightweight Geo-material

Lightweight Geo-material (Figure 1.13) was developed to reduce residual and uneven settlement, decrease earth pressure, prevent lateral displacement and improve

Figure 1.10 High pressure injection (http://www.kajima.co.jp/news/digest/jul_2010/searching/index).

Figure 1.11 Hybrid of mechanical and high pressure injection (by the coutesy of Fudo Tetra Corporation).

earthquake resistance of port and airport facilities in which dredged soil is mixed with binder and either air foam or expanded polystyrene (EPS) beads of 1 to 3 mm in diameter (Tsuchida *et al.*, 1996; Tsuchida and Egashira, 2004; Coastal Development Institute of Technology, 2008c). The density of the stabilized soil can be controlled from 6 to 15 kN/m³ by changing the amount of air foam or polystyrene beads, and water content of soil. In the execution, the mixture manufactured in a mixing plant is transported and placed at the designated area by means of a tremie pipe. The tremie pipe is usually used in order not to entrap seawater into the mixture during the placement.

As clayey soil to be stabilized has a relatively high water content, the mixture has high liquidity at the mixing stage and then loses liquidity quickly. The earth pressure of the stabilized soil is very small by the effect of its relatively large strength as well

(a) Mixing by shear type dumper.

(b) Stabilized soil conveyed to reclaimed site.

Figure 1.12 Premixing Method (by the coutesy of Dr. Yamazaki).

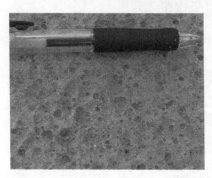

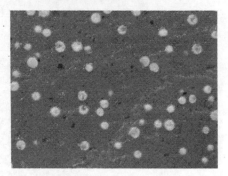

(a) Air foam stabilized soil.

(b) Expanded polystirol beads stabilized soil.

Figure 1.13 Lightweight Geo-material (by the courtesy of SGM Lightweight Treated Soil Method Association).

as light-weight characteristics, which can downsize superstructures such as concrete caisson and sheet pile wall. The lightweight geo-material has been used in backfill behind a new quay wall, reinforcement of existing structure, and embankment on soft ground (Figure 1.14).

4.3.3 Dewatered stabilized soil

In order to produce extremely high strength and compacted stabilized soil with low water content, a dewatering stabilized soil method was developed in which the soil is mixed with binder such as cement, lime and magnesium powder and dewatered at high compressive pressure of the order of 1 to $4\,MN/m^2$ by a compressor as shown in Figure 1.15(a). By the procedure, the stabilized soil with a cone resistance, q_c of 400 to $600\,kN/m^2$ can be obtained. The stabilized and compressed soil is crushed to granular material as shown in Figure 1.15(b) and used for the sand drain method or the sand

Figure 1.14 Application of Lightweight Geo-material for backfill at Kobe Port (by the courtesy of SGM Lightweight Treated Soil Method Association).

(a) High pressure compressor.

(b) Dewatered Stabilized soil.

Figure 1.15 Dewatered stabilized soil.

compaction pile method. In addition, the mixture may be used for making structural materials such as beams by molding.

4.3.4 Pneumatic flow mixing method

The Pneumatic Flow Mixing Method was developed for land reclamation, in which dredged soil is mixed with a relatively small amount of cement in the transporting pipe line (Kitazume and Satoh, 2003, 2005; Coastal Development Institute of Technology, 2008d). The soil-binder mixture forms several separated mud plugs in the transporting pipe, and is thoroughly mixed by means of the turbulent flow in the pipe. The mixture transported and placed at reclamation site gains relatively large strength rapidly so that no additional soil improvement is required. This method is expected to provide an economical and rapid construction for land reclamation. Figure 1.16 shows a group of barges for one kind of the Pneumatic Flow Mixing Method, which consists of a pneumatic barge, a binder supplier barge and a placement barge. The dredged clay in the soil transport barge is loaded into the hopper on the pneumatic barge at first, and is transported by the help of compressed air through the binder supplier barge to the reclamation site. The binder is injected to the soil on the binder supplier barge and

Figure 1.16 Group of barges for Pneumatic Flow Mixing Method.

they are thoroughly mixed in the pipeline during transport. The binder in slurry or dry form may be added to the soil, while slurry form is common practice. There are two types in the location of binder injection; *compressor addition type* and *line addition type*. In the former type, the binder is injected to the soil before the compressed air is injected into the pipeline. In the latter type, the binder is injected to the soil after the air injection. The soil mixture is transported and placed at a reclamation site through a cyclone on the placement barge, which functions to release the air pressure transporting the soil plugs. A tremie pipe is usually used to place the soil mixture under seawater not to entrap seawater within the soil-binder mixture, which can cause considerable decrease of the strength of stabilized soil. Several variations in the binder injection techniques and transporting techniques were put into practice by construction firms for the Pneumatic Flow Mixing (Coastal Development Institute of Technology, 2008d).

5   SCOPE OF THE TEXT

The deep mixing method was developed in Japan and put into practice in the middle of the 1970s. The total volume of stabilized soil by the deep mixing method from 1977 to 2010 reached 72.3 million m^3 for the wet method and 32.1 million m^3 for the dry method. The text is aimed to provide researchers and practitioners with the latest State of Practice of the deep mixing method based on the researches done in Japan and experience accumulated by numerous projects since the mid-1970s to 2010 in Japan. The organization of the current book is as follows.

Chapter 1 overviewed various ground improvement techniques and explained the deep mixing as a technique in the category of admixture stabilization by lime or cement as a binder. The chapter also showed that a wide range of admixture stabilization techniques including in-situ and ex-situ stabilizations gained popularity in Japan. Also the diversity of the deep mixing execution system worldwide was shown, which were designed to accomplish the local needs.

Chapter 2 discusses the influence of various factors on the strength increase by lime and cement. The information compiled in the chapter is basically applicable to all the admixture stabilization and useful in evaluating the feasibility of admixture stabilization to a specific soil, in the selection of appropriate binder, and in interpreting the laboratory or field test results.

Chapter 3 describes the engineering characteristics of stabilized soil by cement and lime. Correlation between unconfined compressive strength of stabilized soil and other engineering characteristics will benefit the understanding of stabilized soil regardless the execution system and type of application. The characteristics of in-situ stabilized soil such as the relation between the average field strength and laboratory strength and variability of field strength are important information for the geotechnical design, which, however, was discussed based only upon the experience gained by the Japanese execution system. This is because the quality of in-situ stabilized soil heavily depends upon the mixing process. The chapter concluded the necessity and responsibility of contractors to accumulate information on the quality of in-situ stabilized soil produced by their proprietary system.

Chapter 4 describes the column installation patterns and typical applications in Japan which will help project owner and geotechnical designer judge the applicability of deep mixing to their project at hand. The applications and pattern of column installation exemplified in the chapter involve the necessity of reliable overlap operation and/or reliable contact with the underlying stiff soil layers, which is not always accomplished by all the execution system. The designer should be aware of the capability of locally available execution system before deciding the pattern of applications.

Chapter 5 describes the construction, quality control and quality assurance during production. The chapter concentrates on the relevant issues on the Japanese execution systems that include dry and wet methods utilizing the mixing tool with vertical rotary shaft and mixing blades at the end of each shaft. Also the high pressure injection mixing was discussed.

Chapter 6 describes the geotechnical design procedure currently employed in Japan. The geotechnical design is an act to determine the required quality of stabilized soil and required geometry of stabilized ground as a composite system of stiffer stabilized soil and un-stabilized soft soils based on the simplified assumptions on the behavior of the composite system. Therefore the chapter starts with two introductory sections. One is to provide the state of the art on the engineering behavior of the ground improved by deep mixing. The information is necessary for the geotechnical engineer to understand the limitation of the currently adopted design procedure. The other is to provide the geotechnical engineer the importance of understanding the capability of the execution system to avoid the unrealistic requirements on the in-situ stabilized soil or the composite system.

Chapter 7 focuses upon the quality control and quality assurance for deep mixing. The concept of QC/QA described in the chapter is generally applicable to all the admixture stabilization. However, quality control procedures during production differs for different mixing process and also the laboratory mix test program as a pre-production QA differs for different mixing process. The current chapter focuses on the mechanical mixing by vertical rotary shaft and mixing blades. The quality of stabilized soil depends upon a number of factors as discussed in Chapter 2, which include the type and condition of original soil, the type and amount of binder, and the execution process. QC/QA of deep mixing, however, cannot be achieved only through process control during production. The chapter emphasizes the importance of collaboration among owner, designer and contractor along with the deep mixing project by explaining the pre-production, during production and post-production activities related to QC/QA.

The Appendix includes the standard laboratory mix test procedure in Japan with visual examples.

REFERENCES

American Society of Civil Engineers (1997) Ground improvement, Ground reinforcement, Ground treatment – Developments 1987–1997. *Proceedings edited by Schaefer*, V.R., ASCE Geotechnical Publication. No. 69.

American Society of Civil Engineers Committee on Grouting (1995) Verification of geotechnical grouting. *ASCE Geotechnical Special Publication.* No. 57, 177p.

Bruce, D.A., Bruce, M.E. & DiMillio, A.F. (2000) Deep mixing: QA/QC and verification methods. *Proc. of the 4th International Conference on Ground Improvement Geosystems.* pp. 11–22.

Cement Deep Mixing Method Association (1999) *Cement Deep Mixing Method (CDM), Design and Construction Manual.* (in Japanese).

Coastal Development Institute of Technology (2002) *The Deep Mixing Method – Principle, Design and Construction.* A.A.Balkema Publishers. 123p.

Coastal Development Institute of Technology (2003) *The Premixing Method – Principle, Design and Construction.* A.A.Balkema Publishers. 140p.

Coastal Development Institute of Technology (2008a) *Technical Manual of Deep Mixing Method for Marine Works.* 289p. (in Japanese).

Coastal Development Institute of Technology (2008b) *Technical Manual of Premixing Method.* 215p. (in Japanese).

Coastal Development Institute of Technology (2008c) *Technical Manual of Light-weight Geomaterial.* 370p. (in Japanese).

Coastal Development Institute of Technology (2008d) *Technical Manual of Pneumatic Flow Mixing Method.* 187p. (in Japanese).

Furudoi, T. (2005) Second phase construction project of Kansai International Airport – Large-scale reclamation works on soft deposits. *Proc. of the 16th International Conference on Soil Mechanics and Geotechnical Engineering.* Vol. 1. pp. 313–322.

http://www.chemico.co.jp/02improvement/category01/pdf/mr001.pdf.

JACSMAN Association (2011) *Technical Data for JACSMAN, Ver. 6.* 21p. (in Japanese).

Japan Jet Grouting Association (2011) *Technical Manual of Jet Grouting Method, Ver. 19.* 82p. (in Japanese).

Japanese Geotechnical Society (2000) *Practice for Making and Curing Stabilized Soil Specimens without Compaction. JGS 0821-2000.* Japanese Geotechnical Society. Vol. 1. (in Japanese).

Japanese Geotechnical Society (2009) *Practice for Making and Curing Stabilized Soil Specimens without Compaction. JGS 0821-2009.* Japanese Geotechnical Society. Vol. 1. pp. 426–434 (in Japanese).

Japanese Society of Soil Mechanics and Foundation Engineering (1988) *Ground Improvement Techniques – From Soil Investigation to Design and Construction.* Japanese Society for Soil Mechanics and Foundation Engineering. 383p. (in Japanese).

Katagiri, M., Terashi, M., Kaneko, A. & Uezono, A. (1996) Consolidation settlement of pump-dredged clay suspension – Analysis of a case record. *Proc. of the International Workshop on Technology Transfer for Vacuum-Induced Consolidation: Engineering and Practice.* pp. 51–65.

Kawasaki, T., Niina, A., Saitoh, S. & Babasaki, R. (1978) Studies on engineering characteristics of cement-base stabilized soil. *Takenaka Technical Research Report*. Vol. 19. pp. 144–165 (in Japanese).

Kawasaki, T., Niina, A., Saitoh, S., Suzuki, Y. & Honjyo, Y. (1981a) Deep mixing method using cement hardening agent. *Proc. of the 10th International Conference on Soil Mechanics and Foundation Engineering*. Vol. 3. pp. 721–724.

Kawasaki, T., Niina, A., Suzuki, Y., Saito, S., Suzuki, Y. & Babasaki, R. (1981b) On the deep chemical mixing method using cement hardening agent. *Takenaka Technical Research Report*. Vol. 26. pp. 13–38.

Kitazume, M. (2005) *The Sand Compaction Pile Method*. Taylor & Francis. 232p.

Kitazume, M. & Satoh, T. (2003) Development of pneumatic flow mixing method and its Application to Central Japan International Airport construction. *Journal of Ground Improvement*. Vol. 7. No. 3. pp. 139–148.

Kitazume, M. & Satoh, T. (2005) Quality control in Central Japan International Airport construction. *Journal of Ground Improvement*. Vol. 9. No. 2. pp. 59–66.

Maeda, S. (1989) Research on settlement and stability management system of soft grounds improved by sand drain for large-scale offshore artificial island. *Doctoral thesis* (in Japanese).

Ministry of Transport (1986) Improvement of soft soils (II). *Internal Report of the Ministry of Transport*. (in Japanese).

Niina, A., Saitoh, S., Babasaki, R., Miyata, T. & Tanaka, K. (1981) Engineering properties of improved soil obtained by stabilizing alluvial clay from various regions with cement slurry. *Takenaka Technical Research Report*. Vol. 25. pp. 1–21 (in Japanese).

Niina, A., Saitoh, S., Babasaki, R., Tsutsumi, I. & Kawasaki, T. (1977) Study on DMM using cement hardening agent (Part 1). *Proc. of the 12th Annual Conference of the Japanese Society of Soil Mechanics and Foundation Engineering*. pp. 1325–1328 (in Japanese).

Noto, S., Kuchida, N. & Terashi, M. (1983) Case histories of the deep mixing method. *Proc. of the Journal of Japanese Society of Soil Mechanics and Foundation Engineering, Tsuchi To Kiso*. Vol. 31. No. 7. pp. 73–80 (in Japanese).

Ogawa, F. & Matsumoto, K. (1978) Correlation between engineering coefficients of soils in the port and harbour regions. *Report of the Port and Harbour Research Institute*. Vol. 17. No. 3. pp. 3–8 (in Japanese).

Okumura, T. & Terashi, M. (1975) Deep-lime-mixing method of stabilization for marine clays. *Proc. of the 5th Asian Regional Conference on Soil Mechanics and Foundation Engineering*. Vol. 1. No. 1. pp. 69–75.

Okumura, T., Mitsumoto, T., Terashi, M., Sakai, T. & Yoshida, T. (1972a) Deep-lime-mixing method for soil stabilization (1st Report). *Report of the Port and Harbour Research Institute*. Vol. 11. No. 1. pp. 67–106 (in Japanese).

Okumura, T., Terashi, M., Mitsumoto, T., Sakai, T. & Yoshida, T. (1972b) Deep-lime-mixing method for soil stabilization (2nd Report). *Report of the Port and Harbour Research Institute*. Vol. 11. No. 4. pp. 103–121 (in Japanese).

Okumura, T., Terashi, M., Mitsumoto, T., Yoshida, T. & Watanabe, M. (1974) Deep-lime-mixing method for soil stabilization (3rd Report). *Report of the Port and Harbour Research Institute*. Vol. 13. No. 2. pp. 3–44 (in Japanese).

Public Works Research Center (2004) *Technical Manual on Deep Mixing Method for On Land Works*. 334p. (in Japanese).

Saitoh, S. (1988) Experimental study of engineering properties of cement improved ground by the deep mixing method. *Doctoral thesis, Nihon University*. 317p. (in Japanese).

Saitoh, S., Suzuki, Y. & Shirai, K. (1985) Hardening of soil improved by the deep mixing method. *Proc. of the 11th International Conference on Soil Mechanics and Foundation Engineering*. Vol. 3. pp. 1745–1748.

Shibasaki, M. (1996) State of the art grouting in Japan. *Proc. of the 2nd International Conference on Ground Improvement Geosystems.* pp. 851–867.

Terashi, M. (1983) Problems and research orientation of the deep mixing method. *Monthly Journal of Japanese Society of Soil Mechanics and Foundation Engineering, Tsuchi To Kiso.* Vol. 31. No. 8. pp. 75–83 (in Japanese).

Terashi, M. & Tanaka, H. (1981a) Ground improved by deep mixing method. *Proc. of the 10th International Conference on Soil Mechanics and Foundation Engineering.* Vol. 3. pp. 777–780.

Terashi, M. & Tanaka, H. (1981b) On the permeability of cement and lime treated soils. *Proc. of the 10th International Conference on Soil Mechanics and Foundation Engineering.* Vol. 4. pp. 947–948.

Terashi, M. & Tanaka, H. (1983) Settlement analysis for deep mixing method. *Proc. of the 8th European Regional Conference on Soil Mechanics and Foundation Engineering.* Vol. 2. pp. 955–960.

Terashi, M., Kitazume, M. & Nakamura, T. (1988a) External forces acting on a stiff soil mass improved by DMM. *Report of the Port and Harbour Research Institute.* Vol. 27. No. 2. pp. 147–184 (in Japanese).

Terashi, M., Kitazume, M. & Nakamura, T. (1988b) Failure mode of treated soil mass by deep mixing method. *Technical Note of the Port and Harbour Research Institute.* No. 622, 18p. (in Japanese).

Terashi, M., Kitazume, M. & Yajima, M. (1985) Interaction of soil and buried rigid structure. *Proc. of the 11th International Conference on Soil Mechanics and Foundation Engineering.* Vol. 3. pp. 1757–1760.

Terashi, M., Okumura, T. & Mitsumoto, T. (1977) Fundamental properties of lime-treated soils. *Report of the Port and Harbour Research Institute.* Vol. 16. No. 1. pp. 3–28 (in Japanese).

Terashi, M., Tanaka, H. & Kitazume, M. (1983b) Extrusion failure of ground improved by the deep mixing method. *Proc. of the 7th Asian Regional Conference on Soil Mechanics and Foundation Engineering.* Vol. 1. pp. 313–318.

Terashi, M., Tanaka, H. & Okumura, T. (1979) Engineering properties of lime-treated marine soils and D.M.M. method. *Proc. of the 6th Asian Regional Conference on Soil Mechanics and Foundation Engineering.* Vol. 1. pp. 191–194.

Terashi, M., Tanaka, H., Mitsumoto, T., Honma, S. & Ohhashi, T. (1983a) Fundamental properties of lime and cement treated soils (3rd Report). *Report of the Port and Harbour Research Institute.* Vol. 22. No. 1. pp. 69–96 (in Japanese).

Terashi, M., Tanaka, H., Mitsumoto, T., Niidome, Y. & Honma, S. (1980) Fundamental properties of lime and cement treated soils (2nd Report). *Report of the Port and Harbour Research Institute.* Vol. 19. No. 1. pp. 33–62 (in Japanese).

Tsuchida, T. & Egashira, K. (2004) *The Lightweight Treated Soil method – New Geomaterials for Soft Ground Engineering in Coastal Areas.* A. A. Balkema Publishers. 120p.

Tsuchida, T., Yokoyama, Y., Mizukami, J., Shimizu, K. and Kasai, J. (1996) Field test of lightweight geo-materials for harbor structures: *Technical Note of the Port and Harbour Research Institute.* No. 833, 30p. (in Japanese).

Watanabe S. (1974) Relations between soil properties. *Railway Technical Research Report.* No. 933. pp. 1–34 (in Japanese).

Yanase, S. (1968) Stabilization of marine clays by quicklime. *Report of the Port and Harbour Research Institute.* Vol. 7. No. 4. pp. 85–132 (in Japanese).

Zen, K., Yamazaki, H., Watanabe, A., Yoshizawa, H. & Tamai, A. (1987) Study on a reclamation method with cement-mixed sandy soils – Fundamental characteristics of treated soils and model tests on the mixing and reclamation. *Technical Note of the Port and Harbour Research Institute.* No. 579. 41p. (in Japanese).

Chapter 2

Factors affecting strength increase

1 INTRODUCTION

The strength increase of lime and cement stabilized soils is influenced by a number of factors, because the basic strength increase mechanism is closely related to the chemical reaction between soil and binder. The factors can be roughly divided into four categories: I. Characteristics of binder, II. Characteristics and conditions of soil, III. Mixing conditions, and IV. Curing conditions, as shown in Table 2.1 (Terashi *et al.*, 1983; Terashi, 1997).

The characteristics of binder mentioned in Category I strongly affect the strength of stabilized soil. Therefore, the selection of an appropriate binder is an important issue. There are many types of binder available on the Japanese market (Japan Lime Association, 2009, Japan Cement Association, 2007). The basic mechanisms of admixture stabilization using quicklime or cement were extensively studied by highway engineers many years ago. This is because lime or cement stabilized soils have been used for sub-base or sub-grade materials in road constructions (*e.g.* Ingles and Metcalf, 1972). The stabilizing mechanisms of various binders have been investigated further by geotechnical engineers (e.g. Babasaki *et al.*, 1996). The factors in Category II (characteristics and conditions of soil) are inherent characteristics of each soil and the way that it has been deposited. It is usually quite difficult to change these conditions at the site to perform deep improvement. Thompson (1966) studied the influence of the properties of Illinois soils on the lime reactivity of a compacted lime-soil mixture and concluded that the major influential factors were acidity (*pH*) and organic carbon content of the original soil. Japanese research groups have also performed similar studies on lime and cement stabilized soils manufactured without compaction (Okumura *et al.*, 1974; Kawasaki *et al.*, 1978, 1981, Terashi *et al.*, 1977, 1979, 1980; Saitoh, 1988). Their valuable works have provided engineers with good qualitative information. The factors in Category III (mixing condition) are easily altered and controlled on site during execution based on the judgment of engineers responsible for the execution. The factors in Category IV (curing conditions) can be controlled easily in laboratory studies but cannot be controlled at work sites in most cases.

The influences of various factors on the strength of the lime and the cement stabilized soils are shown in Sections 2 and 3 respectively, where the unconfined compressive strength, q_u of stabilized soil is mainly used as an index representing the stabilizing effect. The test specimens for the unconfined compression test are, in principle, prepared in the laboratory by the procedure standardized by the Japanese

Table 2.1 Factors affecting strength increase (Terashi *et al.*, 1983; Terashi, 1997).

I.	Characteristics of binder	1. Type of binder
		2 . Quality
		3. Mixing water and additives
II.	Characteristics and conditions of soil (especially important for clays)	1. Physical, chemical and mineralogical properties of soil
		2. Organic content
		3. potential Hydrogen (*pH*) of pore water
		4. Water content
III.	Mixing conditions	1. Degree of mixing
		2. Timing of mixing/re-mixing
		3. Quantity of binder
IV.	Curing conditions	1. Temperature
		2. Curing period
		3. Humidity
		4. Wetting and drying/freezing and thawing, *etc.*
		5. Overburden pressure

Geotechnical Society (formerly Japanese Society of Soil Mechanics and Foundation Engineering). The test procedure was originally proposed by Terashi *et al.* (1980) and welcomed by Japanese researchers and engineers. Essentially the same procedure was adopted by the Japanese Society of Soil Mechanics and Foundation Engineering in 1981 as its Draft Standard JSF: T31-81T (Japanese Society of Soil Mechanics and Foundation Engineering, 1982). The draft standard was later officially standardized by the Japanese Society of Soil Mechanics and Foundation Engineering in 1990 (Japanese Society of Soil Mechanics and Foundation Engineering, 1990) and experienced a minor revision by the Japanese Geotechnical Society in 2000 and 2009 (Japanese Geotechnical Society, 2000, 2009). The laboratory test procedure is described in the Appendix.

2 INFLUENCE OF VARIOUS FACTORS ON STRENGTH OF LIME STABILIZED SOIL

2.1 Mechanism of lime stabilization

When mixed with soil, quicklime (CaO) absorbs moisture in the soil corresponding to 32% of the weight of quicklime to become hydrated lime ($Ca(OH)_2$) as shown in Equation (2.1).

$$CaO + H_2O = Ca(OH)_2 + 15.6 \, kcal/mol \qquad (2.1)$$

This hydration reaction is rapid and generates a large amount of heat. During the process, quicklime doubles in volume. The water content of the soil is reduced by the chemical absorption, accompanied by a slight increase in shear strength. For the soil this process is a kind of consolidation successfully applied to the "quicklime pile method of soil improvement" (Tanaka and Tobiki, 1988).

With the existence of sufficient pore water, hydrated lime dissolves into the water and increases the calcium ion, Ca^{2+} and hydroxyl ion, OH^{-1} contents. Then Ca^{2+} ion exchanges with cations on the surface of the clay minerals. The cation exchange

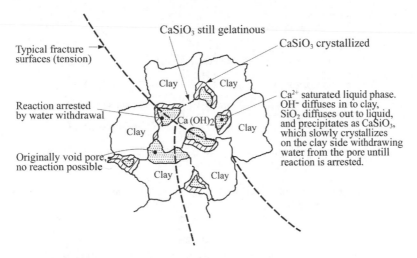

Figure 2.1 Basic lime stabilization mechanism (Ingles and Metcalf, 1972).

reaction alters the characteristics of water films on the clay minerals. In general, the plastic limit (w_p) of soil increases, reducing the plasticity index, I_p, as shown later in Figure 3.7 in Chapter 3. Furthermore, under high concentration of hydroxyl ions (under high pH condition), silica and/or aluminum in the clay minerals dissolve into the pore water and react with calcium to form a tough water-insoluble gel of calcium-silicate and/or calcium-aluminate. The reaction, so-called pozzolanic reaction, proceeds as long as the high pH condition is maintained and calcium ion exists in excess. The strength increase of lime stabilized soil is attributed mainly to the pozzolanic reaction product, which solidifies the clay particles together. The basic mechanism of lime stabilization is shown schematically by Ingles and Metcalf (1972) in Figure 2.1.

As is described above, the strength increase of lime stabilized soil relies solely upon the chemical reaction between clay minerals and lime. The formation of cementation material commences after the attack of lime on clay minerals. Therefore, thorough mixing of soil and lime is absolutely necessary to increase their contact areas.

2.2 Characteristics of lime as a binder

The four types of lime have been used for soil stabilization in Japan: calcium oxide (quicklime), calcium hydroxide (slaked lime, hydrated lime), wet hydroxide and lime-based special binder, as shown in Figure 2.2 (Japan Lime Association, 2009). The chemical constituents of calcium oxide and calcium hydroxide are specified by Japanese Industrial Standards (JIS) as shown in Table 2.2 (Japan Industrial Standard, 2006a). Wet hydroxide is a mixture of hydrated lime and 20 to 25% of water. It can be effectively applied to stabilization in a dust-proof requested site or to stabilization of low water content soil.

Lime-based special binders are a mixture of quicklime or hydrated lime as a mother material and gypsum, micro powder of slag, alumina or fly ash. They are specifically

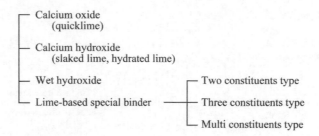

Figure 2.2 Types of lime for stabilization (Japan Lime Association, 2009).

Table 2.2 Specified proportions of chemical constituents of quicklime and hydrated lime (Japan Industrial Standard, 2006a).

	grade	CaO (%)	CaO and MgO (%)	CO₂ (%)	fineness (residue) 600 μm (%)	150 μm (%)
quicklime	special grade	>=94.0	–	<=2	–	–
	grade 1	>=90.0	–	–	–	–
	grade 2	>=80.0	–	–	–	–
hydrated lime	special grade	>=72.5	–	<=1.5	0	<=5.0
	grade 1	>=70.0	–	–	0	–
	grade 2	>=65.0	–	–	0	–

Table 2.3 Chemical constituents of lime-based special binders (Japan Lime Association, 2009).

	Chemical constituents (%) CaO	SiO_2	Al_2O_4	SO_4
two constituents type	60–95	1–20	2–25	0–20
three constituents type	50–85	1–40	1–40	2–40
multi constituents type	50–95	1–25	0–20	0–25

manufactured for stabilizing various soils which neither quicklime nor hydrated lime can stabilize effectively. Typical chemical components of lime-based special binders are tabulated in Table 2.3 (Japan Lime Association, 2009). They are divided into three categories depending on the number of materials combined together.

Figure 2.3 shows the influence of quicklime and lime-based special binder on the unconfined compressive strength, q_u of stabilized soil (Japan Lime Association, 2009). Although the q_u values increase with the amount of binder, the stabilizing effect is highly dependent upon the type of binder and the characteristics of soil.

2.2.1 Influence of quality of quicklime

Figure 2.4 shows the influence of the quality of quicklime on the strength of stabilized soil, in which four types of quicklime (named A to D) with different chemical activities

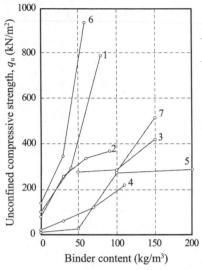

No.	site	soil type	natural water content (%)	wet density (g/cm³)	dry density (g/cm³)
1	Tokyo	sandy soil	67.8	1.563	0.931
2	Tokyo	sandy soil	25.4	1.891	1.508
3	Tokyo	sandy soil (silt)	47.1	1.725	1.173
4	Tokyo	sandy soil (clay)	49.4	1.711	1.145
5	Chiba	ornagic cohesive soil	204.2	1.198	0.394
6	Kanagawa	sandy soil	24.7	1.972	1.582
7	Nagano	volcanic cohesive soil	143.7	1.294	0.531

(a) Quicklime.

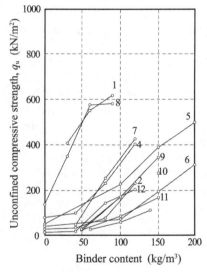

No.	site	soil type	natural water content (%)	wet density (g/cm³)	dry density (g/cm³)
1	Tokyo	sandy soil	42.4	1.744	1.225
2	Tokyo	volcanic cohesive soil	58.6	1.601	1.009
3	Tokyo	organil cohesive soil	46.4	1.722	1.176
4	Chiba	volcanic cohesive soil	125.3	1.321	0.586
5	Chiba	sand	47.6	1.659	1.124
6	Chiba	sand	59.9	1.593	0.996
7	Chiba	sandy soil	30.9	1.868	1.427
8	Kanagawa	sandy soil	24.7	1.972	1.582
9	Tochigi	volcanic cogesive soil	89.6	1.451	0.765
10	Tochigi	volcanic cogesive soil	122.5	1.323	0.595
11	Tochigi	cohesive soil	57.8	1.618	1.025
12	Gunma	volcanic cohesive soil	65.8	1.565	0.944

(b) Lime-based special binder.

Figure 2.3 Relationship between binder content and q_u value for quicklime and lime-based special binder (Japan Lime Association, 2009).

were mixed with the Honmoku marine clay (w_L of 92.4%, w_p of 46.9% and w_i of 120%) (Okumura *et al.*, 1974). In Figure 2.4(b), the w_i and aw represent the initial water content of the original soil and the binder factor, respectively. The binder factor, aw is defined by a ratio of the dry weight of binder to the dry weight of original

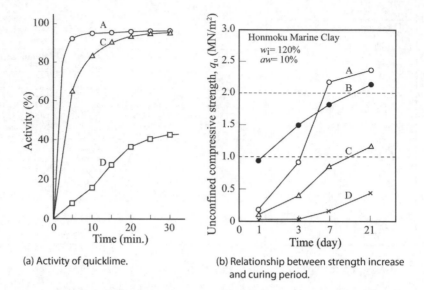

(a) Activity of quicklime.

(b) Relationship between strength increase and curing period.

Figure 2.4 Influence of activity of quicklime on unconfined compressive strength (Okumura *et al.*, 1974).

soil. The chemical activity is an index to represent the rate of hydration reaction of quicklime. The two types of quicklime, A and B, are ones with high activity burned at comparatively low temperature (about 1,000°C), while the two types of quicklime, C and D, are ones with low activity burned at comparatively higher temperature. As shown in Figure 2.4(a), the quicklime A has the highest activity among them while the quicklime D has the lowest. Figure 2.4(b) shows that the unconfined compressive strength, q_u of the quicklime stabilized soils is highly influenced by the activity of quicklime. The strength increases of the two types of quicklime A and B are much larger than those of the two types of quicklime C and D. This emphasizes that an appropriate type of quicklime should be selected to obtain high strength increase in lime stabilization.

2.3 Characteristics and conditions of soil

2.3.1 Influence of soil type

Figure 2.5 shows the influence of soil type on the unconfined compressive strength of the hydrated lime stabilized soils at 91 days curing (Saitoh, 1988). The soils are not organic clays whose plasticity index and initial water content range from 23 to 69 and 39 to 175% respectively. The figure clearly shows that the strength increase is highly influenced by the type of soil irrespective of marine and on-land soils. By examining the mineralogical properties of each soil, it was found that the soil having low crystallinity and fine clay minerals shows high strength, high pozzolanic reaction. Also showed that the allophane, halloysite and hydrated halloysite have high pozzolanic reactivity, but kaolinite, chlorite and illite have quite low pozzolanic reactivity.

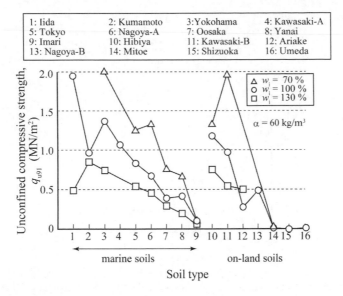

Figure 2.5 Influence of soil type on unconfined compressive strength at 91 days curing (Saitoh, 1988).

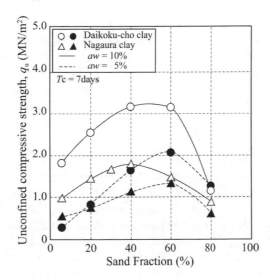

Figure 2.6 Influence of grain size distribution on strength of quicklime stabilized soil (Terashi *et al.*, 1977).

2.3.2 *Influence of grain size distribution*

As already shown in Figure 2.5, the q_u value of stabilized soils is highly influenced by the characteristics of soil. Figure 2.6 shows the influence of the grain size distribution of soil on the strength of quicklime stabilized soil (Terashi *et al.*, 1977). In the tests, a certain amount of the Toyoura standard sand (D_{50} of about 0.2 mm) was mixed with two different clays, the Daikoku-cho clay (w_L of 96.1% and w_p of 47.8%) and

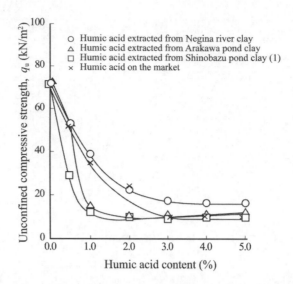

Figure 2.7 Influence of humic acid content on unconfined compressive strength (Okada *et al.*, 1983).

the Nagaura clay (w_L of 86.0% and w_p of 45.0%) so as to obtain artificial soils with different sand fractions. These artificial soils were stabilized with quicklime of the same binder factor, aw of 5 and 10%. In the figure, the unconfined compressive strength, q_u at 7 days curing, is shown on the vertical axis. The figure shows that the unconfined compressive strength is influenced by the amount of sand fraction of the soil and has a peak value at around 40 to 60%.

2.3.3 Influence of humic acid

Figure 2.7 shows the influence of humic acid of original soil on the unconfined compressive strength (Okada *et al.*, 1983). Four kinds of artificial soil were prepared by mixing various amounts of humic acid with the Kaolin clay (w_L of 50.6 %), in which three kinds of humic acid extracted from Japanese clays and one obtained on the market were mixed. These artificial soils with an initial water content of 60.6% were stabilized with aw of 5% of hydrated lime. The figure clearly shows the strength decreases quite rapidly with increasing amount of humic acid irrespective of the type of humic acid. The amount of humic acid should be one of the critical influence factors on the strength of hydrated lime stabilized soil, because some types of clay and sludge at marine and on-land areas contain a few percent order of humic acid.

2.3.4 Influence of potential Hydrogen (pH)

Figure 2.8 is the test results obtained on the compacted lime stabilized Illinois soils (Thompson, 1966). These two figures directly or indirectly explain the influence of soil acidity on the strength. The influence of *pH* of the original soil on the unconfined compressive strength, q_u is shown in Figure 2.8(a). The figure shows the tendency of decreasing strength with decreasing *pH*. Figure 2.8(b) shows the relationship between

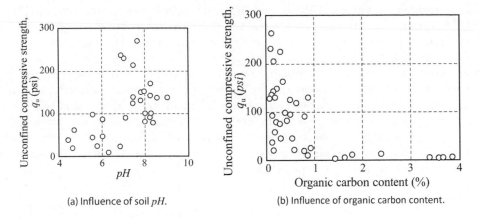

(a) Influence of soil *pH*. (b) Influence of organic carbon content.

Figure 2.8 Influence of soil *pH* and organic carbon content on strength of compacted hydrated lime stabilized soil (Thompson, 1966).

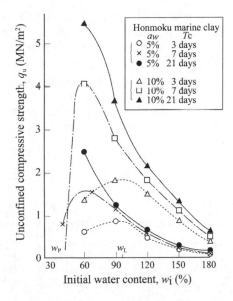

Figure 2.9 Influence of initial water content on strength of quicklime stabilized soil (Terashi *et al.*, 1977).

the q_u and the organic carbon content of original soil. The strength is very much different depending on the soil type as long as the organic carbon content of original soil is less than about 1%, but becomes negligibly small when the organic carbon content of original soil exceeds about 1%.

2.3.5 Influence of water content

The influence of the water content of original soil on the unconfined compressive strength, q_u is shown in Figure 2.9 (Terashi *et al.*, 1977). A marine clay excavated at

Honmoku Wharf, Yokohama Port was prepared to different initial water contents, w_i, and was stabilized with quicklime of two different binder factors and cured for 3, 7 and 21 days until the unconfined compression test. Figure 2.9 shows that the maximum strengths of the stabilized soil are achieved at around the liquid limit of the original soil, w_L for short term strength at 3 days curing. With increasing curing period, the water content providing the maximum strength shifts toward the dry side. The strength decreases considerably with increasing initial water content when it exceeds the liquid limit, w_L. In the cases of marine construction in Japan, this phenomenon might not cause serious problems because the natural water content of normally consolidated Japanese marine clay is close to its liquid limit in most cases. Care, however, should be taken on-land reclamation areas with pump dredged clay whose water content is usually much higher than its liquid limit.

2.4 Mixing conditions

2.4.1 Influence of amount of binder

Figure 2.10 shows the relationship between the binder factor, aw and the unconfined compressive strength, q_u, in which two different marine soils were stabilized (Terashi *et al.*, 1977). In the case of the Yokohama reclaimed soil (w_L of 78.8% and w_p of 49.1%), the unconfined compressive strength increases almost linearly with the amount of quicklime, irrespective of the curing period. In the case of the Honmoku marine clay (w_L of 92.4% and w_p of 46.9%), however, a clear peak strength can be seen and the amount of binder at the peak strength becomes larger with longer curing period. A similar phenomenon was found in the Haneda marine soil (w_L of 99.1% and w_p of 49.7%) (Terashi *et al.*, 1977).

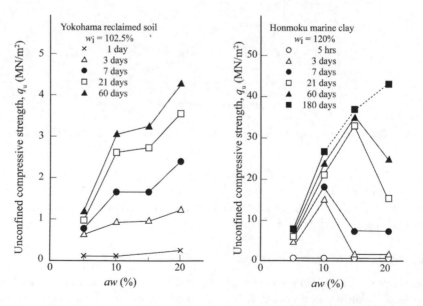

Figure 2.10 Influence of amount of binder in quicklime stabilization (Terashi *et al.*, 1977).

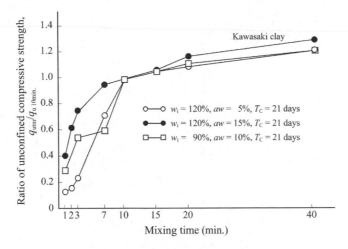

Figure 2.11 Influence of mixing time on strength (Terashi *et al.*, 1977).

2.4.2 Influence of mixing time

Figure 2.11 shows the influence of mixing time on the unconfined compressive strength, q_u by changing the mixing time of soil mixer in the preparation of laboratory specimens (Terashi *et al.*, 1977). In the tests, the Kawasaki clay (w_L of 87.8% and w_p of 49.7%) with various initial water contents, w_i were stabilized with quicklime. The vertical axis of Figure 2.11 shows the strength ratio, which is defined by the ratio of strength of stabilized soil prepared with arbitrary mixing time to those with mixing time of 10 min. The strength ratio decreases considerably when the mixing time is shorter than 10 min., especially for the case of small binder factor. When the mixing time exceeds 10 min., the strength ratio increases only slightly with the mixing time. A similar phenomenon was found on the cement stabilized soils by Nakamura *et al.* (1982), as shown later in Figure 2.30.

In the above description, the mixing time is an index to represent how sufficiently the mixing of soil and binder has been achieved. The degree of mixing depends not only on the mixing time but also on the type of mixer and the characteristics of original soil to be stabilized in the laboratory. Based on the past experiences of Japanese alluvial clays with water content around the liquid limit, Terashi *et al.* (1977) proposed a mixing time of 10 min. and use of the recommended soil mixer. In running laboratory mix tests with different types of soil and mixer, the responsible engineer should confirm the appropriate mixing time. The laboratory mix test procedure standardized by the Japanese Geotechnical Society (Japanese Society of Soil Mechanics and Foundation Engineering, 1980, Japanese Geotechnical Society, 2000, 2009) prescribes "sufficient mixing" in the main text and suggests 10 min. in the commentary (see Appendix).

2.5 Curing conditions

2.5.1 Influence of curing period

Figure 2.12 shows the influence of curing period on the unconfined compressive strength, q_u of various kinds of clay stabilized by quicklime with the same binder

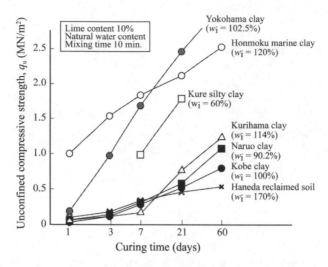

Figure 2.12 Influence of curing period in quicklime stabilization (Terashi *et al.*, 1977).

factor of 10% (Terashi *et al.*, 1977). In the figure, the curing period is plotted in logarithmic scale along the horizontal axis. The strength increase is much dependent upon the type of clay even if the amount of binder is the same, but the strengths of all the stabilized soils increase almost linearly with the logarithm of curing period. The strength increase of stabilized soils for more than 10 years will be shown later in Figure 3.31 in Chapter 3, in which the strength also increases almost linearly with the logarithm of curing period for longer term.

3 INFLUENCE OF VARIOUS FACTORS ON STRENGTH OF CEMENT STABILIZED SOIL

3.1 Mechanism of cement stabilization

The types of cement used as a binder are usually ordinary Portland cement and blast furnace slag cement type B in Japan. Ordinary Portland cement is manufactured by adding gypsum to cement clinker and grinding it to powder. Cement clinker is formed by minerals; $3CaO \cdot SiO_2$, $2CaO \cdot SiO_3$, $3CaO \cdot Al_2O_3$ and $4CaO \cdot Al_2O_3 \cdot Fe_2O_3$. A cement mineral, $3CaO \cdot SiO_2$, for example, reacts with water in the following way to produce cement hydration products.

$$2(3CaO \cdot SiO_2) + 6H_2O = 3CaO \cdot 2SiO_2 \cdot 3H_2O + 3Ca(OH)_2 \qquad (2.2)$$

During the hydration of cement, calcium hydroxide is released. The cement hydration product has high strength, which increases as it ages, while calcium hydroxide contributes to the pozzolanic reaction as in the case of lime stabilization. Blast furnace slag cement is a mixture of Portland cement and blast furnace slag. Finely powdered

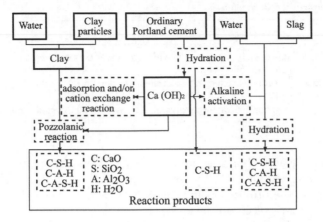

Figure 2.13 Chemical reactions between clay, cement, slag and water (Saitoh *et al.*, 1985).

Table 2.4 Chemical components of Japanese cements (Japanese Industrial Standard, 2009a, 2009b).

	CaO (%)	SiO$_2$ (%)	Al$_2$O$_4$ (%)	Fe$_2$O$_4$ (%)	SO$_4$ (%)	others
ordinary Portland cement	64–65	20–24	4.8–5.8	2.5–4.6	1.5–2.4	MgO, Na$_2$O,
high-early-strength Portland cement	64–66	20–22	4.0–5.2	2.4–4.4	2.5–4.4	K$_2$O, MnO,
blast furnace slag cement type B	52–58	24–27	7.0–9.5	1.6–2.5	1.2–2.6	P$_2$O$_5$

blast furnace slag does not react with water but has the potential to produce pozzolanic reaction products under high alkaline condition. In blast furnace slag cement, silicon dioxide, SiO$_2$ and aluminum oxide, Al$_2$O$_3$ contained in slag are actively released by the stimulus of the large quantities of Ca^{2+} and SO$_4^{2-}$ released from the cement, so that fine hydration products abounding in silicates are formed rather than cement hydration products, and the long-term strength is enhanced. The rather complicated mechanism of cement stabilization is simplified and schematically shown in Figure 2.13 for the chemical reactions between clay, pore water, cement and slag (Saitoh *et al.*, 1985).

3.1.1 Characteristics of binder

In Japan, ordinary Portland cement (OPC) and blast furnace slag cement type B have often been used as a binder for stabilizing clay and sand, whose chemical components are specified by Japanese Industrial Standard (Japanese Industrial Standard, 2009a, 2006b) as tabulated in Table 2.4. In addition to the two types of cement, cement-based special binders have been on the Japanese market as shown in Table 2.5 (Japan Cement Association, 2007).

Cement-based special binders are specially manufactured for the specific purpose of stabilizing soil or similar material by reinforcing certain constituents of the ordinary cement, by adjusting Blaine fineness or by adding ingredients effective for particular

Table 2.5 Cement-based special binders.

Type	Characteristics
for soft soils	appropriate for soft soils with high water content, e.g. sand, silt, clay and volcanic soil
for problematic soils	to reduce leaching of Hexavalent chromium (chromium VI) from stabilized soil
for organic soils	appropriate for highly organic soils, e.g. humus, organic soil, sludge

soil types. They are actually a mixture of cement as a mother material and gypsum, micro powder of slag, alumina or fly ash. The chemical components of cement-based special binders are the proprietary information of cement manufactures and are not specified by the Japanese Industrial Standard.

As shown in Table 2.5, cement-based special binders are designed for high water content soil, high organic soil and for reducing the leaching of Cr^{6+} from stabilized soil. The improvement effect in organic soils is said to be affected by the composite ratio, $((SiO_2 + Al_2O_3)/CaO)$, of the constituent elements in cement and cement-based special binders (Hayashi *et al.*, 1989).

Other than those special binders, "delayed stabilizing" or "long-term strength control" type binders are available by which the rate of strength increase can be controlled. They are obtained by adjusting the quantities of ingredients such as gypsum or lime. These binders react slowly with soil and exhibit smaller strength in the short term, but result in sufficiently high strength in the long term in comparison with ordinary Portland cement or blast furnace slag cement type B. These binders are useful for cases where the rate of strength increase has to be controlled, for example, for the convenience of the overlapping execution.

3.1.2 *Influence of chemical composition of binder*

An example of the effects of chemical compounds, CaO, SO_3 and Al_2O_3, on the strength is shown in Figure 2.14 (Japan Cement Association, 2009). In the test, a dredged clay (w_L of 60.7%, w_p of 29.1% and I_p of 31) was stabilized with a mixture of several types of cement and cement-based special binders so that the effects of the chemical compounds can be highlighted. After four weeks curing, the stabilized soils were subjected to unconfined compression test. The unconfined compressive strength, q_u is compared with the content of chemical compounds in the binder. In the effect of CaO, Figure 2.14(a), the strength remains almost constant irrespective of the amount of CaO as far as the amount of binder is about $80 \, kg/m^3$. When the amount of binder is increased to 140 and $200 \, kg/m^3$, however, the strength decreases with the content of CaO. In the effect of SO_3, Figure 2.14(b), the strength is almost constant irrespective of the amount of binder as far as the amount of SO_3 remains lower than about 8%. However, when the amount of SO_3 becomes about 9%, the strength rapidly increases. In the effect of Al_2O_3, Figure 2.14(c), the strength remains almost constant irrespective of the amount of Al_2O_3 as far as the amount of binder is about $80 \, kg/m^3$. When the amount of binder becomes 140 and $200 \, kg/m^3$, however, the strength increases almost linearly with the content of Al_2O_3.

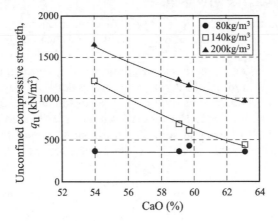

(a) Effect of content of CaO.

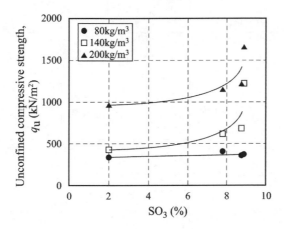

(b) Effect of content of SO_3.

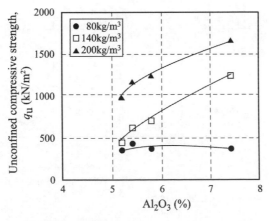

(c) Effect of content of Al_2O_3.

Figure 2.14 Effect of chemical compound on strength of cement stabilized soil (Japan Cement Association, 2009).

3.1.3 Influence of type of binder

Figure 2.15 shows the influence of the type of cement on the strength of stabilized soil in which ordinary Portland cement and blast furnace slag cement type B were compared at the curing period, t_c of 28 days to 5 years (Saitoh, 1988). The tests were conducted on two different sea bottom sediments; the Yokohama Port clay (w_L of 95.4%, w_p of 42.4% and w_i of 97.9%) and the Osaka Port clay (w_L of 79.4%, w_p of 40.2% and w_i of 94.9%). For each clay three different amounts of cement, α of 100 to 300 kg/m³ were mixed. The binder content, α is defined as a dry weight of cement added to 1 m³ of original soil. The horizontal axes of the figures show the curing period, t_c. The vertical axis of the upper figures for each clay is the unconfined

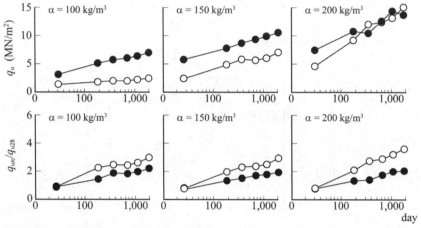

(a) Influence of binder type on Yokohama Port clay.

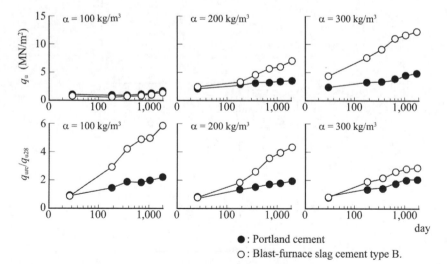

● : Portland cement
○ : Blast-furnace slag cement type B.

(b) Influence of binder type on Osaka Port clay.

Figure 2.15 Influence of cement type on unconfined compressive strength (Saitoh, 1988).

compressive strength, q_u of the stabilized soil, while the axis of the lower figures is the normalized unconfined compressive strength at arbitrary curing period, t_c by that of 28 days strength: q_{utc}/q_{u28}. In the case of the Yokohama Port clay which exhibits high pozzolanic reactivity, ordinary Portland cement is much more effective than blast furnace slag cement type B. Whereas in the case of the Osaka Port clay with lower pozzolanic reactivity than the Yokohama Port clay, blast furnace slag cement type B is much more effective. These test results suggest that the appropriate selection of the type of cement may be made if the pozzolanic reactivity of soil is evaluated beforehand. An evaluation method for the pozzolanic reaction of soil was proposed where the pozzolanic reactivity of natural soil may be judged by stabilizing the soil with hydrated lime as shown in Figure 2.3(a) (Saitoh, 1988). It is interesting to see the q_{utc}/q_{u28} is higher for blast furnace slag cement type B than for ordinary Portland cement, irrespective to the difference of soil type.

Figure 2.16 shows the influence of various cement-based special binders on the strength of various types of organic soil. The physical and chemical properties of the soils are tabulated in Table 2.6. The letters along the horizontal axis of the figures represent the types of binder. The chemical components of some binders are shown in Table 2.7. The figures show that cement-based special binders are effective in general but that the most effective binder for a particular soil is not always the best binder for the other type of organic soil. For these difficult soils, the selection of appropriate binder by laboratory test is important. A similar phenomenon on the strength of stabilized organic soils will be shown in Figure 2.19.

The overlapping execution is required for the block, wall and grid type improvement, as shown later in Chapter 4. The overlapping execution is carried out by cutting the side surface of a previously stabilized soil column during penetration and create new one during retrieval. In order to achieve tight overlapping, the low initial strength of stabilized soil is desirable, while ensuring the design strength in long term. For ease of overlapping execution, some cement-based special binders have been developed for retarding the short-term strength gain. Figure 2.17 shows the effect of one of the special binders on the strength of laboratory stabilized soils (Kuwahara et al., 2000). In the figure, the strengths of the stabilized soils with blast furnace slag cement type B are plotted together for comparison. The strength of stabilized soils with the special binder remains lower than those with blast furnace slag cement type B within about a couple of days curing, while the strengths increase with the curing period and they are almost same as those with the blast furnace slag cement type B at 28 days curing. These special binders were applied successfully to several construction projects.

3.1.4 Influence of type of water

Table 2.8 shows the influence of the type of water for preparing binder slurry on the strength of stabilized soil, where the clay excavated at Tokyo Port (w_L of 94.1% and w_p of 45.8%) was stabilized with ordinary Portland cement (Kawasaki et al., 1978). The cement slurry was prepared by two types of water: tap water and seawater obtained at Tokyo Port. The table shows that the strength of the stabilized soil with the tap water is slightly smaller than that with the sea water but the influence of the water type on the strength is negligibly small from the practical point of view.

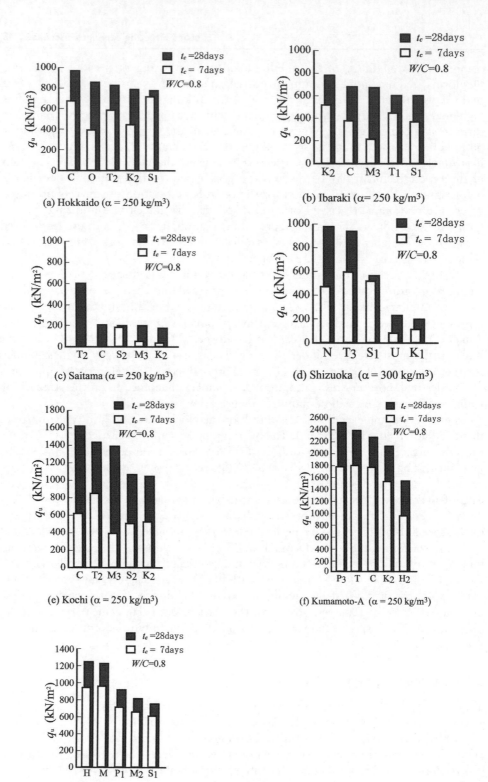

Figure 2.16 Unconfined compressive strength of organic soils stabilized with cement-based special binders.

Table 2.6 Physical and chemical properties of soils.

Depth (m)	Hokkaido −0.5 to −1.0	Ibaraki −0.5 to −1.0	Saitama −0.5 to −1.0	Shizuoka −3.0 to −4.0	Kochi −1.0 to −1.5	Kumamoto -A −5.0 to −7.5	Kumamoto -B −0.5 to −1.0
Physical properties							
Grain size distribution							
gravel (%)	–	–	–	–	0.0	0	0.0
sand (%)	–	–	–	–	0.0	0	2.9
silt (%)	–	–	–	–	71.8	40.5	42.0
clay (%)	–	–	–	–	28.2	59.5	55.1
Consistency limits							
liquid limit, w_L (%)	–	251.2	–	–	271.6	174.8	181.4
plastic limit, w_p (%)	–	92.7	–	–	69.1	76.2	47.4
plasticity index, I_p	–	158.5	–	–	202.5	97.6	144.0
Particle density	1.969	1.688	2.099	1.700	2.249	2.279	1.572
Natural condition							
water content, w (%)	492	246	940	840	295	156.4	159
density, ρ_c (g/cm³)	1.11	1.16	1.04	1.045	1.14	1.400	1.26
Chemical properties							
Organic content							
Ignition loss test (%)	55.2	47.7	67.4	70.5	24.8	22.2	24.0
Dichromate test (%)	42.4	25.2	59.0	–	17.6	–	11.5
Humus content (%)	8.1	15.2	28.6	17.2	4.1	–	7.4
pH	4.9	4.7	4.5	–	4.0	6.7	5.0

Table 2.7 Chemical components of binders.

Binder	SiO_2	Al_2O_3	Fe_2O_2	CaO	MgO	SO_3	Na_2O	K_2O
C	21.5	7.8	1.6	51.7	2.5	9.4	0.2	0.4
H	20.8	8.3	2.0	53.0	3.1	9.7	0.3	0.3
M	17.3	4.9	2.5	59.9	1.8	8.4	0.1	0.1
N	19.8	7.3	1.8	53.0	2.6	12.9	0.1	0.1
O	17.6	4.5	2.9	57.8	1.4	11.3	0.4	0.5

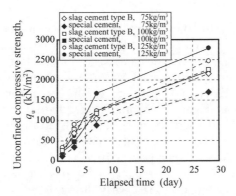

Figure 2.17 Strength increase of stabilized soil with special cement (Kuwahara et al., 2000).

3.2 Characteristics and conditions of soil

3.2.1 Influence of soil type

In order to investigate the influential factors on cement stabilization, Babasaki *et al.* (1996) collected 231 test results on soils taken from 69 locations in Japan from the

Table 2.8 Influence of type of water of cement slurry on strength of stabilized soil (Kawasaki *et al.*, 1978).

initial water content, w_i (%)	binder factor, aw (%)	curing period (day)	unconfined compressive strength, q_u (kN/m²)		strength ratio
			tap water	sea water	
79.9	14.1	7	2400	2640	0.91
	14.1	28	4500	4700	0.95
85.1	14.1	7	2080	2090	0.99
	14.1	28	4090	2980	1.04

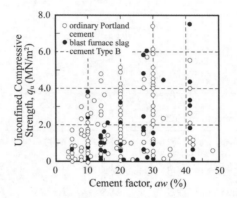

Figure 2.18 Relationship between unconfined compressive strength, q_u and binder factor, a_w (Babasaki *et al.*, 1996).

fourteen literatures published during 1981 to 1992 in Japan. For deducing the influence of soil type from the test data conducted by different laboratories, the other factors listed in Table 2.1 should be kept constant. Regarding to the characteristics of binder, only the test data for ordinary Portland cement and blast furnace slag cement type B were compared. The mixing and curing conditions except for the binder factor were the same for all the tests. Figure 2.18 compares the binder factor, *aw* and the 28 day unconfined compressive strength, q_u of various soils. Even for the same value of *aw*, the q_u varies considerably according to the type of soil tested.

It is well known that the strength of a particular soil stabilized by cement increases with increasing binder factor as shown later in Figures 2.28 and 2.29. The large variation of strength found in Figures 2.18 clearly shows that the strength gain by cement stabilization heavily depends upon the type and properties of soil.

The influence of soil type on the unconfined compressive strength, q_u is also shown in Figure 2.19, in which a total of 21 different soils were stabilized by ordinary Portland cement with binder factor, *aw* of 20% (Niina *et al.*, 1981). In the figure, various physical and chemical properties of the original soils are shown. The figure indicates that the humic acid content and *pH* of original soil are the most dominant factors influencing the strength.

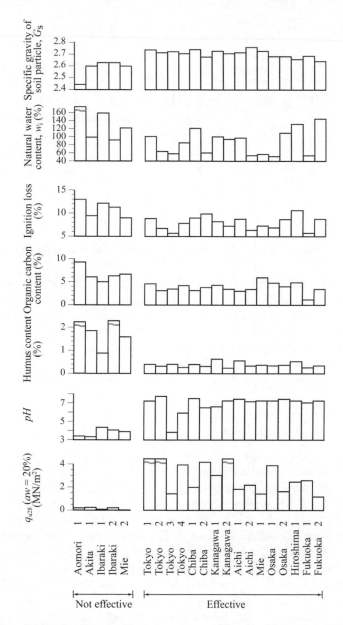

Figure 2.19 Influence of soil type in cement stabilization (Niina *et al.*, 1981).

3.2.2 *Influence of grain size distribution*

Figure 2.20 shows the influence of the grain size distribution of soil on the unconfined compressive strength, q_u of cement stabilized soil (Niina *et al.*, 1977). Two artificial soils B and C were prepared by mixture of two natural soils, the Shinagawa alluvial clay (w_L of 62.6% and w_p of 24.1%), named A and the Ooigawa sand, named D,

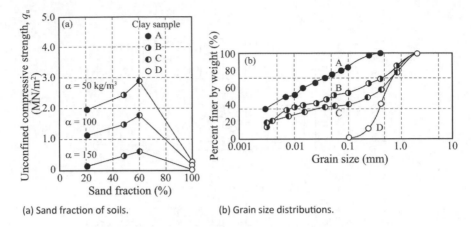

(a) Sand fraction of soils.

(b) Grain size distributions.

Figure 2.20 Influence of grain size distribution in cement stabilization (Niina et al., 1977).

whose grain size distributions are shown in Figure 2.20(b). The soils were stabilized with ordinary Portland cement with three magnitudes of binder content, α. Unconfined compression tests were carried out on the stabilized soils after 28 days curing. Similar to the lime stabilized soil as already shown in Figure 2.6, the unconfined compressive strength, q_u is dependent upon the sand fraction and the highest improvement effect can be achieved at around 60% of sand fraction irrespective of the amount of cement. This amount of sand fraction is quite close to that found for the lime stabilized soil.

3.2.3 Influence of humic acid

Figure 2.21 shows the influence of humic acid content on the unconfined compressive strength of cement stabilized soil (Okada et al., 1983). Artificial soil samples were prepared by mixing various amount of humic acid with the Kaolin clay (w_L of 50.6%), in which three kinds of humic acid extracted from Japanese clays and a commercially available humic acid were mixed. These artificial soils having the same initial water content of 60% were stabilized with aw of 5% of ordinary Portland cement. The figure clearly shows the influence of the humic acid depends on its characteristics: the acid extracted from the Negina River clay gives negligible influence on the strength, while the acid extracted from Shinobazu Pond clay gives considerably large influence on the strength.

Figure 2.22 also shows the influence of humic acid content of soil on the unconfined compressive strength (Miki et al., 1984). Artificial soil samples were prepared by adding various amounts of humic acid extracted from the clay at Arakawa Pond to the Kaolin clay, in which the humic acid content was 0 to 5% of the dry weight of the Kaolin clay. In the tests, these artificial soils were stabilized by nine types of binder whose chemical compositions are shown in Figure 2.22(a). Figure 2.22(b) shows the relationship between the unconfined compressive strength, q_u and the humic acid content. The unconfined compressive strength of the stabilized soil is highly dependent upon the binder, but decreases considerably with increasing humic acid content

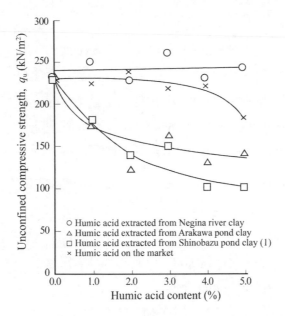

Figure 2.21 Influence of humic acid content on unconfined compressive strength (Okada *et al.*, 1983).

irrespective of the type of binder. The strength decreases to about one third when the humic acid content is about 5%.

3.2.4 Influence of ignition loss

The same data on stabilized soils from 69 locations as explained in Figure 2.18 are used to examine the influence of organic matter content. Ignition loss is a simple measure to estimate the organic matter content although it contains the loss due to inorganic matter. The relationship between the ignition loss and the unconfined compressive strength, q_u is shown in Figure 2.23 (Babasaki *et al.*, 1996). Type of binder and binder factor for each test result can be identified by the legend.

When the ignition loss is smaller than 15%, higher strength can be generally achieved with larger binder factor. For the soils with ignition loss exceeding 15% the unconfined compressive strength, q_u remains low value even with *aw* exceeding 20%, which means that high strength can't be achieved within a practical amount of binder. The soils encircled in the half-tone dot mesh do not exhibit strength increase, despite the increase in binder factor. In these soils the ignition loss is lower than 15% but the proportion of humus in the soil exceeds 0.9%, which is a higher figure than that for usual soils. Although there are some exceptions, the ignition loss is a convenient index to determine the stabilizing effect of various soils.

3.2.5 Influence of pH

Figure 2.24 shows the relationship between the *pH* of original soil and the unconfined compressive strength, q_u (Babasaki *et al.*, 1996). As the figure shows, most of the soils

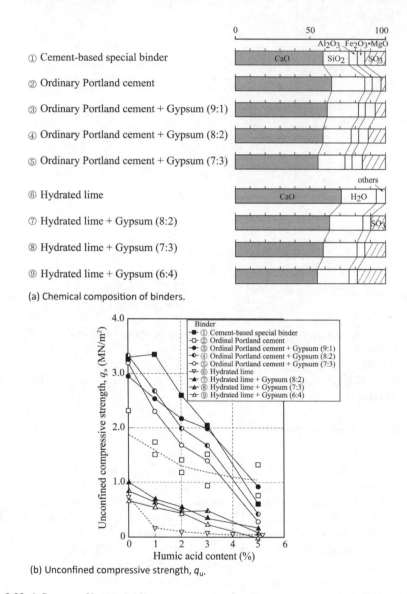

(a) Chemical composition of binders.

(b) Unconfined compressive strength, q_u.

Figure 2.22 Influence of humic acid content on unconfined compressive strength (Miki *et al.*, 1984).

with *pH* lower than 5 show a smaller strength increase compared with those with *pH* higher than 5 for the same binder content. Although there are some soils in which the improvement effect is not low even with low *pH* value, the *pH* value is a convenient and effective indicator to evaluate the effectiveness of soil improvement.

The relationship between the *pH* of original soil and the unconfined compressive strength, q_u of stabilized soil is proposed by Nakamura *et al.* (1980). In Figure 2.25, the test results of the five different soils are plotted, where their major characteristics

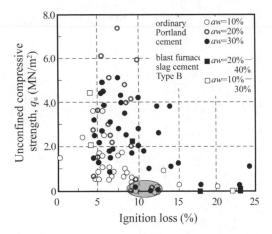

Figure 2.23 Relationship between unconfined compressive strength, q_u and ignition loss (Babasaki *et al.*, 1996).

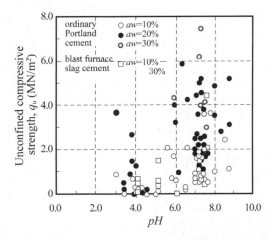

Figure 2.24 Relationship between unconfined compressive strength, q_u and pH (Babasaki *et al.*, 1996).

are tabulated in the attached table. On the horizontal axis of the figure, a parameter, F is plotted to incorporate the influence of pH, which is defined by Equation (2.3). The figure shows that the q_u value is roughly proportional to the F and the relationship between the q_u and the F is found as Equation (2.4).

$$\left. \begin{array}{ll} F = Wc/(9 - pH) & \text{for } pH < 8 \\ F = Wc & \text{for } pH > 8 \end{array} \right\}$$

(2.3)

$$q_u = 32.5 \cdot F - 1.625$$

(2.4)

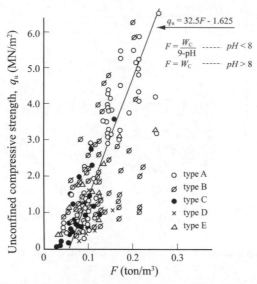

Type	Wet unit weight, γ_t (g/cm³)	Natural water content, w_n (%)	Liquid limit w_L (%)	Plastic limit w_p (%)	Grain size distribution (%)			pH (H₂O)	Ignition loss
					Sand, gravel fraction	Silt fraction	Clay fraction		
A	1.38 ~ 1.76	55 ~ 144	51 ~ 121	2 ~ 18	30 ~ 47	33 ~ 50	29 ~ 52	8.1 ~ 8.7	2.0 ~ 7.0
B	1.28 ~ 1.70	38 ~ 160	27 ~ 204	3 ~ 42	18 ~ 76	27 ~ 70	16 ~ 54	7.3 ~ 8.9	3.8 ~ 12.9
C	1.50 ~ 1.76	42 ~ 86	49 ~ 110	3 ~ 43	22 ~ 66	36 ~ 54	12 ~ 55	5.5 ~ 7.9	4.0 ~ 12.0
D	1.10 ~ 1.40	114 ~ 740	–	–	–	–	–	5.5 ~ 6.0	19.0 ~ 64.0
E	1.49 ~ 1.97	25 ~ 56	–	49 ~ 86	–	9 ~ 35	2 ~ 2	5.4 ~ 9.3	3.3 ~ 15.2

Figure 2.25 Effects of pH on cement stabilized soil (Nakamura *et al.*, 1980).

where
  F : parameter
  Wc : dry weight of cement added to original soil of 1 m³.

3.2.6 Influence of water content

The influence of the initial water content of soil on the unconfined compressive strength, q_u is shown in Figure 2.26 (Coastal Development Institute of Technology, 2008). In the tests, two kinds of marine clay were stabilized with either ordinary Portland cement or blast furnace slag cement type B. The unconfined compressive strength decreases almost linearly with increasing initial water content irrespective of the type of soil and the type of cement.

Figure 2.27 shows the relationship between the water content, w_t, in terms of the total water (including pore water and mixing water) and the q_u of stabilized soil with binder factor, aw of 10, 20, 30 and 35% (Babasaki *et al.*, 1996). The figure shows that the strength of stabilized soils decreases rapidly with the total water content. For soils with water content, w_t higher than 200%, increase of binder factor does not lead

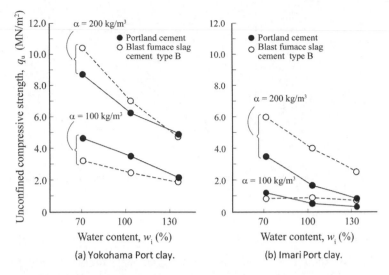

Figure 2.26 Influence of initial water content on strength (t_c of 91 days) (Coastal Development Institute of Technology, 2008).

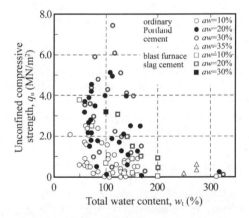

Figure 2.27 Relationship between unconfined compressive strength, q_u and total water content, w_t (Babasaki et al., 1996).

to greater strength. Such soils here with high water content are sludge, marshy soil, and surplus soil left after construction work, and are special soils from the viewpoint of admixture stabilization. For a specific soil, the lower the water content, w_t, and the higher the content of binder, aw, the greater the strength, q_u. But as can be seen in the figure, even when the water content, w_t and the binder factor, aw remain the same, the difference in soil characteristics leads to large differences in the improvement effect. There are some soils which are difficult to improve even when their water contents are lower than 200%. These soils usually contain high amount of organic material, or are acidic soils with low pH value.

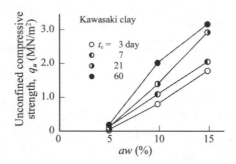

Figure 2.28 Influence of amount of cement on strength (Terashi *et al.*, 1980).

3.3   Mixing conditions

3.3.1   Influence of amount of binder

Figure 2.28 shows the influence of the amount of cement, *aw* on the unconfined compressive strength, q_u, in which the Kawasaki clay with an initial water content of 120% was stabilized with ordinary Portland cement, and tested at four curing periods (Terashi *et al.*, 1980). The unconfined compressive strength increases almost linearly with the amount of cement. The figure also shows that a minimum amount of cement of about 5% is necessary irrespective of curing period to obtain an improvement effect for this particular soil.

A similar phenomenon for organic soils is shown in Figure 2.29, in which the horizontal axis is the binder content, α, the dry weight of cement per 1 m^3 of original soil (Babasaki *et al.*, 1980). The strength is relatively small in the organic soils, but it increases with the binder content. The figure clearly shows that there exists a minimum binder content to achieve appreciable strength increase. The minimum binder content for these organic soils is around 50 kg/m^3.

3.3.2   Influence of mixing time

Figure 2.30 shows the relationship between the mixing time and the unconfined compressive strength, q_u in laboratory mix tests (Nakamura *et al.*, 1982). The laboratory mix tests were conducted as the same manner as the standardized procedure (Japanese Society of Soil Mechanics and Foundation Engineering, 1990) except for the mixing time. In the tests, the Narashino clay (w_i of 68%) was stabilized with ordinary Portland cement in either dry form or slurry form with a water to cement ratio, W/C of 100%. The unconfined compressive strength decreases with decreasing mixing time, similarly to that of quicklime stabilization as already shown in Figure 2.11. The figure also shows that the strength deviation increases with decreasing mixing time.

3.3.3   Influence of time and duration of mixing and holding process

In the laboratory mix test, a test specimen is produced by the following steps: 1) disaggregation and homogenization of original soil, 2) preparation of binder-water slurry at prescribed water/binder ratio, 3) mixing of soil and binder-slurry to prepare

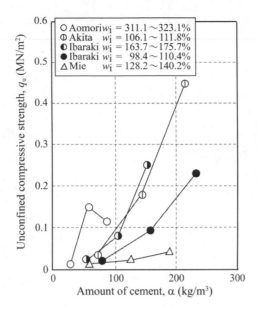

Figure 2.29 Influence of amount of cement on strength of stabilized organic soils (Babasaki *et al.*, 1980).

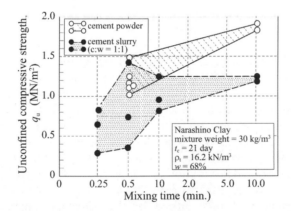

Figure 2.30 Influence of mixing time on strength and deviation of cement stabilized soil (Nakamura *et al.*, 1982).

uniform soil-binder mixture (about 10 min), 4) rest time before the molding, 5) filling the soil-binder mixture into the prescribed number of molds. A chemical reaction between binder and water starts when water is added to the binder at step 2. The chemical reaction between binder and soil continues when the binder slurry is added to the soil at step 3. As these chemical reactions progress with time, the time duration in steps 2) to 4) may influence the test results. For example, if the time for mixing binder slurry and soil and/or the time until molding is unnecessarily long, the chemical reaction products in the early phase may be broken during the molding procedure. Also

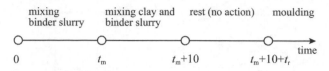

Figure 2.31 Process chart of mixing and moulding.

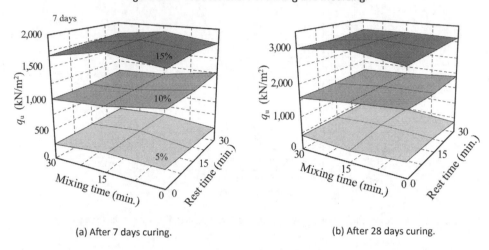

(a) After 7 days curing.

(b) After 28 days curing.

Figure 2.32 Influence of cement-slurry mixing time and rest time after clay-cement mixing on strength (Kitazume and Nishimura, 2009).

anticipated is the change of fluidity of soil-binder mixture may invite the difficulty of molding. The time duration of steps 2) to 4) is shown in Figure 2.31.

Although the time of mixing soil and binder slurry is not clearly specified in the Japanese standard test procedure, 10 min. mixing is the de facto standard in Japan (Japanese Geotechnical Society, 2009). The other time duration are considered to vary considerably on lab. to lab. basis, depending on the number of lab. technicians and number of specimens prepared from a batch of soil binder mixture. Any delay in the test procedure may cause deterioration of stabilized soil specimens' properties.

Figure 2.32 shows the effects of the mixing time, t_m of binder-slurry and the rest time, t_r on the strength of stabilized soil (Kitazume and Nishimura, 2009). The rest time is defined as the time period between the end of mixing and the start of molding. In the tests, the Kawasaki clay (w_L of 54.1%, w_p of 24.0% and w_i of 65%) was stabilized with ordinary Portland cement slurry of the W/C ratio of 100%, in which the binder factor, aw was changed 5, 10 and 15%. In the tests, t_m and t_r were changed. The case $t_m = 0$ corresponds to the situation where the binder and water are simultaneously added to the soil or to the test condition for the dry method of deep mixing. The unconfined compressive strengths measured at 7 and 28 days curing are shown in Figures 2.32(a) and 2.32(b) respectively. The standard deviation of q_u in each condition (three tests) was 2.6 to 2.9% in the average. The results indicate little influence of the time after mixing the binder and water, t_m and the time after mixing the soil and binder slurry, t_r on the q_u. The unit weight of a specimen exhibits little variability, being correlated more to the initial water content of the batches.

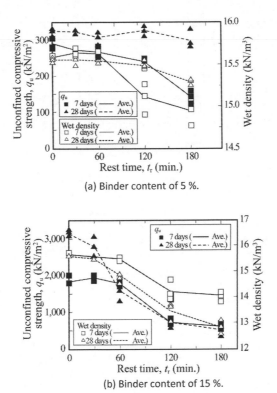

(a) Binder content of 5 %.

(b) Binder content of 15 %.

Figure 2.33 Influence of cement-slurry mixing time and rest time after clay-cement mixing on strength (Kitazume and Nishimura, 2009).

Figure 2.33 shows additional test results with extended rest time after mixing, t_r to identify the limit beyond which the soundness of specimen preparation is compromised (Kitazume and Nishimura, 2009). The test results reveal that it is t_r exceeding 40 min. that the specimen quality starts being affected by the soil-binder's reduced fluidity, and hence by the difficulty in 'compacting' through tapping actions. Longer t_r resulted in inclusions of numerous voids in the completed specimens, and lower unit weight, is closely related to q_u.

3.4 Curing conditions

3.4.1 Influence of curing period

Figure 2.34 shows the strength increase of cement stabilized soil with the curing period (Kawasaki *et al.*, 1981). In the tests, four types of soil excavated at Tokyo, Chiba, Kanagawa and Aichi were stabilized with ordinary Portland cement of *aw* of 10, 20 and 30%. The unconfined compressive strength, q_u increases with the curing period irrespective of the soil type, and the strength increase with time is more dominant for the stabilized soil with a larger amount of binder. Similar test results were obtained for

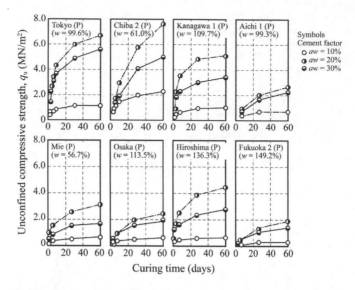

Figure 2.34 Strength increase with curing period (Kawasaki et al., 1981).

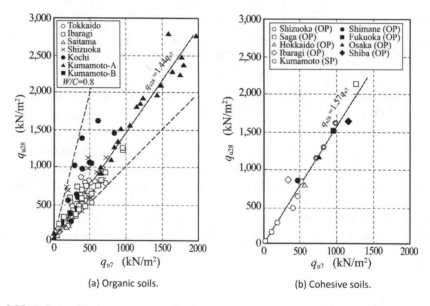

(a) Organic soils.   (b) Cohesive soils.

Figure 2.35 Relationship between unconfined compressive strength at 28 days curing and that at 7 days curing (Cement Deep Mixing Method Association, 1999).

the stabilized soils with either ordinary Portland cement or blast furnace slag cement type B (Saitoh, 1988).

The relationships between the strength of stabilized soil at two different curing periods have been studied. Figure 2.35 shows two typical examples of the relationship for organic soils and cohesive soils respectively (Cement Deep Mixing Association,

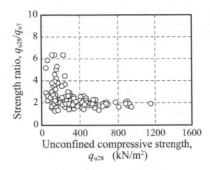

Figure 2.36

Figure 2.36 Relationship between unconfined compressive strength and strength ratio (Coastal Development Institute of Technology, 2008).

Table 2.9 Effect of curing period on unconfined compressive strength (Cement Deep Mixing Method Association, 1999).

	Ordinary Portland cement	Blast furnace slag cement type B
q_{u7}–q_{u28}	$q_{u28} = 1.49\ q_{u7}$	$q_{u28} = 1.56\ q_{u7}$
q_{u7}–q_{u91}	$q_{u91} = 1.97\ q_{u7}$	$q_{u91} = 1.95\ q_{u7}$
q_{u28}–q_{u91}	$q_{u91} = 1.44\ q_{u28}$	$q_{u91} = 1.20\ q_{u28}$

1999). In Figure 2.35(a), the q_{u28}/q_{u7} ranges within 1 to 4 with mean value of around 1.44 for the stabilized organic soils. For the cohesive soils (Figure 2.35(b)), on the other hand, the mean value of q_{u28}/q_{u7} is 1.57. A similar relationship q_{u28}/q_{u7} of 1.4 to 2.3, q_{u91}/q_{u7} of 1.8 to 5.9, and q_{u91}/q_{u28} of 1.2 to 2.1 for the clay and sand was reported by Saitoh (1988). The strength ratio, q_{u28}/q_{u7}, depends on the soil type, the type and amount of binder.

Figure 2.36 shows the relationship between the strength ratio, q_{u28}/q_{u7}, and q_{u28} on the 14 laboratory stabilized clays with blast furnace slag cement type B (Coastal Development Institute of Technology, 2008). The figure shows the strength ratio ranges about 2 to 6 as far as the q_{u28} is lower than about 400 kN/m², but the ratio decreases rapidly to around 2 when the q_{u28} is higher than about 400 kN/m².

Other examples of the relationship between the q_{u7}, q_{u28} and the q_{u91} on laboratory cement stabilized soil are tabulated in Table 2.9 (Cement Deep Mixing Method Association, 1999).

3.4.2 *Influence of curing temperature*

The influence of curing temperature is shown in Figure 2.37, in which the stabilized soils, the Yokohama clay (w_L of 95.4% and w_p of 42.4%) and the Osaka clay (w_L of 79.4% and w_p of 40.2%) were cured at various temperatures up to four weeks (Saitoh *et al.*, 1980). In the figure, the strength of stabilized soil cured at arbitrary temperature is normalized by the strength of the stabilized soil cured at 20°C. The figure shows that larger strength can be achieved at a higher curing temperature. This influence of curing temperature is more dominant on the short-term strength but it becomes less dominant as the curing period becomes longer.

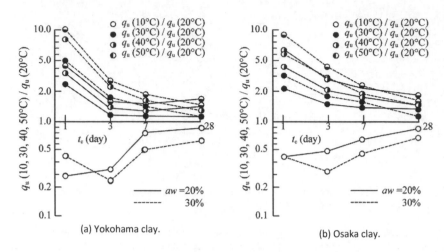

Figure 2.37 Effects of curing temperature on strength of cement stabilized soils (Saitoh et al., 1980).

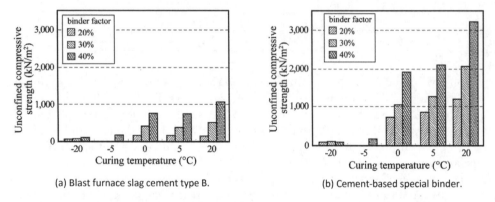

Figure 2.38 Relationship of 28 days for curing unconfined compressive strength and curing temperature (Kido et al., 2009).

Figure 2.38 shows the relationship between the unconfined compressive strength of laboratory stabilized peat and the curing temperature (Kido *et al.*, 2009). The peat was excavated in Hokkaido, whose natural water content, density of soil particle and ignition loss were 550%, 1.854 g/cm^3 and 66% respectively. The peat was stabilized with either blast furnace slag cement type B or cement-based special binder with three different binder factors, *aw* of 20, 30 and 40%. In Figure 2.38(a), the unconfined compressive strength, q_u of the stabilized soil with blast furnace slag cement type B and cured at 0 and 5°C are about 60% of those cured at 20°C. The strengths of the stabilized soils cured at −20 and −5°C are a quite low value of around 50 kN/m^2 even in the case of 40% in binder factor, and they are lower than one third of those at 20°C. For stabilization with the cement-based special binders, Figure 2.38(b), the unconfined compressive strength cured at 0, 5 and 20°C are comparatively higher than those of the stabilized soil with blast furnace slag cement type B, which demonstrates the high

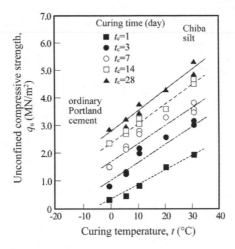

Figure 2.39 Relationship between unconfined compressive strength, q_u and curing temperature, T_c (Enami *et al.*, 1985).

applicability of the cement-based special binder. The strengths of the stabilized soils cured at 0 and 5°C are about 60% of that at 20°C, which are almost the same ratio as that for the stabilized soil with blast furnace slag cement type B. The strength cured at −20 and −5°C are quite low value, where negligible improvement effect is found in the both binders.

Figure 2.39 shows the influence of the curing temperature on the strength of stabilized soil for various curing periods (Enami *et al.*, 1985). At the same curing period, the higher the curing temperature the larger the soil strength. Looking at the same curing temperature, the strength increases with the curing period.

3.4.3 *Influence of maturity*

In concrete engineering, the influence of the curing temperature and the curing period on the strength is often explained by the Maturity index. The Maturity is a concept to combine the effects of time and temperature. Equation (2.5) shows four definitions of Maturity proposed by the previous studies (M_1: general definition for cement-concrete, M_2: Nakama *et al.* (2004), M_3: Åhnberg and Holm (1984), and M_4: Babasaki *et al.* (1996)). The correlation between the strength of stabilized soil and the logarithm of Maturity, expressed differently, means that temperatures as an environmental condition does not have a significant effect on the long-term strength but has a considerable effect on the short term strength.

$$M_1 = \Sigma(T_c - T_{c0}) \cdot t_c \tag{2.5a}$$

$$M_2 = 2.1^{(T_c - T_{c0})/10} \cdot t_c \tag{2.5b}$$

$$M_3 = \{20 + 0.5 \cdot (T_c - 20)\}^2 \cdot \sqrt{t_c} \tag{2.5c}$$

$$M_4 = 2 \cdot \exp\left(\frac{T_c - T_{c0}}{10}\right) \cdot t_c \tag{2.5d}$$

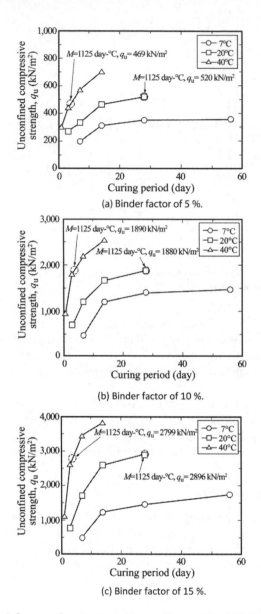

Figure 2.40 Influence of curing period on q_u (Kitazume and Nishimura, 2009).

where

M : maturity

T_c : curing temperature (°C)

T_{c0} : reference temperature (−10°C)

t_c : curing period (day).

The variations of q_u with curing period observed for the Kawasaki clay are shown in Figures 2.40(a) to 2.40(c) for various binder factors and curing temperatures

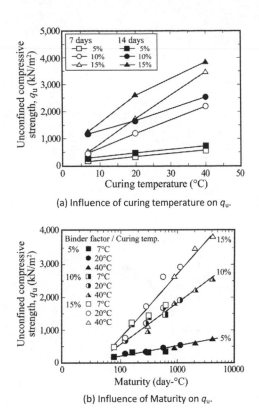

(a) Influence of curing temperature on q_u.

(b) Influence of Maturity on q_u.

Figure 2.41 Influences of curing temperature and Manurity on q_u (Kitazume and Nishimura, 2009).

(Kitazume and Nishimura, 2009). The observed effects of these factors follow the patterns, with higher curing temperature and longer curing period giving higher strength. In the figure, the Maturity, M, is shown, which is defined by M_4 (Equation (2.5d)).

The unconfined compressive strength, q_u is plotted against temperature in Figure 2.41(a) and the M_4 in Figure 2.41(b) (Kitazume and Nishimura, 2009). Use of the M_4 brings the q_u data points broadly along unique lines, each of which represents different binder contents. One potential application of this result is to estimate the standard 28 days, and 20°C strength from shorter-term tests at higher temperature. Equation (2.5d) implies that the 4.8 days at 40°C is equivalent to 28 days at 20°C in terms of the Maturity. For the particular clay tested, the strengths at these two conditions match well. It should be noted, however, that the curing at low temperature (7°C) expressed by the square in Figure 2.41(b) exhibited very small long-term gains in strength, as indicated by the concaved shape of the q_u and M relationships. It therefore seems difficult to estimate the long-term strength at very low temperature through extrapolation of short-term strength obtained for moderate to high curing temperature.

Figure 2.42 shows another test results on the relationship between the q_u and the M_4 on five different types of soil: silt, peat (w_n of 456.9%), fine sand, loam (w_n of

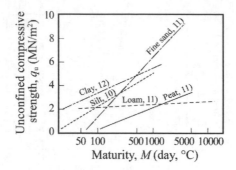

Figure 2.42 Relationship between unconfined compressive strength, q_u and maturity, M (Babasaki et al., 1996).

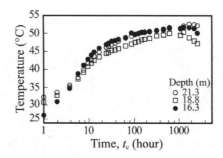

Figure 2.43 On-site measurements of hydration-generated heat in stabilized ground (Omura et al., 1981).

109.9%) and clay (Enami et al., 1985, Horiuchi et al., 1984, Babasaki et al., 1984). At any curing temperature and curing period, the strength, q_u is expressed as Equation (2.6) (Babasaki et al., 1996), but the magnitude of the parameters are quite different depending on the type of soil.

$$q_u = A \cdot \log M_4 + B \tag{2.6}$$

The curing temperature of stabilized soil in the field is affected by the ground temperature, but the heat generation brought about by the hydration of the binder also affects the curing temperature. The actual temperature change with the process of hydration is determined by the amount of heat generated through hydration of the binder, the specific heat of soil, the thermal capacity, the size and the geometry of the stabilized soil, and the ground temperatures as a background. The greater the bulk of stabilized soil, the greater the content of binder and the higher the background temperature, the higher the temperature will become. Figure 2.43 shows the change of ground temperature with time after stabilization (Omura et al., 1981). The temperature was measured at various depths within a large block of stabilized soil mass at Yokohama Port. As shown in the figure, a high temperature of the order of 50°C is maintained over several months. The prediction of the temperature in the stabilized soil mass is possible by the thermal analysis (Babasaki et al., 1984).

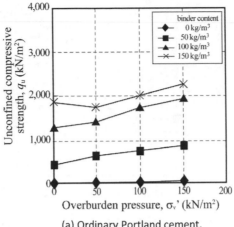

(a) Ordinary Portland cement.

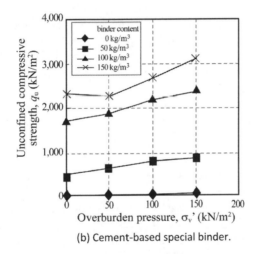

(b) Cement-based special binder.

Figure 2.44   Relationship between unconfined compressive strength, q_u and overburden pressure, σ'_v (Yamamoto *et al.*, 2002).

3.4.4   Influence of overburden pressure

Field stabilized soils are subjected to an overburden pressure due to the weight of soil during the curing period. Figure 2.44 shows the effect of the overburden pressure during the curing on the strength of the cement stabilized soil, where the Ube clay (w_L of 45.4%, w_p of 20.1% and Fc of 61.0%) was stabilized with either ordinary Portland cement or cement-based special binder (SiO_2 of 15 to 20%, Al_2O_4 of more than 4.5%, CaO of 40 to 70%, SO_4 of more than 4.0%) (Yamamoto *et al.*, 2002). Figure 2.44 shows the relationship between the unconfined compressive strength at 7 days curing with the overburden pressure, σ'_v (Yamamoto *et al.*, 2002). The figure clearly shows that the strength increases almost linearly with the overburden pressure irrespective of the type and amount of binder.

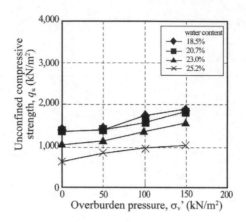

Figure 2.45 Relationship between unconfined compressive strength, q_u and overburden pressure, σ_v' (Yamamoto *et al.*, 2002).

Figure 2.45 shows a similar relationship on the stabilized sandy soil, (D_{max} of 4.8 mm, w_L of 46.6%, w_p of 22.4%, w_n of 16.4% and Fc of 18.9%). The figure shows the strength of the stabilized sandy soils having different initial water contents. The strength increases almost linearly with the overburden pressure, as similar to those on the stabilized clays.

The effect of various loading patterns of overburden pressure, loading time, loading period, stepwise loading, *etc.* were discussed in detail by Yamamoto *et al.* (2002) and Suzuki *et al.* (2005).

4 PREDICTION OF STRENGTH

In a deep mixing project, the strength of in-situ stabilized soil should be predicted and confirmed at various stages of planning, testing, design, and implementation. There are many proposed formulas to predict the laboratory strength and field strength of stabilized soil, which incorporate various factors for the improvement effect. The general formula may be written as:

$$q_{ul} = \text{function (soil type, binder, } C/W_t, O_c, F_c, T_c, \text{etc.)} \tag{2.7}$$

$$q_{uf} = \text{function}(q_{ul}, T_c, t_c, \text{mixedness, environment, machine, procedure)} \tag{2.8}$$

where
  C/W_t : ratio of the weight of the binder to that of total weight of water including mixing water
  Fc : fine grain content (may be substituted by the amount of soluble silica and alumina)
  Oc : organic matter content (may be substituted by pH or ignition loss)
  q_{uf} : unconfined compressive strength of in-situ stabilized soil (kN/m²)
  q_{ul} : unconfined compressive strength of stabilized soil manufactured in the laboratory (kN/m²)

t_c : curing period (day)

T_c : curing temperature (°C).

The formula for predicting q_{ul} is presented above by much simpler form than that for q_{uf}, because the laboratory test can be conducted according to standardized test procedures which reduce the number of factors. Many papers have proposed a simplified version of the above formula for predicting q_{ul} and compared them with laboratory test results. One of such proposals is Equation (2.4). However, we are not yet at the stage where we can predict the laboratory strength with a reasonable level of accuracy.

There is no widely applicable formula for estimating the field strength which incorporates all the relevant factors, because the strength of in-situ stabilized soil is also influenced by the mixing and curing conditions, which differ from one machine to another and according to specific site conditions. Because of this, most predictions are now made by performing the laboratory mix test and then estimating the field strength on the basis of laboratory test results and past experience. In large scale projects, laboratory test results are often confirmed by a field trial installation of a stabilized soil column at the construction site. For small scale work, reference is made to previous soil improvement work done in similar areas.

Nevertheless, the information compiled in the present chapter is extremely valuable in planning the deep mixing work and also interpreting the laboratory test results if properly used by the experienced engineer.

REFERENCES

Åhnberg, H. & Holm, G. (1984) On the influence of curing temperature on the strength of lime and cement stabilised soils. *Swedish Geotechnical Institute Report*. Vol. 30. pp. 93–146 (in Swedish).

Babasaki, R, Terashi, M., Suzuki, T., Maekawa, A., Kawamura, M. & Fukazawa, E. (1996) Japanese Geotechnical Society Technical Committee Reports: Factors influencing the strength of improved soil. *Proc. of the 2nd International Conference on Ground Improvement Geosystems*. Vol. 2. pp. 913–918.

Babasaki, R., Kawasaki, T., & Niina, A. (1980) Study of the deep mixing method using cement hardening agent (Part 9). *Proc. of the 15th Annual Conference of the Japanese Society of Soil Mechanics and Foundation Engineering*. pp. 713–716 (in Japanese).

Babasaki, R., Saito, S. & Suzuki, Y. (1984) Temperature characteristics of cement improved soil and temperature analysis of ground improved using the deep cement mixing method. *Proc. of the Symposium on strength and deformation of composite ground*. pp. 33–40 (in Japanese).

Cement Deep Mixing Method Association (1999) *Cement Deep Mixing Method (CDM), Design and Construction Manual* (in Japanese).

Coastal Development Institute of Technology (2008) *Technical Manual of Deep Mixing Method for Marine Works*. 289p. (in Japanese).

Enami, A., Yoshida, M., Hibino, S., Takahashi, M. & Akitani, K. (1985) In situ measurement of temperature in soil cement columns and influence of curing temperature on unconfined compressive strength of soil cement. *Proc. of the 20th Annual Conference of the Japanese Society of Soil Mechanics and Foundation Engineering*. pp. 1737–1740 (in Japanese).

Hayashi, H., Noto, S. & Toritani, N. (1989) Cement improvement of Hokkaido peat. *Proc. of the Symposium on High Organic Soils*. pp. 101–106 (in Japanese).

Horiuchi, N., Ito, M., Morita, T., Yoshihara, S., Hisano, T., Hanazono, H. & Tanaka, T. (1984) Strength of soil mixture under lower temperatures. *Proc. of the 19th Annual Conference of the Japanese Society of Soil Mechanics and Foundation Engineering.* pp. 1609–1610 (in Japanese).

Ingles, O.G. & Metcalf, J.B. (1972) *Soil Stabilization, Principles and Practice.* Butterworth.

Japan Cement Association (2007) *Soil Improvement Manual using Cement Stabilizer (3rd edition).* Japan Cement Association. 387p. (in Japanese).

Japan Cement Association (2009) *Committee Report on Soil Stabilization of Dredged Soil.* Internal Report of Japan Cement Association. 57p. (in Japanese).

Japan Lime Association (2009) *Technical Manual on Ground Improvement using Lime.* Japan Lime Association. 176p. (in Japanese).

Japanese Geotechnical Society (2000) *Practice for Making and Curing Stabilized Soil Specimens without Compaction. JGS 0821-2000.* Japanese Geotechnical Society (in Japanese).

Japanese Geotechnical Society (2009) *Practice for Making and Curing Stabilized Soil Specimens without Compaction. JGS 0821-2009.* Japanese Geotechnical Society. Vol. 1. pp. 426–434 (in Japanese).

Japanese Industrial Standard (2006a) *Industrial lime. JIS R 9001:2006* (in Japanese).

Japanese Industrial Standard (2006b) *Portland Blast-furnace Slag Cement, JIS R 5211: 2006* (in Japanese).

Japanese Industrial Standard (2009) *Portland Cement, JIS R 5210: 2009* (in Japanese).

Japanese Society of Soil Mechanics and Foundation Engineering (1990) *Practice for Making and Curing Stabilized Soil Specimens without Compaction. JGS T 821-1990.* Japanese Society of Soil Mechanics and Foundation Engineering (in Japanese).

Kawasaki, T., Niina, A., Saitoh, S. & Babasaki, R. (1978) Studies on engineering characteristics of cement-base stabilized soil. *Takenaka Technical Research Report.* Vol. 19. pp. 144–165 (in Japanese).

Kawasaki, T., Niina, A., Saitoh, S., Suzuki, Y. & Honjyo, Y. (1981) Deep mixing method using cement hardening agent. *Proc. of the 10th International Conference on Soil Mechanics and Foundation Engineering.* Vol. 3. pp. 721–724.

Kido, Y., Hishimoto, S., Hayashi, H. & Hashimoto, H. (2009) Effects of curing temperatures on the strength of cement-treated peat. *Proc. of the International Symposium on Deep Mixing and Admixture Stabilization.* pp. 151–154.

Kitazume, M. & Nishimura, S. (2009) Influence of specimen preparation and curing conditions on unconfined compression behaviour of cement-treated clay. *Proc. of the International Symposium on Deep Mixing and Admixture Stabilization.* pp. 155–160.

Kuwahara, S., Nishi, S., Endo, T. & Fukumitsu, K. (2000) Overlap execution in DJM method – Development of special binder for retarding the short-term stress gain. *Monthly Journal of Kisoko.* pp. 84–86 (in Japanese).

Miki, H., Kudara, K. & Okada, Y. (1984) Influence of humin acid content on ground improvement (part 2). *Proc. of the 49th Annual Conference of the Japan Society of Civil Engineers.* Vol. 4. pp. 407–408 (in Japanese).

Nakamura, M, Akutsu, H. & Sudo, F. (1980) Study of improved strength based on the deep mixing method (Report 1). *Proc. of the 15th Annual Conference of the Japanese Society of Soil Mechanics and Foundation Engineering.* pp. 1773–1776 (in Japanese).

Nakamura, M., Matsuzawa, S. & Matsushita, M. (1982) Studies on mixing efficiency of stirring wings for deep mixing method. *Proc. of the 17th Annual Conference of the Japanese Society of Soil Mechanics and Foundation Engineering.* Vol. 2. pp. 2585–2588 (in Japanese).

Niina, A., Saitoh, S., Babasaki, R., Miyata, T. & Tanaka, K. (1981) Engineering properties of improved soil obtained by stabilizing alluvial clay from various regions with cement slurry. *Takenaka Technical Research Report.* Vol. 25. pp. 1–21 (in Japanese).

Niina, A., Saitoh, S., Babasaki, R., Tsutsumi, I. & Kawasaki, T. (1977) Study on DMM using cement hardening agent (Part 1). *Proc. of the 12th Annual Conference of the Japanese Society of Soil Mechanics and Foundation Engineering*. pp. 1325–1328 (in Japanese).

Okada, Y., Kudara, K. & Miki, H. (1983) Effect of humic acid on soil stabilization. *Proc. of the 53rd Annual Conference of the Japan Society of Civil Engineering*. pp. 467–468 (in Japanese).

Okumura, T., Terashi, M., Mitsumoto, T., Yoshida, T. & Watanabe, M. (1974) Deep-lime-mixing method for soil stabilization (3rd Report). *Report of the Port and Harbour Research Institute*. Vol. 13. No. 2. pp. 3–44 (in Japanese).

Omura, T., Murata, M. & Hirai, N. (1981) Site measurement of hydration-generated temperature in ground improved by deep mixing method and effect of curing temperature on improved soil. *Proc. of the 36th Annual Conference of the Japan Society of Civil Engineering*. pp. 732–733 (in Japanese).

Saitoh, S. (1988) Experimental study of engineering properties of cement improved ground by the deep mixing method. *Doctoral thesis, Nihon University*. 317p. (in Japanese).

Saitoh, S., Niina, A. & Babasaki, R. (1980) Effect of curing temperature on the strength of treated soils and consideration on measurement of elastic modules. *Proc. of the Symposium on Testing of treated Soils, Japanese Society of Soil Mechanics and Foundation Engineering*. pp. 61–66 (in Japanese).

Saitoh, S., Suzuki, Y. & Shirai, K. (1985) Hardening of soil improved by the deep mixing method. *Proc. of the 11th International Conference on Soil Mechanics and Foundation Engineering*. Vol. 3. pp. 1745–1748.

Tanaka, H. & Tobiki, I. (1988) Properties of soils treated by the quick lime pile method. *Report of the Port and Harbour Research Institute*. Vol. 27. No. 4. pp. 201–223 (in Japanese).

Terashi, M. (1997) Theme Lecture: Deep mixing method – Brief State of the Art. *Proc. of the 14th International Conference on Soil Mechanics and Foundation Engineering*. Vol. 4. pp. 2475–2478.

Terashi, M., Fuseya, H. & Noto, S. (1983) Outline of the deep mixing method. *Proc. of the Journal of Japanese Society of Soil Mechanics and Foundation Engineering, Tsuchi To Kiso*. Vol. 31. No. 6. pp. 57–64 (in Japanese).

Terashi, M., Okumura, T. & Mitsumoto, T. (1977) Fundamental properties of lime-treated soils. *Report of the Port and Harbour Research Institute*. Vol. 16. No. 1. pp. 3–28 (in Japanese).

Terashi, M., Tanaka, H. & Okumura, T. (1979) Engineering properties of lime-treated marine soils and D.M.M. method. *Proc. of the 6th Asian Regional Conference on Soil Mechanics and Foundation Engineering*. Vol. 1. pp. 191–194.

Terashi, M., Tanaka, H., Mitsumoto, T., Niidome, Y. & Honma, S. (1980) Fundamental properties of lime and cement treated soils (2nd Report). *Report of the Port and Harbour Research Institute*. Vol. 19. No. 1. pp. 33–62 (in Japanese).

Thompson, R. (1966) Lime reactivity of Illinois soils. *Proc. of the American Society of Civil Engineering*. 92 (SM-5).

Yamamoto, T., Suzuki, M., Okabayashi, S., Fujino, H., Taguchi, T. & Fujimoto, T. (2002) Unconfined compressive strength of cement-stabilized soil cured under an overburden pressure. *Journal of Geotechnical Engineering*. pp. 387–399 (in Japanese).

Chapter 3

Engineering properties of stabilized soils

I INTRODUCTION

The engineering properties of lime or cement stabilized soils have been extensively studied by highway engineers since the 1960s. However, the purpose of their stabilization was to improve sub-base or sub-grade materials and the stabilization was characterized by the low water content of the original soil and a small amount of binders. Mixing a few percent of binder with respect to the dry weight of soil is enough to change the physical properties of soil in order to enable efficient compaction that follows the mixing.

Soils to be stabilized by the deep mixing method in Japan are very soft dredged clay, organic soil, and soft alluvial soil which usually have a water content nearly equal to or exceeding their liquid limit. Compaction of a nearly saturated soil-binder mixture is ineffective and practically impossible to carry out at depth. The purpose of stabilization is to manufacture strong stabilized columns, walls or blocks in situ and expect them to transfer the external loads to a reliable deeper stratum. Due to these differences in manufacturing process and in expected function of stabilized soils, the fundamental engineering properties of lime or cement stabilized clays and sands have been studied in detail in Japan.

Although the magnitude of strength gain by stabilization is influenced by various factors including the type of binders (Chapter 2), the engineering properties of cement stabilized soils and lime stabilized soils are quite similar. Each property of stabilized soils will be described in this Chapter without distinction of the type of binder unless noted otherwise.

The descriptions in this chapter are mostly based on the researches done in Japan or on accumulated experience on Japanese soils and machines. The soil properties introduced here are not directly applied in the other parts of the world.

2 PHYSICAL PROPERTIES

2.1 Change of water content

Water content is altered by the hydration of binder, as described in Chapter 2. The hydration of quicklime, CaO, is expressed as Equation (3.1).

$$CaO + H_2O = Ca(OH)_2 + 15.6\,Kcal/mol \qquad (3.1)$$

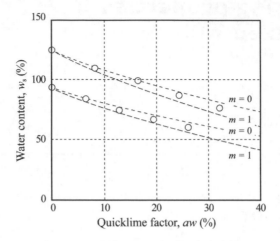

Figure 3.1 Change of water content by laboratory quicklime stabilization (Japan Lime Association, 2009).

The ratio of molecular weight of calcium oxide, CaO, water and calcium hydroxide, $Ca(OH)_2$ is 1:0.32:1.32. The water content of quicklime stabilized soil is calculated by Equation (3.2). The ratio of generated heat for evaporating water in soil, m ranges from 0 to 1: $m = 0$ for no water evaporated due to generated heat, and $m = 1$ for water evaporated due to generated heat. Its magnitude depends upon the type and conditions of the original soil.

$$w_s = \frac{w_0 - (\lambda_{w/CaO} + m \cdot \eta) \cdot aw}{100 + \lambda_{Ca(OH)_2/CaO} \cdot aw} \times 100 \qquad (3.2)$$

where,

aw	: binder factor (%)
m	: ratio of generated heat for evaporating water in soil
w_o	: water content of original soil (%)
w_s	: water content of stabilized soil (%)
$\lambda_{w/CaO}$	: weight ratio of water to CaO (0.32)
$\lambda_{Ca(OH)_2/CaO}$	: weight ratio of $Ca(OH)_2$ to CaO (1.32)
η	: amount of water evaporated due to heat by unit weight of CaO (0.478 g/g).

Figure 3.1 shows the relationship between the water content of laboratory stabilized volcanic cohesive soil with quicklime and the binder factor (Japan Lime Association, 2009). The measured water contents plotted by open circles decrease with increasing binder factor. In the figure, the estimated values by Equation (3.2) for $m = 0$ and 1 are also plotted. The measured values are plotted between the estimated values of $m = 0$ and 1. A similar phenomenon was also found by Terashi *et al.* (1977).

The water contents of in-situ stabilized soils with quicklime are shown in Figure 3.2 (Kamata and Akutsu, 1976). In the field tests, eight types of clay were stabilized with quicklime with binder factor, aw of 10 to 25%. In the figure, estimated water content derived by Equation (3.2) with m of 0 is also shown. It can be seen that the measured data almost coincide with the estimation.

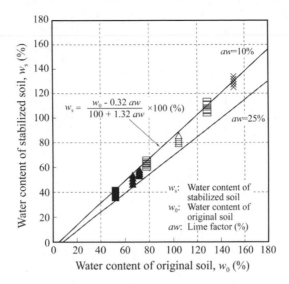

Figure 3.2 Change of water content by in-situ quicklime stabilization (after Kamata and Akutsu, 1976).

Shimomura (2001) proposed that the magnitude of m can be assumed 0 or 1 for the soil with a fine content, Fc, higher than 80% or Fc lower than 30% respectively. However, according to Figures 3.1 and 3.2, m can be assumed as 0 irrespective of the type of soil in the case of the deep mixing method.

For cement stabilization, more complicated chemical reactions take place. The water content of cement stabilized soil after cement hydration can be estimated by Equation (3.3). The required amount of water for cement hydration, λ, is dependent upon the type and composition of cement, but can be assumed about 0.25 to 0.28 of the dry weight of cement.

$$w_t = \frac{w_o + (\beta - \lambda) \cdot aw}{100 + (1 + \lambda) \cdot aw} \times 100 \qquad (3.3)$$

where
  aw : binder factor (%)
  w_o : water content of original soil (%)
  w_s : water content of stabilized soil (%)
  β : water binder ratio (%)
  λ : ratio of required water for cement hydration (0.25 to 0.28).

Figure 3.3 shows the water content of the cement stabilized soils, in which the Shinagawa clay (w_L of 62.6%, w_p of 23.1% and w_i of 76.5%) was stabilized with ordinary Portland cement with binder content, aw of 5, 10, 15 and 20% (Kawasaki *et al.*, 1978). The water content of the stabilized soil decreases gradually with the binder content. In the figure, estimated values by Equation (3.3) with λ of 0.25 and β of 0 are also plotted. The estimated values coincide with the measured values very well.

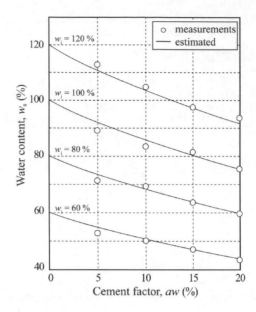

Figure 3.3 Change of water content by cement stabilization (Kawasaki *et al.*, 1978).

2.2 Change of unit weight

The saturated density of quicklime stabilized soil can be calculated by Equation (3.4), in which the volume change of lime due to hydration is considered. The increment of density by stabilization can be roughly estimated about 5, 10 and 15% for *aw* of 10, 20 and 30% respectively.

$$\rho_s = \frac{100 + w_o + aw}{\dfrac{100}{G_s} + \dfrac{w_o - \lambda_{Ca(OH)_2/CaO} \cdot aw}{G_w} + \dfrac{\lambda_{Ca(OH)_2/CaO} \cdot aw}{G_{Ca(OH)_2}}} \times \rho_w \qquad (3.4)$$

where
aw : binder factor (%)
G_s : specific gravity of soil particle
G_w : specific gravity of water
$G_{Ca(OH)_2}$: specific gravity of $Ca(OH)_2$
w_o : water content of original soil (%)
ρ_w : density of water (g/cm^3)
ρ_s : density of stabilized soil (g/cm^3).

Figure 3.4 shows the change of density due to quicklime stabilization without any compaction (Kamata and Akutsu, 1976). Although the increment of density is estimated about 10% according to Equation (3.4), the actual change of density is relatively small.

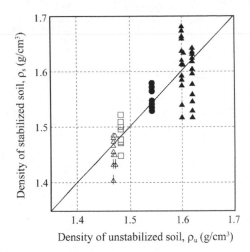

Figure 3.4 Change of density by in-situ quicklime stabilization (Kamata and Akutsu, 1976).

The saturated density of cement stabilized soil can be calculated by Equation (3.5).

$$\rho_s = \frac{100 + w_o + (1 + \beta) \cdot aw}{\dfrac{100}{G_s} + \left(\dfrac{100}{G_c} + \dfrac{100\beta}{G_w}\right) \cdot aw + \dfrac{w_o}{G_w}} \times \rho_w \qquad (3.5)$$

where
G_c : specific gravity of binder
G_s : specific gravity of soil particle
G_w : specific gravity of water
w_o : water content of original soil (%)
β : water binder ratio
ρ_s : density of stabilized soil (g/cm^3)
ρ_w : density of water (g/cm^3).

Figure 3.5 shows the density of the cement stabilized soils, in which the Kawasaki clay (w_L of 62.6% and w_p of 23.1%) was stabilized with ordinary Portland cement with cement factor, aw of 5, 10, 15 and 20% (Kawasaki *et al.*, 1978). The densities of the stabilized soils increase gradually with the cement factor. In the figure, estimated values by Equation (3.5) are also plotted. The estimated values coincide with the measured values very well.

Figure 3.6 shows the change of density of in-situ cement stabilized soil without compaction (Japan Cement Association, 2007). In the figure, the ratio of density of stabilized soil to that of the original soil is plotted against the cement content, α. The wet density increases by cement stabilization in the case of the dry method and its increment becomes larger for a larger cement content. In the case of the wet method, on the other hand, the change of density is negligibly small even if the cement factor is increased.

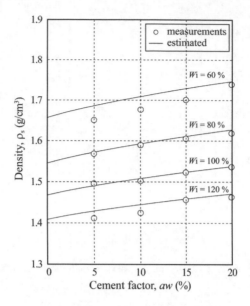

Figure 3.5 Change of density by cement stabilization (Kawasaki *et al.*, 1978).

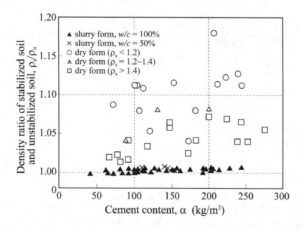

Figure 3.6 Change of density by in-situ cement stabilization (Japan Cement Association, 2007).

2.3 Change of consistency of soil-binder mixture before hardening

The water content decreases in many cases due to the hydration of quicklime and cement. At the same time, the consistency of the soil-binder mixture changes from that of the original soil due to ion exchange. Figure 3.7 shows the effect of the quicklime stabilization on the consistency of the soil-binder mixture measured at three hours after mixing (Japan Lime Association, 2009). The liquid limit, w_L, decreases with increasing quicklime content, while the plastic limit, w_p, increases. As a result, the plasticity index, I_p, sharply decreases with increasing quicklime content.

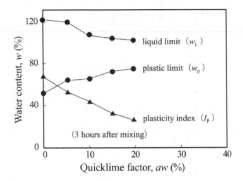

Figure 3.7 Change of consistency by quicklime stabilization (Japan Lime Association, 2009).

3 MECHANICAL PROPERTIES (STRENGTH CHARACTERISTICS)

3.1 Stress–strain curve

Figure 3.8 shows the stress–strain curves on quicklime stabilized clay which are obtained in the consolidated undrained (CU) tests (Terashi *et al.*, 1980). The Kawasaki clay (w_L of 87.8% and w_p of 39.7%) was stabilized with quicklime of 7.5% in *aw*, whose unconfined compressive strength was 1,300 kN/m². The stabilized soil was allowed to isotropically consolidate under various consolidation pressure, σ ranging from 0 to 8,100 kN/m² and was subjected to undrained compression. The test data of σ of 0 kN/m² corresponds to the unconfined compressive strength, q_u. The change of modulus of elasticity, Young's modulus, and the peak strength due to the change of consolidation pressure are negligibly small as far as the consolidation pressure remains

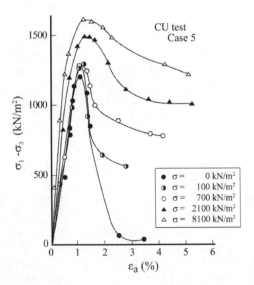

Figure 3.8 Stress and strain curves of quicklime stabilized soils (Terashi *et al.*, 1980).

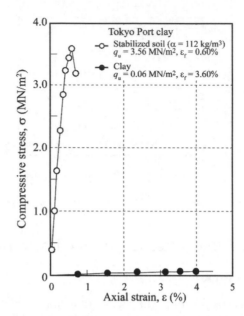

Figure 3.9 Stress–strain of in-situ cement stabilized soil (Sugiyama *et al.*, 1980).

lower than about q_u of the stabilized soil, but they increase with the consolidation pressure when it exceeds about q_u. The deviator stress, $\sigma_1 - \sigma_3$ sharply decreases after the peak in the case of unconfined compression, σ of $0\,kN/m^2$, but the reduction in the deviator stress becomes smaller with the consolidation pressure, σ.

A stress–strain curve of in-situ cement stabilized soil in an unconfined compression test is shown in Figure 3.9, in which the Tokyo Port clay (w_L of 93.1% and w_p of 35.8%) was stabilized with ordinary Portland cement with cement content, α of $112\,kg/m^3$ (Sugiyama *et al.*, 1980). In the figure, the stress-strain curve of the original clay is plotted together. The figure clearly shows that the stress–strain curve of the stabilized soil is characterized by very high strength and small axial strain at failure, while the original soil is characterized by small strength and large axial strain at failure.

Figure 3.10(a) shows the stress–strain curves on the laboratory cement stabilized clay in consolidated undrained (CU) tests on the cement stabilized clay together with the stress strain curves for unconfined compression tests. The Tokyo Bay clay (w_L of 100% and w_p of 46%) was remolded with an initial water content of 120% and then stabilized with ordinary Portland cement whose binder factor, aw, is 14%. After about 4 weeks curing, triaxial compression tests were carried out on the specimen where the consolidation pressure is changed from 0 to $686\,kN/m^2$ (about 0 to 85% of q_u) (Tatsuoka and Kobayashi, 1983). The residual strength of stabilized soil is about 20% in the case of unconfined compression. But even under small confining pressure of the order of a couple of percentages of q_u, the residual strength of stabilized soil is increased to almost 80% of the unconfined compressive strength, q_u. The elastic modulus of the stabilized soil is almost the same irrespective of the type of test condition, undrained and drained shear. In the CU test, the consolidation pressure gives negligible influence on the peak deviator stress, $\sigma_1 - \sigma_3$, but considerably influence the

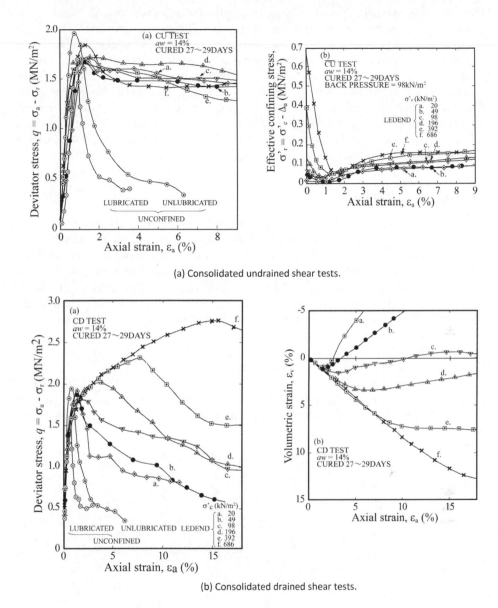

(a) Consolidated undrained shear tests.

(b) Consolidated drained shear tests.

Figure 3.10 Consolidation and shear tests on laboratory cement stabilized soils (Tatsuoka and Koayashi, 1983).

stress–strain curve after the peak. The effective confining pressure quickly decreases with the axial strain at first, which indicates that the stabilized soil shows the negative dilation phenomenon. After then, the effective confining pressure increases slightly and keeps an almost constant value for further axial strain.

In the consolidated drained (CD) test, Figure 3.10(b), the stress–strain curves show almost same phenomenon before the peak deviator stress irrespective of the magnitude of consolidation pressure, but is considerably influenced after the peak.

When the consolidation pressure is quite small in magnitude, the deviator stress, $\sigma_1 - \sigma_3$ sharply decreases to a quite small residual stress after the peak. But when the consolidation pressure increases, the deviator stress doesn't decrease sharply and the residual strength increases with increasing consolidation pressure. When the consolidation pressure exceeds the unconfined compressive strength, q_u, the deviator stress still increases after the peak stress, and shows the strain hardening phenomenon. The volumetric strain also indicates the above phenomenon, where the volumetric strain turns negative in the case of low consolidation pressure but increases continuously in the case of higher consolidation pressure. This shows that the stabilized soil behaves like heavily over-consolidated clay.

3.2 Strain at failure

As shown in Figures 3.8 and 3.9, the axial strain at failure of stabilized soil is quite small compared to that of the original soil. Figure 3.11 shows the relationship between the axial strain at failure, ε_f and the unconfined compressive strength, q_u of stabilized soils (Terashi *et al.*, 1980). In the tests, marine clay excavated at Kawasaki Port (w_L of 87.7% and w_p of 39.7%) and at Kurihama Port (w_L of 70.9% and w_p of 30.8%) were stabilized with either hydrated lime, quicklime or ordinary Portland cement in a laboratory. The soil samples were subjected to the unconfined compression test. In the unconfined compression, the magnitude of axial strain at failure, ε_f is of the order of a few percent and markedly smaller than that of unstabilized clay. The axial strain at failure decreases with the unconfined compressive strength, q_u.

In the case of confining conditions as shown in Figure 3.10(a), the magnitude of ε_f is negligibly influenced by the consolidation pressure in the undrained shear condition, but considerably influenced in the drained condition. In the case of the drained shear

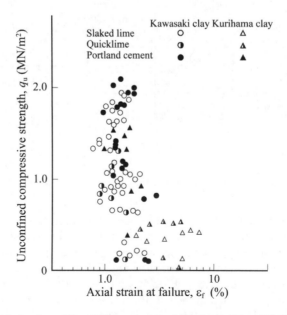

Figure 3.11 Strain at failure of laboratory stabilized soils (Terashi *et al.*, 1980).

condition, the axial strain at failure increases with the consolidation pressure when it exceeds about the unconfined compressive strength.

3.3 Modulus of elasticity (Yong's modulus)

The modulus of elasticity of stabilized soils is plotted in Figure 3.12 against the unconfined compressive strength, q_u (Terashi *et al.*, 1977). The two types of clays excavated at Honmoku Wharf (w_L of 92.3% and w_p of 46.9%) and Kawasaki Port (w_L of 87.7% and w_p of 39.7%) were stabilized with various amounts of quicklime in a laboratory and subjected to the unconfined compression test. The modulus of elasticity, E_{50} is defined by the secant modulus of elasticity in a stress–strain curve at half of the unconfined compressive strength, q_u. The magnitude of E_{50} exponentially increases with the q_u and is 75 to 200 × q_u for the Honmoku stabilized clay and 200 to 1000 × q_u for the Kawasaki stabilized clay.

A similar relationship is shown in Figure 3.13 on laboratory cement stabilized soils, where a total of 16 clays and sandy silts were stabilized with ordinary Portland cement with *aw* of 10, 20 or 30% (Niina *et al.*, 1981). The E_{50} almost linearly increases with the q_u and is 350 to 1,000 × q_u.

3.4 Residual strength

As already shown in Figures 3.8 and 3.10, the stress–strain curves and residual strength of stabilized soils are heavily influenced by the confining pressure, which show that the deviator stress, $\sigma_1 - \sigma_3$, sharply decreases after the peak in the case of unconfined compression, but the reduction in the deviator stress becomes smaller with the confining pressure, σ'_c. Figure 3.14 shows the relationship between the strength ratio of the

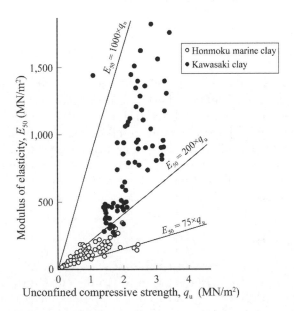

Figure 3.12 Modulus of elasticity, E_{50}, of quicklime stabilized soils stabilized in a laboratory (Terashi *et al.*, 1977).

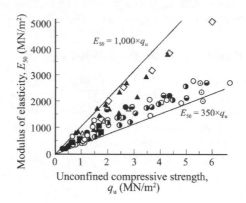

Figure 3.13 Modulus of elasticity, E_{50} of cement stabilized soils stabilized in laboratory (Niina et al., 1981).

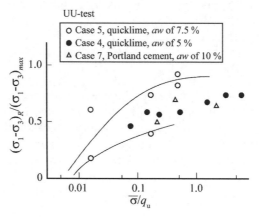

Figure 3.14 Relationship between residual strength and confining pressure (Terashi et al., 1980).

residual strength against the peak strength, $(\sigma_1 - \sigma_3)_R/(\sigma_1 - \sigma_3)_{max}$ and the confining pressure ratio (σ'/q_u) which is obtained in UU tests on quicklime and cement stabilized clays having a q_u value of 600 to 1,300 kN/m^2 (Terashi et al., 1980). The strength ratio increases with the confining pressure ratio, and the strength ratio is about 50 to 80% for the confining pressure ratio exceeding about 0.1 irrespective of the type of binder (Terashi et al., 1980). In the case of the CU test as already shown in Figure 3.10(a), the residual strength of stabilized soil is about 80% of the unconfined compressive strength, q_u even under a small confining pressure of the order of a couple percentages of the q_u (Tatsuoka and Kobayashi, 1983).

3.5 Poisson's ratio

The Poisson's ratio, μ of in-situ cement stabilized soils is shown in Figure 3.15 against the unconfined compressive strength, q_u, in which the unconfined compression tests were carried out on small scale specimens of 50 mm in diameter (Niina et al., 1977). In the tests, the Shinagawa clay (w_L of 77.9% and w_p of 32.5%) was stabilized with either quicklime, hydrated lime or cement. The Poisson's ratio was calculated by measurements of longitudinal and radial strains in the unconfined compression tests, and

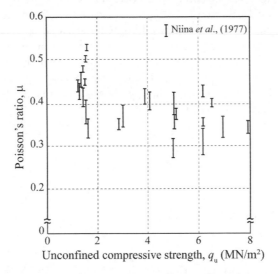

Figure 3.15 Poisson's ratio of in-situ stabilized soils (Niina *et al.*, 1977).

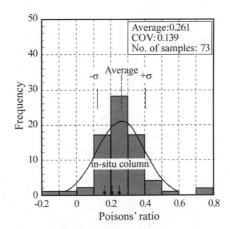

Figure 3.16 Frequency of measured Poisson's ratio of stabilized soils (The Building Center of Japan, 1997).

those for a shear stress lower than 70% of q_u are plotted in the figure. Although there is a relatively large scatter in the test data, it can be seen that the Poisson's ratio is around 0.28 to 0.45, irrespective of the unconfined compressive strength, q_u.

The Poisson's ratio on the large size stabilized sands was measured in the unconfined compression tests, whose diameter and height are about 1.0 to 1.2 m and 1.5 to 2.4 m respectively (Hirade *et al.*, 1995). In the tests, the Poisson's ratio was obtained by the measured longitudinal and radial strains, and the Poisson's ratio ranging about 0.2 to 0.3 irrespective of the strength of stabilized soil.

Figure 3.16 shows the frequency distribution of Poisson's ratio measured on various types of laboratory stabilized soils, sand, loam, silt, organic soil and Shirasu (deposits of volcanic ash and sand). The Poisson's ratio in the figure is calculated by the measured axial and volumetric strains as Equation (3.6) in the consolidated drained triaxial compression (CD) tests (The Building Center of Japan, 1997). In the figure, Poisson's ratio measured by an unconfined compression test on a full-scale stabilized soil column by the wet mixing method are also plotted by down arrows. The measured Poisson's ratio ranges between 0.19 and 0.30, and the average is 0.26. The Poisson's ratio is not so much dependent upon the type of soils, and is almost same irrespective of laboratory and field stabilized soils.

$$\mu = \frac{\varepsilon_f - \varepsilon_{vf}}{2\varepsilon_f} \tag{3.6}$$

where

ε_f : axial strain at failure (%)

ε_{vf} : volumetric strain at failure (%)

μ : Poisson's ratio.

3.6 Angle of internal friction

Figure 3.17 shows the relationship between the consolidation pressure and the undrained shear strength in the isotropically consolidated undrained shear (CIU) tests (Terashi *et al.*, 1980). In Cases 1 to 3, the Kawasaki clay (w_L of 87.7% and w_p of 39.7%) having an initial water content of about 120% was stabilized with quicklime of aw of 5, 10 and 15% respectively. In Case 6, the Kawasaki clay having an initial water content of about 200% was stabilized with ordinary Portland cement of aw of 10%. In the figure, the test data on the unstabilized soil are also plotted by open circles. The figure shows that the undrained shear strength, c_u of the stabilized soil is larger than that of the unstabilized soil, and almost constant as long as the consolidation pressure is low. But when the consolidation pressure exceeds the consolidation yield pressure (the pseudo pre-consolidation pressure), p_y the undrain shear strength

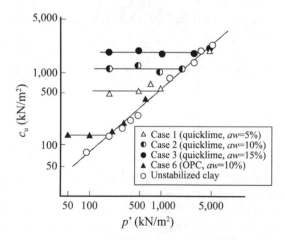

Figure 3.17 Relationship between the consolidation pressure and undrained shear strength (Terashi *et al.*, 1980).

increases with increasing consolidation pressure. The phenomenon can be seen irre-spective of type and amount of binder. The increasing ratio in the c_u of the stabilized soil is almost same as that of the unstabilized soil. According to the figure, the angle of internal friction, ϕ' of stabilized soil is almost zero as far as the consolidation pressure is lower than the consolidation yield pressure and the same as that of the unstabilized soil when the consolidation pressure is higher than the yield pressure.

3.7 Undrained shear strength

As shown in Figure 3.17, the undrained shear strength, c_u obtained by an isotropi-cally consolidated undrained compression test (CIU test) is almost constant as long as the consolidation pressure does not exceed the consolidation yield pressure (the pseudo pre-consolidation pressure), p_y. The undrained shear strength, c_u, increases with increasing consolidation pressure when the consolidation pressure exceeds the consolidation yield pressure, p_y and the increment ratio of c_u is equivalent to that of the original clays consolidated to the same stress level.

3.8 Dynamic property

Figure 3.18 shows the secant shear modulus, G_{sec} and the equivalent shear modulus, G_{eq} against the shear strain, γ and pulsating shear strain, γ_{SA} of the in-situ cement stabilized sandy soil (Shibuya et al., 1992). The slurry of cement, sandy soils and water was prepared by 1,177 kg of the Sengenyama sand (D_{50} of 0.3 mm), 80 kg of cement, 110 kg of mudstone powder and 520 kg of sea water per cubic meters. The pre-mixed slurry were casted through a tremie pipe into a huge ship-building dock filled with sea water. After about three weeks, the specimens were carefully sampled in the blocks. After four to seven months curing period, a series of monotonic and cyclic loading tests were performed under both undrained and drained conditions on the isotropically consolidated samples. The test results are shown in the figure together with the shear modulus obtained from the in-situ shear wave velocity by the cross-hole method. The term "local" and "external" mean that the measurement of strain was by the local deformation transducer (LDT) and by the conventional displacement transducer respectively. They advised to apply the former to measure

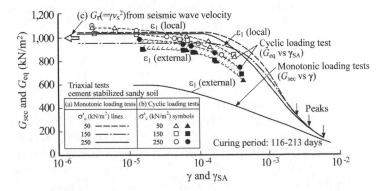

Figure 3.18 Relationship between shear moduli and shear strain (Shibuya *et al.*, 1992).

the axial strain with a high accuracy. The shear moduli measured in the three tests are practically identical for a small strain range between 10^{-6} and 10^{-5}. The G_{max} value is scarcely affected by the confining pressure and the shear stress level in the test condition.

Figure 3.19(a) shows the relationship between the equivalent shear modulus, G_{eq} and the shear strain which were measured in the dynamic triaxial tests on the stabilized soils (Enami et al., 1993). The Toyoura sand (D_{50} of about 0.1 to 0.2 mm) was stabilized with cement-based special binder whose binder content, α was 250, 300 or 350 kg/m^3. The unconfined compressive strengths of the stabilized soil were 800, 1,100 and 1,400 kN/m^2 for α of 250, 300 and 350 kg/m^3 respectively. Figure 3.19(b) shows the effect of confining pressure, σ'_c where the equivalent shear modulus, G_{eq} increases with σ'_c. Figure 3.19(c) shows the effect of the confining pressure on the initial shear modulus, G_0. In the figure, the initial shear modulus, G_0 defined as G_{eq} at γ of 10^{-6} is plotted against the normalized confining pressure, σ'_c/q_u. The G_0 increases with the confining pressure, while the binder content doesn't give a large effect on the G_0 value.

The damping ratio, h_{eq} of the stabilized sand is plotted in Figure 3.20 against the shear strain, γ (Enami et al., 1993). In the test, Toyoura sand (D_{50} of about 0.1 to 0.2 mm) was stabilized with cement-based special binder whose binder content, α was 250, 300 or 350 kg/m^3. The unconfined compressive strength of the stabilized soil were 800, 1,100 and 1,400 kN/m^2 for α of 250, 300 and 350 kg/m^3 respectively. The damping ratio increases with the shear strain as shown in Figure 3.20(a), while the relationship isn't influenced so much by the confining pressure, σ'_c. Figure 3.20(b) shows the effect of the confining pressure on the damping ratio. The h_{eq} slightly decreases with increasing confining pressure, irrespective of the binder content.

Figure 3.21 shows the relationship between the initial shear modulus, G_0 at the shear strain of 10^{-6} and the q_u of the cement stabilized clays (Tanaka and Terashi, 1986). For the laboratory stabilized soil, the clay excavated at Kawasaki Port (w_L of 88% and w_p of 44%) having an initial water content of 100 to 150% were stabilized with ordinary Portland cement with aw of 10 to 25%, and the stabilized soils were subjected to the resonant column test. For the field stabilized soil, the clay at Sakai Port (w_L of 93.3% and w_p of 27.3%) was stabilized in situ by a dual mixing shafts type deep mixing machine, where ordinary Portland cement of about 130 kg/m^3 was mixed and cured in situ. At 140 days after the execution, the stabilized soil was retrieved by coring and trimmed for the test. The G_0 almost linearly increases with q_u irrespective of the laboratory and field stabilized soils.

3.9 Creep strength

Figure 3.22 shows the relationship between the strain rate and loading period of cement stabilized clay (Terashi et al., 1983). The Kawasaki clay (w_L of 88% and w_p of 40%) having an initial water content of 150 or 200% was stabilized with ordinary Portland cement of aw of 15 or 20%. The specimen was subjected to a constant load, q_{cr} whose magnitude was changed from 0.52 to 0.91 q_u. The strain rate decreases almost linearly on the double-logarithmic graph with the time duration. The decreasing ratio is almost constant irrespective of the load intensity, q_{cr}/q_u. The figure shows that the stabilized

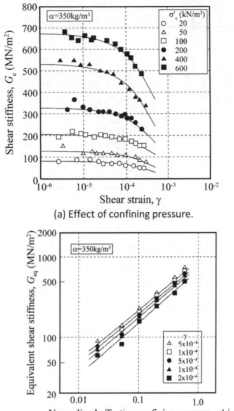

(a) Effect of confining pressure.

(b) Effect of confining pressure on equivalent shear stiffness.

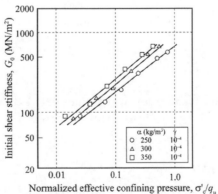

(c) Effect of confining pressure on initial shear stiffness.

Figure 3.19 Relationship between shear modulus and confining pressure (Enami *et al.*, 1993).

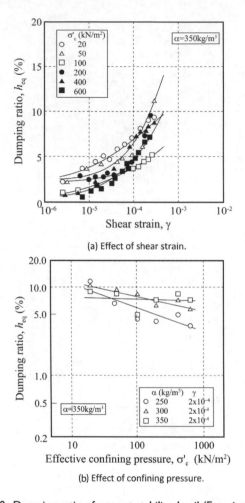

(a) Effect of shear strain.

(b) Effect of confining pressure.

Figure 3.20 Dumping ratio of cement stabilized soil (Enami *et al.*, 1993).

soil subjected to a vertical load q_{cr}/q_u higher than 0.91 exhibits creep failure, but the specimens do not fail as far as the load intensity is lower than about 0.8.

3.10 Cyclic strength

Figure 3.23(a) shows the relationship between the axial strain and the number of loading cycles, N, on the stabilized soils (Terashi *et al.*, 1983). In the tests, the Kawasaki clay having an initial water content of 200% was stabilized with ordinary Portland cement of 15% in *aw*, whose unconfined compressive strength was about 470 kN/m². The stabilized soil was subjected to the cyclic loading whose maximum and minimum pressures were 0.7 q_u and 0 kN/m² respectively. In the figure, the data plotted as ε_l and ε_r show the residual axial strain at σ_{max} loading and σ_{min} loading respectively, and the

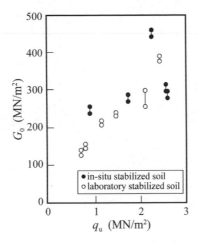

Figure 3.21 Relationship between initial shear modulus and the q_u of the cement stabilized clays (Tanaka and Terashi, 1986).

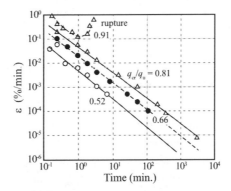

Figure 3.22 Relationship between strain rate and the loading period (Terashi *et al.*, 1983).

$\varepsilon_l - \varepsilon_r$ is also plotted. The axial strains increase gradually with the number of loading cycles, and increases to failure with the loading cycles.

Figure 3.23(b) shows the relationship between the cyclic loading pressure and number of loadings at failure, N_f, on the stabilized soils (Terashi *et al.*, 1983). In the tests, the stabilized soil was subjected to a cyclic loading whose minimum pressure was $0 \, kN/m^2$. When the σ_{max}/q_u decreases, the number of cyclic loadings at failure, log N_f, increases almost linearly.

The relationship between the $(\sigma_{max} - \sigma_{min})/q_u$ and the N_f for the case of σ_{min} being larger than $0 \, kN/m^2$ is shown in Figure 3.23(c), where the stress difference, $(\sigma_{max} - \sigma_{min})/q_u$ is plotted against N_f. In the figure, the range of test results for the σ_{min} of $0 \, kN/m^2$ as shown in Figure 3.23(a) are shown by broken lines together. The test data for σ_{min} higher than $0 \, kN/m^2$ are within those of the σ_{min} of $0 \, kN/m^2$, which reveals that the $(\sigma_{max} - \sigma_{min})/q_u$ governs the cyclic strength rather than σ_{max}.

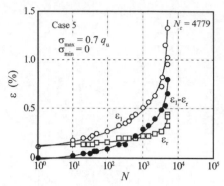

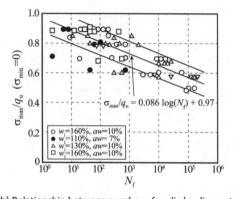

(a) Relationship between the axial strain and the number of loading cycles.

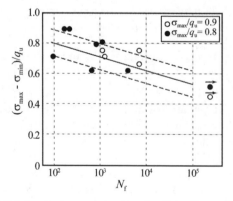

(b) Relationship between number of cyclic loading until failure
and magnitude of loading ($\sigma_{min}=0$).

(c) Relationship between number of cyclic loading until failure
and magnitude of loading ($\sigma_{min}\neq0$).

Figure 3.23 Effect of cyclic loading (Terashi *et al.*, 1983).

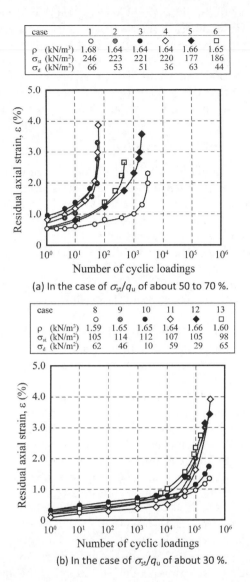

case	1	2	3	4	5	6
	○	◎	●	◇	◆	□
ρ (kN/m³)	1.68	1.64	1.64	1.64	1.66	1.65
σ_{st} (kN/m²)	246	223	221	220	177	186
σ_d (kN/m²)	66	53	51	36	63	44

(a) In the case of σ_{st}/q_u of about 50 to 70 %.

case	8	9	10	11	12	13
	○	◎	●	◇	◆	□
ρ (kN/m³)	1.59	1.65	1.65	1.64	1.66	1.60
σ_{st} (kN/m²)	105	114	112	107	105	98
σ_d (kN/m²)	62	46	10	59	29	65

(b) In the case of σ_{st}/q_u of about 30 %.

Figure 3.24 Relationship between residual axial strain and number of cyclic loading (Kudo *et al.*, 1993).

Figure 3.24 shows the relationship between the residual axial strain and the number of cyclic loadings on the stabilized clay (Kudo *et al.*, 1993). The alluvial clay (w_L of 78.8% and w_p of 34.0%) having an initial water content of 60% was stabilized with a cement-based special binder of 6% in *aw*. The unconfined compressive strength at 28 days curing was 355 kN/m² in average. The stabilized soil was subjected to an unconfined compressive stress with various magnitudes of the initial vertical stress, σ_{st} and then subjected to cyclic loading with various magnitudes of the half-amplitude, σ_d. As shown in Figure 3.24(a) for the case of the relatively large initial stress level, σ_{st}/q_u of about 50 to 70%, the residual axial strain accumulates gradually as far as the number of cyclic loadings is small, but it accumulates very rapidly to fail for further increase of the number of cyclic loadings. In the case of a relatively small initial vertical stress level, σ_{st}/q_u of about 30% as shown in Figure 3.24(b), the residual strain accumulates gradually with the number of cyclic loadings as far as the number of cyclic loadings is smaller than about 100,000, but accumulates very quickly to failure for further loadings. The figure shows that the number of cyclic loadings at failure is influenced by the initial stress level, σ_{st} and/or the maximum axial stress, $\sigma_{st} + \sigma_d$.

3.11 Tensile and bending strengths

The tensile strength of stabilized soil is evaluated by various tests: split tension test (Brazilian tension test, indirect tension test), simple tension test and bending test. In the split tension test, a disc of the stabilized soil is loaded across a diameter, and the tensile strength is calculated by the compressive load at failure. In the simple tension test, a cylindrical specimen is subjected to direct tensile force. In the bending test, a rectangular shape beam of stabilized soil is bent by load, and the tensile strength (bending strength) is calculated by the tensile stress induced at the bottom surface of the specimen. Here the strengths measured by the three tests are expressed by σ_{ts} (by split tension test), σ_{td} (by simple tension test) and σ_{tb} (by bending test) respectively.

The tensile strength of the stabilized soil was evaluated by the split tension tests and bending tests (Terashi *et al.*, 1980). In the tests, the Kawasaki clay (w_L of 87.8% and w_p of 39.7%) having different initial water contents, w_i was stabilized with either quicklime or ordinary Portland cement to form a disc shape specimen of 100 mm in diameter and 50 mm in height for the former test and a beam with rectangular cross-section of 50 mm in width, 50 mm in height and 250 mm in length for the latter test. The unconfined compressive strength, q_u, was also measured on the reference column shape specimen of 50 mm in diameter and 100 mm in height.

Figure 3.25(a) shows the relationship between the tensile strength, σ_{ts} and the unconfined compressive strength, q_u. The figure shows the tensile strength, σ_{ts} increases almost linearly with unconfined compressive strength, q_u irrespective of the type, amount of binder and initial water content of the soil, but its increment becomes lower with increasing q_u. The tensile strength is about 0.15 of the unconfined compressive strength, q_u. Figure 3.25(b) shows the relationship between the tensile strength measured by the bending test, σ_{tb} and the unconfined compressive strength, q_u. The figure shows the bending strength is around 0.1 to 0.6 of the unconfined compressive strength irrespective of the type of binder and the initial water content of original soil.

The tensile strengths of the cement stabilized soils was obtained from three types of test: split tension test, σ_{ts}, direct tension test, σ_{td}, and bending test, σ_{tb} (Namikawa and Koseki, 2007). In the test, the Toyoura sand (D_{50} of about 0.1 to 0.2 mm) was stabilized with ordinary Portland cement and bentonite for the target strength of 1,800 kN/m². Figure 3.26 compares the tensile strengths measured by different tests and the unconfined compressive strength, in which their test results are referred as "Experiment in this study." They carried out the FEM analyses to simulate the loading tests and present the results referred as "Simulation in this study." Figure 3.26(a) shows the relationship between the tensile strength, σ_{ts} and the unconfined compressive strength, q_u. The tensile strength linearly increases with the q_u. The tensile strength measured by the

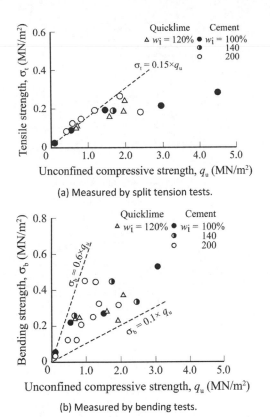

(a) Measured by split tension tests.

(b) Measured by bending tests.

Figure 3.25 Tensile strength of laboratory stabilized soils (Terashi *et al.*, 1980).

direct tension test, σ_{td} as shown in Figure 3.26(b) has a large scatter but also shows the linear relation between the σ_{td} and the q_u. The tensile strength measured by the bending test, σ_{tb}, also shows the linear increase with increasing q_u. The strength ratio of the tensile strength and q_u can be obtained, σ_{ts}/q_u of 0.08 to 0.30, σ_{td}/q_u of 0.07 to 0.20, and σ_{tb}/q_u of 0.15 to 0.51.

Figure 3.27 shows the relationship between the tensile strength and the q_u of laboratory prepared cement stabilized soils and in-situ cement stabilized soils (Saitoh *et al.*, 1996). Figure 3.27(a) shows the relationship between the strength ratio, σ_{ts}/q_u, σ_{td}/q_u, and the q_u of the laboratory stabilized soils. The tensile strength measured by the split tension test, σ_{ts} gives an almost constant value of about 0.1 irrespective of the unconfined compressive strength and the type of soil. The tensile strength by the simple tension test, σ_{td} is larger than the σ_{ts} and is highly influenced by the q_u, in which the strength ratio, σ_{td}/q_u decreases almost linearly with the q_u irrespective of the type of soil. Figure 3.27(b) shows the relationship between the strength ratio, σ_{ts}/q_u and the water content of in-situ stabilized soils. In the figure, the test results on the stabilized clays and stabilized sands are plotted together. There is a large scatter in the data, but the strength ratio is in the approximate range of 0.06 to 0.2 irrespective of the type of soil.

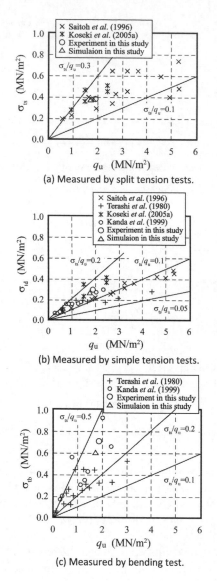

Figure 3.26 Tensile strength of laboratory stabilized soils (Namikawa and Koseki, 2007).

3.12 Long term strength

The deep mixing method is adopted in hundreds of projects annually in Japan alone. On each project a laboratory mix test is carried out to determine the strength increase with time. After the construction the strength increase is confirmed by verification testing at the actual construction site. Numerous data, however, are based on samples aged less than a month or two. Long term strength in years or decades has been studied by a limited number of research groups.

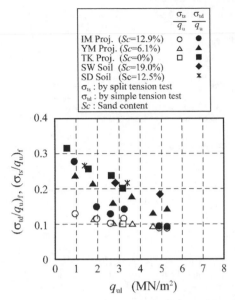

(a) Tensile strength ratios of laboratory stabilized soils.

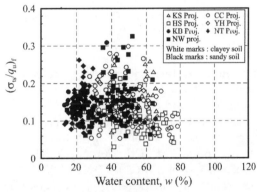

(b) Relationship between tensile strength and water content.

Figure 3.27 Tensile strength ratio (Saitoh *et al.*, 1996).

There are two aspects when the long term strength of stabilized soil is concerned. One is the strength increase with time at the core portion of the stabilized soil column which is negligibly influenced by the surrounding conditions and the other is the possible strength decrease with time due to deterioration in the periphery of the stabilized soil column, as shown in Figure 3.28.

3.12.1 Strength increase

Figures 3.29 is an example of strength increase with time confirmed by a laboratory study (Coastal Development Institute of Technology, 2008). The figure shows

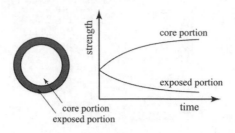

Figure 3.28 Image of long term strength of stabilized soil.

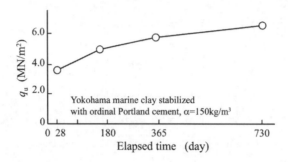

Figure 3.29 Relationship between q_u and elapsed time (Coastal Development Institute of Terchnology, 2008).

the relationship between the unconfined compressive strength, q_u and the elapsed time on laboratory manufactured stabilized soil, in which the Yokohama marine clay was stabilized with ordinary Portland cement of α of 150 kg/m³. Laboratory specimens, prepared in accordance with the Japanese Geotechnical Society standard, were wrapped in high polymer film to avoid contact with the environment and prevent a change of water content, and stored in the humid chamber until testing, which correspond to the core portion shown earlier in Figure 3.28.

Figure 3.30 shows test results on the relationship, in which the influence of the type and amount of cement on unconfined compressive strength were investigated (Saitoh, 1988). In the tests, two marine clays, the Yokohama Port clay (w_L of 95.4% and w_p of 32.3%) and the Osaka Port clay (w_L of 79.3% and w_p of 30.2%) were stabilized with either ordinary Portland cement or blast furnace slag cement type B. The strengths of stabilized soils increase with the elapsed time irrespective of the type of soil and the type and amount of binder, while a larger strength increment with elapsed time is found in the blast furnace slag cement type B rather than ordinary Portland cement. A similar phenomenon has been obtained by Kitazume et al. (2003).

Long term strength increase has been also studied on the in-situ stabilized soils (Niina et al., 1981; Terashi and Kitazume, 1992; Niigaki et al., 2001; Hayashi et al., 2001; Ikegami et al., 2002a, 2002b, 2005; Kitazume and Takahashi, 2009). In these studies the test specimens are retrieved from the in-situ stabilized soil column by core boring and subjected to an unconfined compression test in a laboratory. Figures 3.31 shows the relationships between the unconfined compressive strength, q_u and the

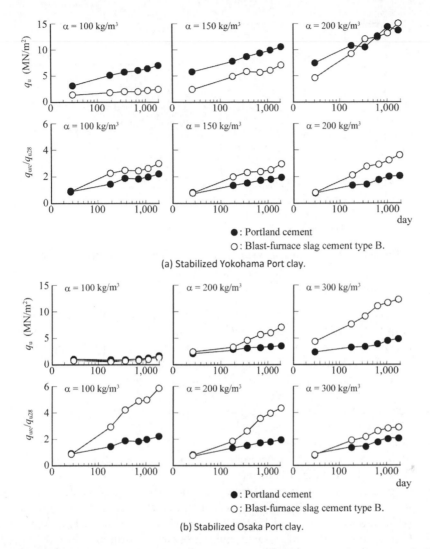

**: Portland cement
○: Blast-furnace slag cement type B.

(a) Stabilized Yokohama Port clay.

**: Portland cement
○: Blast-furnace slag cement type B.

(b) Stabilized Osaka Port clay.

Figure 3.30 Influence of type and amount of cement on unconfined compressive strength (Saitoh, 1988).

elapsed time, in which several types of soil were stabilized with various types and amounts of binder. The age of stabilized soil varies from 3 to 20 years. The strength of stabilized soil is highly dependent upon the type of soil, and the type and amount of binder. However, the strength of stabilized soil increases almost linearly with the logarithm of elapsed time irrespective of the type of soil, and the type and amount of binder.

According to the accumulated data, it can be concluded that the strength of stabilized soil at the core part increases almost linearly with the logarithm of elapsed time, irrespective of laboratory prepared/in-situ stabilized soil, and the soil type and the type and amount of binder.

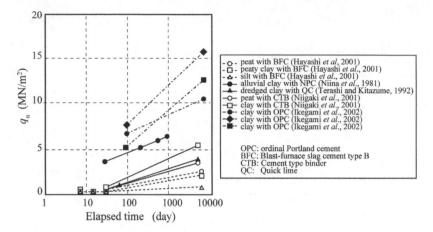

Figure 3.31 Relationship between q_u and elapsed time (in-situ stabilized soils).

3.12.2 Strength decrease

The possibility of strength decrease at the periphery of in-situ stabilized soil was studied by Terashi *et al.* (1983) and Saitoh (1988) based on laboratory tests. In these studies, laboratory mixed specimens are subjected to different exposure conditions such as direct contact with seawater or tap water, contact with saturated clay and compared with specimens wrapped with sealant. The findings in these studies are that the deterioration (strength reduction) starts at the outer surface first and progress inward, the depth of deterioration from the surface (or rate of progress) differs with different exposure conditions, the deterioration is a slow process and that the leaching of Ca^{2+} from the stabilized soil may be one of the reasons for the strength decrease. These initial studies also emphasized the importance of long term observation of actual stabilized soil columns in the real life environment.

Two separate research projects focusing upon the long term strength of in-situ stabilized soils were started in 2001 which included the detailed investigation of the periphery of stabilized soils aged 17 and 20 years (Hayashi *et al.*, 2003; Ikegami *et al.*, 2005). A series of experiments to determine the deterioration on laboratory prepared samples (Kitazume *et al.*, 2003) and efforts to numerically simulate the ion migration from the periphery of a stabilized soil were also conducted (Nishida *et al.*, 2003).

3.12.2.1 Strength distribution

Figure 3.32 shows the strength profile of laboratory prepared cement stabilized soil along the distance from exposure surface (Kitazume *et al.*, 2003). In the test, the Kawasaki clay (w_L of 83.4% and w_p of 38.6%) having an initial water content of 160% was stabilized with ordinary Portland cement of cement factor, aw of 30%. After two weeks curing under the 20°C and 95% relative humidity condition, one surface of the specimen was exposed to either tap water, seawater or clay. A specimen wrapped with sealant was also prepared and cured for reference. At the prescribed time, the strength profile of the stabilized soil was measured by the needle penetration

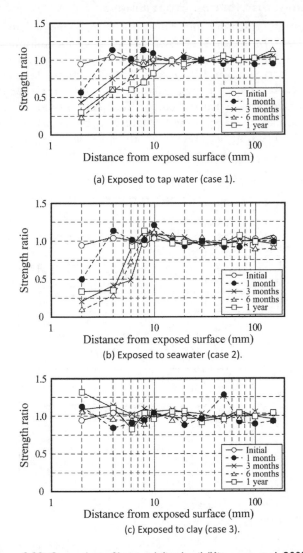

(a) Exposed to tap water (case 1).

(b) Exposed to seawater (case 2).

(c) Exposed to clay (case 3).

Figure 3.32 Strength profile in stabilized soil (Kitazume *et al.*, 2003).

test. In the figure, the strength ratio defined as the ratio of strength at each measuring point to that at the non-deteriorated portion is shown.

In the case of exposure to tap water (Figure 3.32(a)), the initial strength distribution is almost constant within the specimen. But the strength of the soil close to the exposure surface decreases very rapidly and the deterioration progresses gradually inward with time. A similar phenomenon can be seen in the case of exposure to seawater (Figure 3.32(b)). However, in the case of exposure to clay (Figure 3.32(c)), there is negligible strength decrease in the specimen even after twelve months exposure.

3.12.2.2   Calcium distribution in specimens

Figure 3.33 shows the calcium content distribution in terms of CaO for all test cases (Kitazume *et al.*, 2003). In case 1, exposed to tap water, the amount of calcium oxide is almost constant at the non-deteriorated portion (core portion) irrespective of the elapsed time, but decreases gradually toward the exposure surface expect at the immediate vicinity of the exposure surface. A similar phenomenon can be seen in case 2, exposed to seawater. These distributions of the calcium oxide measured in cases 1 and 2 are quite similar in shape to the strength distribution as already shown in Figures 3.32(a) and 3.32(b). In case 3, exposed to clay (Figure 3.33(c)), the distribution of calcium oxide is almost constant in the portion close to the exposure surface.

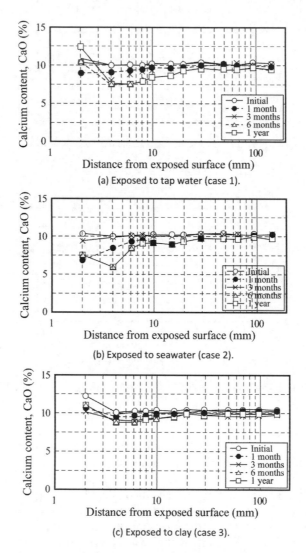

(a) Exposed to tap water (case 1).

(b) Exposed to seawater (case 2).

(c) Exposed to clay (case 3).

Figure 3.33 Distribution of calcium content in stabilized soil (Kitazume *et al.*, 2003).

It is also generally known in concrete engineering that the dissolution of the calcium ion of concrete is one of the major causes of strength decrease. At the periphery of stabilized soil, which is affected by the surrounding conditions (exposure conditions), the calcium ion, Ca^{2+} dissolute gradually from the stabilized soil. The dissolution speed of Ca^{2+} is highly dependent upon the exposure condition of stabilized soil. The extent of the deteriorated portion is anticipated to become large with elapsed time. However, as far as focusing upon cases 1 and 2, exposed to tap water or seawater, it can be concluded that the dissolution of calcium is one of the major causes of strength decrease for cement stabilized soil.

In order to examine the deterioration of in-situ stabilized soil, the stabilized foundation ground at the T2 berth of Daikoku Pier, Yokohama Port was investigated in detail (Ikegami et al., 2002a, 2002b, 2005). The original ground at the T2 berth is a thick alluvial clay from the sea bottom at $-12\,m$ down to $-50\,m$ underlain by a diluvial clay and the bed rock appears at $-70\,m$. The ground was improved by the wet method of deep mixing to a depth of $-49\,m$ as a massive block type column installation pattern. The alluvial soil layer is further divided into three layers, upper, intermediate and lower layers based on physical properties such as grain size distribution and water content. The binder was ordinary Portland cement and the binder content adopted in production was $180\,kg/m^3$ with a water to cement ratio, W/C of 60% throughout the improvement depth irrespective of the three layers mentioned above. In 2001, after 20 years form construction, undisturbed stabilized soil in contact with unstabilized soil at the side surface of the massive stabilized soil block were retrieved by core borings inclined 45 degrees from vertical.

Figure 3.34 shows the strength and calcium content distribution in the cement stabilized soil of the upper layer after 20 years curing in the ground (Ikegami et al., 2002a, 2002b). The horizontal axis of the figure is the horizontal distance from the exposure surface in logarithmic scale. The strength in terms of unconfined compressive strength shown in the upper half of the figure was estimated based on the needle penetration test. The calcium content shown in the lower half of the figure was measured on sliced core samples by means of atomic adsorption spectrometry. The overall pattern of strength and Ca content distributions are in good agreement each other except for the large Ca content found at 5 to 10 mm from the exposure surface. A similar pattern of Ca content distributions is also found in the laboratory exposure test in Figure 3.33. The depth of deterioration in 20 years in this case is 30 to 50 mm. "*Average of inside (1981)*" is the average of Ca contents measured 20 years ago.

In the upper half of the figure, two levels of strength are shown. "*Average of inside (2001)*" is $10200\,kN/m^2$, which is the average unconfined compressive strength of the upper layer measured on core samples that are sufficiently far from the exposure surface. The average unconfined compressive strength of the same layer at 93 days after production was $5785\,kN/m^2$. "*Design strength*" is $2256\,kN/m^2$ for this ground improvement project. While the deterioration progressed at the periphery to 30 or 50 mm, the strength inside the column shows 2.1 times increase.

Figure 3.34(b) shows the calcium content distribution across the exposure surface between stabilized soil and the original soil (Ikegami et al., 2005). The calcium content in the stabilized soil decreases toward the exposure surface and that in the original soil increases toward the exposure surface. The overall pattern of

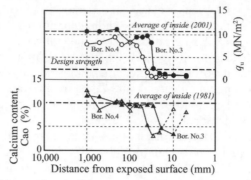

(a) Strength and calcium content distribution in stabilized soil.

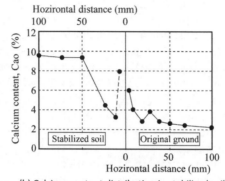

(b) Calcium content distribution in stabilized soil.

Figure 3.34 Long term strength and calcium content in stabilized soil (Ikegami *et al.*, 2002a, 2002b, 2005).

calcium content suggests that the calcium leaching from the stabilized soil to the unstabilized soil is the dominating phenomenon which caused the deterioration at the periphery.

3.12.2.3   Depth of deterioration

Figure 3.35 compares the depth of deterioration and time (Ikegami *et al.*, 2002a, 2002b). In the figure, the field data at Daikoku Pier and the results of the laboratory exposure tests by Terashi *et al.* (1983), Saitoh (1988), Kitazume *et al.* (2003) and Hayashi *et al.* (2004) are plotted together. The strengths, q_{u28} shown as references are the unconfined compressive strength of the stabilized soil specimen after 28 days curing under the sealed condition. The progress of the deterioration depth in logarithmic scale is almost linear to logarithmic time, and the slopes in all the test cases are about 1/2 irrespective of the strength of specimens and the exposure conditions, that means the rate of deterioration was proportional to the square root of time. The same relation between the depth of deterioration and time was also obtained by a numerical simulation proposed by Nishida *et al.* (2003) that assumed ions migration primarily based on

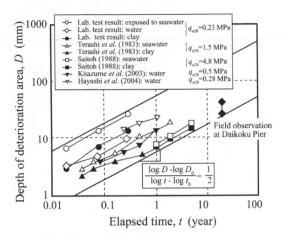

Figure 3.35 Estimation of extension of deterioration with elapsed time (Ikegami *et al.*, 2005).

the diffusion by Ca concentration gradient. Judging from the results of laboratory tests and the numerical analysis, it may be possible to predict long-term deterioration by extrapolation of the short-term result of the exposure test assuming the deterioration progress is in proportion to the square root of time. The general tendency found in Figure 3.35 is that the larger the strength the smaller the depth of deterioration and seawater exposure gives rise to larger depth of deterioration compared to tap water exposure or contact to original soil.

4 MECHANICAL PROPERTIES (CONSOLIDATION CHARACTERISTICS)

4.1 Void ratio – consolidation pressure curve

Figure 3.36 shows $e - \log p$ curves of the laboratory manufactured cement stabilized soils, in which the Tokyo Port clay (w_L of 93.1% and w_p of 35.8%) was stabilized with ordinary Portland cement with two different cement contents, α of 70 and 100 kg/m³ and cured 180 days (Kawasaki *et al.*, 1978). In the laboratory tests, the stabilized soil samples with 20 mm in height and 60 mm in diameter were consolidated one dimensionally up to 12.8 MN/m². The figure shows a sharp bend in the curve. The consolidation pressure at the sharp bend is higher for the larger binder content.

Figure 3.37 shows the $e - \log p$ curves of the laboratory stabilized soils, in which the Kawasaki clay (w_L of 64.8% and w_p of 25.2%) having different initial water contents (about 105% for cases 1–3 to 1–5, and about 140% for cases 1–6 and 1–7) were stabilized with ordinary Portland cement (Takahashi and Kitazume, 2004). This figure also shows a bend in the curve when the amount of binder increases.

These figures show that the shape of $e - \log p$ curves of the stabilized soils are similar to ordinary clay samples, which is characterized by a sharp bend at a

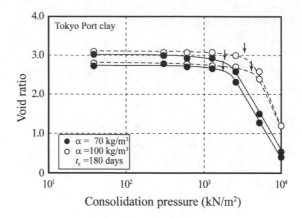

Figure 3.36 e − log p curve (Kawasaki *et al.*, 1978).

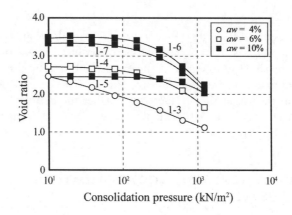

Figure 3.37 e − log p curves of cement stabilized soil (Takahashi and Kitazume, 2004).

pre-consolidation pressure. As the stabilized soil isn't subjected to pre-consolidation pressure, the consolidation pressure at the sharp bend should better be called a consolidation yield pressure, p_y.

4.2 Consolidation yield pressure

The consolidation yield pressure of stabilized soil is closely related to its unconfined compressive strength. Figure 3.38 shows the relationship between the consolidation yield pressure, p_y and the unconfined compressive strength, q_u of the Kawasaki clay (w_L of 87.7% and w_p of 39.7%) and the Kurihama clay (w_L of 70.9% and w_p of 30.8%) stabilized with three different types of binder (Terashi *et al.*, 1980). The figure shows that the consolidation yield pressure, p_y has a linear relationship with the unconfined compressive strength, q_u. The ratio of p_y/q_u of the stabilized soils is between 1.27 and

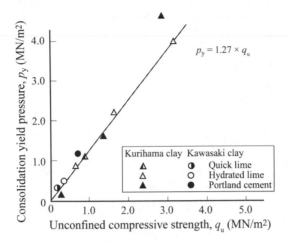

Figure 3.38 Consolidation yield pressure – unconfined compressive strength of laboratory stabilized soils (Terashi *et al.*, 1980).

1.55 for the unconfined compressive strength up to 3 MN/m², irrespective of the type of original soil, and the type of binder.

Figure 3.39(a) shows the relationship between the consolidation yield pressure and the unconfined compressive strength of the Kawasaki clay (w_L of 64.8% and w_p of 25.2%) stabilized with Japanese cement (Takahashi and Kitazume, 2004). In the figure, the test results of the two Finnish clays stabilized with Finnish cement, the Arabianranta clay (w_L of 158.0% and w_p of 24.0%) and the Fallkulla clay (w_L of 67.0% and w_p of 23.2.0%), are also plotted. The figure shows that the relationship between the p_y and the q_u was almost linear irrespective of the type of soil and the type of binder. The ratio of p_y/q_u of the stabilized soils is between 1.27 and 2.0 for the unconfined compressive strength up to 600 kN/m² irrespective of the type of original soil. And the mean value obtained in this study is 1.55.

Figure 3.39(b) plots the relationships in Figures 3.38 and 3.39(a) together. As the stress level of Figure 3.39(a) is quite smaller than that in Figure 3.38, it can be concluded that the ratio of p_y/q_u of the stabilized soils is about 1.3 irrespective of the types of soil and binder.

4.3   Coefficient of consolidation and coefficient of volume compressibility

Figure 3.40 shows the relationship between the coefficient of consolidation of the stabilized clays, c_{vs} and the consolidation pressure, p (Terashi *et al.*, 1980). The coefficient of consolidation of the stabilized soil, c_{vs} is normalized by that of the unstabilized soil, c_{vu} under the same consolidation pressure. The consolidation pressure, p is normalized by the consolidation yield pressure of the stabilized soil, p_y. In the tests, a series of one dimensional consolidation tests on the stabilized and unstabilized soils

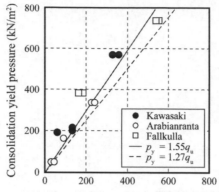

(a) Consolidation yield pressure and unconfined compressive strength
(Takahashi and Kitazume, 2004).

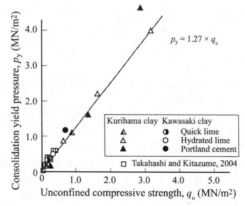

(b) Consolidation yield pressure and unconfined compressive strength
(Terashi *et al.*, 1980; Takahashi and Kitazume, 2004).

Figure 3.39 Consolidation yield pressure and unconfined compressive strength.

were carried out under a wide range of consolidation pressure. Two marine clays, the Kawasaki clay (w_L of 87.7% and w_p of 39.7%) and the Kurihama clay (w_L of 70.9% and w_p of 30.8%), were stabilized with hydrated lime, quicklime or ordinary Portland cement. The size of the soil samples are 20 mm in thickness and 60 mm in diameter. The figure shows the ratio of c_{vs}/c_{vu} is 10 to 100 as long as the normalized consolidation pressure, p/p_y is around 0.1, in a sort of overconsolidated condition, but the c_{vs}/c_{vu} approaches to unity when the p/p_y exceeds 1, in a sort of normally consolidated condition.

Figure 3.41 shows the relationship between the coefficient of volume compressibility of the stabilized soils, m_{vs} and the consolidation pressure, p as a similar manner to Figure 3.40 (Terashi *et al.*, 1980). The figure shows the ratio of m_{vs}/m_{vu} is 0.01

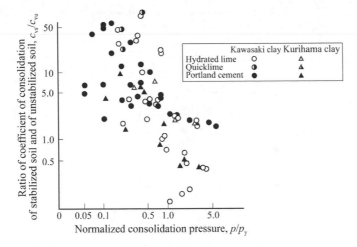

Figure 3.40 Relationship between coefficient of consolidation and consolidation pressure on laboratory stabilized soils (Terashi *et al.*, 1980).

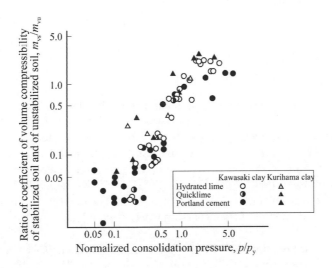

Figure 3.41 Relationship between coefficient of volume compressibility and consolidation pressure on laboratory stabilized soils (Terashi *et al.*, 1980).

to 0.1 as long as the normalized consolidation pressure, p/p_y is around 0.1, but the m_{vs}/m_{vu} approaches to unity when the p/p_y exceeds 1.

These figures indicate that the rate of consolidation of the stabilized soil increases and the compressibility of the soil decreases by lime and cement stabilizations as long as the consolidation pressure is lower than the consolidation yield pressure.

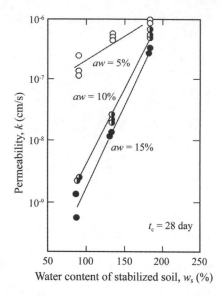

Figure 3.42 Relationship between coefficient of permeability and water content of cement stabilized soils (Terashi *et al.*, 1983).

4.4 Coefficient of permeability

4.4.1 Permeability of stabilized clay

Figure 3.42 shows the coefficient of permeability of the stabilized Kawasaki clay (w_L of 87.7% and w_p of 39.7%) with ordinary Portland cement of 5, 10 and 15% in *aw*, in which the coefficient of permeability is plotted against the water content of the stabilized soils (Terashi *et al.*, 1983). In the tests, the stabilized soil specimen, 20 mm in height and 50 mm in thickness, were subjected to the constant head permeability tests. The figure shows that the coefficient of permeability is dependent upon the water content of stabilized soil and the amount of cement. The coefficient of permeability of the stabilized soil decreases with decreasing water content and with increasing amount of cement.

Figure 3.43 shows the relationship between the coefficient of permeability and the strength of laboratory stabilized soil (Terashi *et al.*, 1983). The coefficient of permeability of the stabilized soil decreases exponentially with increasing strength, q_u.

Figure 3.44(a) shows the relationship between the coefficient of permeability and the void ratio, e, which were obtained by the oedometer tests (Takahashi and Kitazume, 2004). In the tests, the Kawasaki clay (w_L of 64.8% and w_p of 25.2%) having different initial water contents (about 105% for cases 1–3 to 1–5, and about 140% for cases 1–6 and 1–7) were stabilized with ordinary Portland cement. After four weeks curing, the oedometer tests on the soil were carried out. As the accuracy of the coefficient of permeability in the "over-consolidated state" (at the consolidation pressure lower than the consolidation yield pressure) is not high due to the quite small degree of settlement and rapid consolidation process, the coefficients of permeability obtained in the

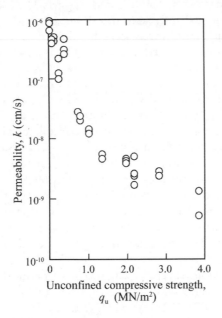

Figure 3.43 Relationship between coefficient of permeability and unconfined compressive strength of stabilized soil (Terashi *et al.*, 1983).

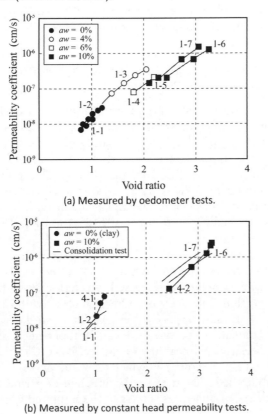

(a) Measured by oedometer tests.

(b) Measured by constant head permeability tests.

Figure 3.44 Coefficient of permeability of stabilized soil (Takahashi and Kitazume, 2004).

"normally consolidated state" are plotted in the figure. The coefficient of permeability on the single logarithmic scale increases almost linearly with increasing void ratio. As shown in the figure, the rate of increase with void ratio is almost the same both for stabilized and unstabilized soils, irrespective of the amount of cement and the initial water content of the original soil.

The solid circles and squares in Figure 3.44(b) show the coefficients of permeability of the original soil ($aw = 0\%$) and the stabilized soils which were measured in the constant head permeability test (Takahashi and Kitazume, 2004). The lines in the figure indicate the test results obtained in the consolidation test, which are shown earlier in Figure 3.44(a). In the permeability tests, the Kawasaki clay having an initial water content of 135% was stabilized with ordinary Portland cement of 10% in aw to form a cylindrical shape specimen of 50 mm in diameter and 100 mm in height. After 26 days curing, the permeability tests on the soils were carried out in a triaxial cell by changing the cell pressure. The measured coefficient of permeability in the constant head permeability test have a linear relationship against the void ratio on the single logarithmic scale graph. The characteristics of the coefficient of permeability in the two tests were similar.

A similar phenomenon where the coefficient of permeability in logarithm scale increases with the void ratio was obtained in the lime and cement stabilized clays (w_L of 133.0% and w_p of 71.4%) (Onitsuka et al., 2003).

From accumulated test data on Japanese clays, it is known that the coefficient of permeability of stabilized soil is equivalent to or lower than that of the unstabilized soils and whose magnitude is of the order of 10^{-9} to 10^{-6} cm/sec (Figures 3.42 to 3.44). Therefore in Japan the stabilized soil is not expected to function as a drainage layer in the current design.

4.4.2 Influence of grain size distribution on the coefficient of permeability of stabilized soil

Figure 3.45 show the influence of grain size distribution on the coefficient of permeability of stabilized soil (Miura et al., 2004). In the tests, five kinds of soil were prepared for the permeability tests, which include the sand excavated in Chiba Prefecture, the cohesive soil excavated at Yokohama Bay, and the mixtures of the sand and the cohesive soil (Chiba sand content of 39.3, 60.0 and 78.6%). The grain size distributions of the soils are shown in Figure 3.45(a). Each soil was stabilized with ordinary Portland cement, whose water to cement ratio, W/C was a constant of 60%. The amount of cement slurry was changed for the test cases, from 100 to 250 kg/m^3. After 28 days curing, a series of permeability tests was carried out on the specimen in a triaxial cell in which an isotropic cell pressure of 137 kN/m^2 was applied.

Figure 3.45(b) shows the relationship between the amount of cement slurry and the coefficient of permeability of the stabilized soil. The figure shows that the coefficient of permeability in logarithm scale decreases almost linearly with the amount of cement slurry irrespective of the soil type. The permeability also decreases with increasing fine grain fraction content irrespective of the amount of cement slurry. On the figure, the measured coefficient of permeability of the stabilized Chiba sand prepared in an unsaturated condition is also plotted, in which the water percolation wasn't performed before the permeability test. The measured coefficient of permeability also decreases with increasing amount of cement as similar to the saturated Chiba sand.

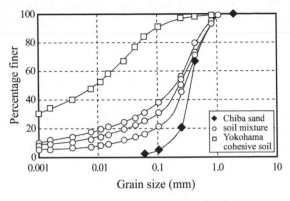

(a) Grain size distributions of soils.

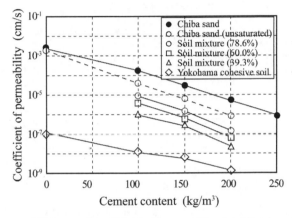

(b) Measured coefficient of permeability of stabilized sand.

Figure 3.45 Coefficient of permeability on stabilized sand (Miura *et al.*, 2004).

5 ENVIRONMENTAL PROPERTIES

5.1 Elution of contaminant

The Soil Contamination Countermeasures Act (Environment Agency, 1975, 2005) was enforced by the Ministry of Environment of the Japanese government in 2005, in order to facilitate the implementation of countermeasures against soil contamination and measures for the prevention of harmful effects on human health, and thereby to protect the health of citizens. In the Act, 26 chemical substances including lead, arsenic, trichloroethylene are designated as "Designated Hazardous Substance" which can bring harmful effects on human health (Table 3.1). The Act designates that not only natural soils but also stabilized soils shall be subjected to the soil contamination investigation to measure the content and elution amount of the substances and report them to the governor. Four regulated values are designated in the Act, of which "soil

Table 3.1 Soil elution criterion and second elution criterion designated by the Soil Contamination Countermeasures Act.

Hazardous substance	Soil elution criterion (mg/l)	Second elution criterion (mg/l)
Cadmium	0.01	0.3
Lead	0.01	0.3
Hexavalent chromium	0.05	1.5
Arsenic	0.01	0.3
Mercury	0.0005	0.005
Selenium	0.01	0.3
Fluorine	0.8	24
Boron	1	30

Table 3.2 Physical properties of soils (Kaneshiro *et al.*, 2006).

	Water content (%)	Density (g/cm³)	pH	Particle size distribution (%)			Classification
				gravel	sand	fine	
sand (1)	34	1.849	3.88	0.0	53.4	46.6	SF
sand (2)	20	1.742	–	0.3	89.4	10.3	S-Cs
clay (1)	61	1.718	7.18	0.6	2.0	97.4	CH
clay (2)	40	1.776	–	0.0	0.0	100.0	CH
volcanic soil	88	1.393	6.16	1.7	5.8	92.5	VH2

elution criterion" and "second elution criterion" are critical concerns for excavation and filling soils. The former is designated by the Minister of Environment for the "Designated Areas." When the situation of contamination by a Designated Hazardous Substance of the soil of the site does not conform to the criteria, the prefectural governor shall designate an area covering such site as an area contaminated by the Designated Hazardous Substance. The soils in the "Designated Areas" should be treated by in-situ in-solubility, in-situ confinement or confinement by impermeable wall.

The effect of stabilization on the leaching of hazardous substances were investigated by a series of laboratory leaching tests, where five soils artificially contaminated by eight chemical reagent designated as "Designated Hazardous Substances" were prepared and stabilized with a cement-based special binder (Kaneshiro *et al.*, 2006). The properties of the soils and the chemical reagents are summarized in Tables 3.2 and 3.3 respectively. The stabilized soil are prepared by the procedure specified by the Japan Cement Association (JCAS L-1: 2006), which is almost the same as the Japanese Geotechnical Society standard. After 7 days curing, the leaching tests were carried out on the specimen according to the testing procedure specified by the Environmental quality standards (Environment Agency, 1975), where the stabilized soil was crushed into pieces, sieved through a 2 mm sieve and naturally dried in advance.

Figure 3.46 shows the leaching test results on eight hazardous substances shown in Table 3.3. For cadmium leaching from the stabilized soil (Figure 3.46(a)), the amount of

Table 3.3 Chemical reagent mixed with soils.

Hazardous substances	Chemical substances	Chemical formula
Cadmium	Cadmium nitrate	$Cd(NO_3)_2 \cdot 4H_2O$
Lead	Lead(II) nitrate	$Pb(NO_3)_2$
Hexavalent chromium	Potassium bichromate	$K_2Cr_2O_7$
Arsenic	Disodium hydrogenarsenate	$Na_2HAsO_4 \cdot 7H_2O$, $KAsO_2$
Mercury	Mercuric chloride	$HgCl_2$
Selenium	Sodium selenate	Na_2SeO_4
Fluorine	Potassium fluoride	$KF \cdot 2H_2O$
Boron	Sodium metaborate	$NaBO_2 \cdot 4H_2O$

cadmium leaching quickly decreases with increasing binder content and becomes lower than the detection limit for all the stabilized soils. For lead leaching (Figure 3.46(b)), the amount of leaching decreases and becomes lower than the detection limit for all the stabilized soils when the binder content is larger than $100\,kg/m^3$ and cured for 28 days. For leaching of hexavalent chromium (Figure 3.46(c)), the amounts of leaching decrease only slightly with the binder content, and the improvement effect varies depending upon the type of soil. For leaching of arsenic (Figure 3.46(d)), the amount of leaching is variable for soil type: decrease by the stabilization for sand (2) and clay (2). For leaching of mercury (Figure 3.46(e)), the amounts of leaching decrease rapidly as far as the binder content is about $100\,kg/m^3$, but increases for further increase of the binder content. This phenomenon can be seen especially for the volcanic soil. For leaching of selenium (Figure 3.46(f)), the amounts of leaching decrease very slightly even if the binder content increases to $300\,kg/m^3$. For leaching of fluorine and boron (Figures 3.46(g) and (h)), the amount of leaching are variable for soil type: decrease by stabilization for sand (1) and clay (1) but slightly decrease for volcanic soil.

According to the test results, the improvement effect by admixture stabilization is variable depending upon the type of soil and type of substances. The high improvement effect is achieved for cadmium and lead where the amount of leaching can be reduced lower than the "Soil Elution Criteria." For the other substances, the effect of stabilization is variable depending upon the type of soil and the amount of binder.

5.2 Elution of Hexavalent chromium (chromium VI) from stabilized soil

Figure 3.47 shows the influence of the type of binder on the elution of hexavalent chromium (chromium VI) from stabilized soils (Hosoya, 2002). In the tests, six soils including two sandy soils, two cohesive soils and two volcanic cohesive soils were stabilized with four types of binder, ordinary Portland cement, blast furnace slag cement type B, two cement-based special binders. The leaching tests were carried out on the stabilized soils according to the testing procedure specified by the Environmental quality standards (Environment Agency, 1975) and the amount of hexavalent chromium was measured by the ultrasonic extraction-diphenylcarbazide colorimetry specified by Japanese Industrial Standard (Japan Industrial Standard, 2010). The time difference

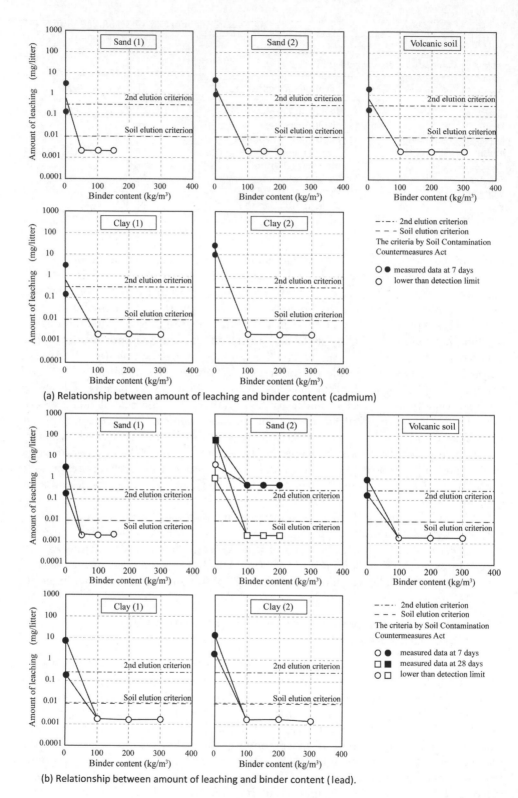

(a) Relationship between amount of leaching and binder content (cadmium)

(b) Relationship between amount of leaching and binder content (lead).

Figure 3.46 Effect of cement stabilization on leaching of hazardous substances (Kaneshiro *et al.*, 2006).

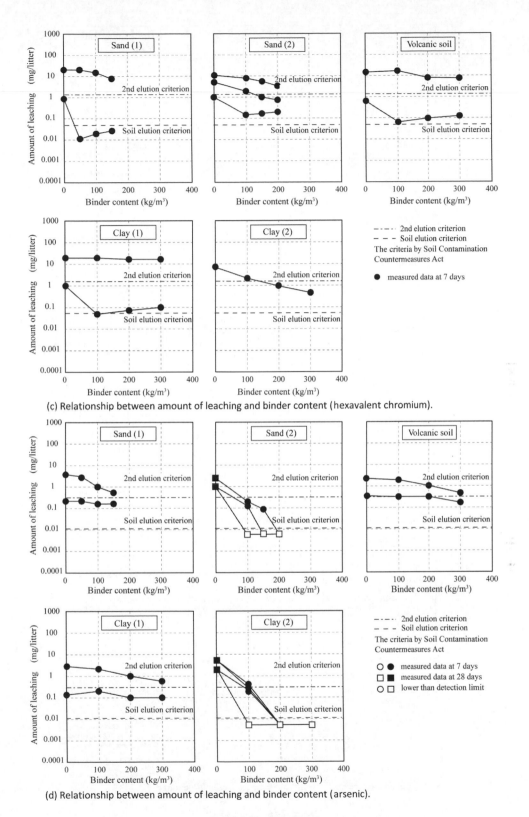

(c) Relationship between amount of leaching and binder content (hexavalent chromium).

(d) Relationship between amount of leaching and binder content (arsenic).

Figure 3.46 Continued.

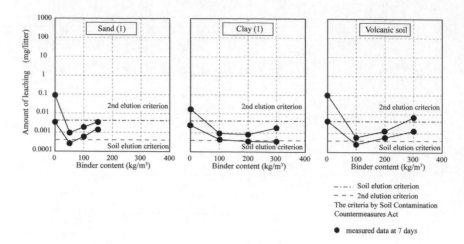

(e) Relationship between amount of leaching and binder content (mercury).

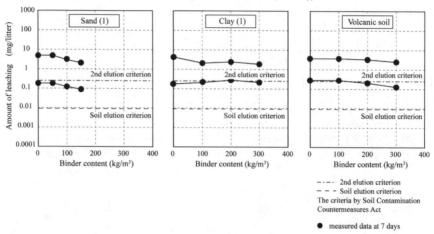

(f) Relationship between amount of leaching and binder content (selenium).

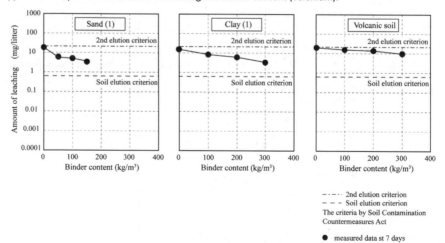

(g) Relationship between amount of leaching and binder content (fluorine).

Figure 3.46 Continued.

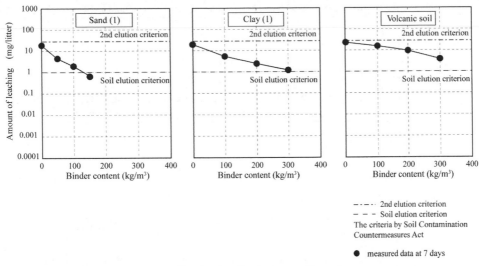

(h) Relationship between amount of leaching and binder content (boron).

Figure 3.46 Continued.

after adding sulfuric acid to diphenylcarbazide is changed either to 1 min. (DC1min) or 5 min. (DC5min). The broken lines in Figure 3.47, 0.05 mg/litter, is the 'Soil Elution Criterion' specified by the Japanese Ministry of Environment. As the measured elution amounts of Cr(VI) are the total amount eluted from not only the original soil but also the binder, the measured value increases with binder content in some cases. The stabilized soils with the cement-based special binders or blast furnace slag cement type B show a lower elution amount of hexavalent chromium, Cr(VI) than that stabilized with ordinary Portland cement. For the effect of the type of soil, the volcanic cohesive soils show a larger elution amount among the soils.

According to the accumulated test results, the leaching phenomenon of hexavalent chromium is prominent in the case where the soil is volcanic soil and in an unsaturated condition, and the binder is ordinary Portland cement. The Ministry of Land, Infrastructure, Transport and Tourism, Japan, notified the legal action on the leaching of hexavalent chromium from stabilized soil in 2000, where a laboratory test should be carried out on the leaching of hexavalent chromium from stabilized soil to assure the amount of leaching should be lower than the criteria designated by the Soil Contamination Countermeasures Act (Table 3.1). Several types of special binder have been available on the Japanese market for mitigating the leaching of hexavalent chromium from stabilized soil.

5.3 Resolution of alkali from stabilized soil

When calcium hydroxide, $Ca(OH)_2$ created by hydration of cement, dissociates in water, the solution shows high alkalinity as shown in Table 3.4 (Japan Cement Association, 2007). The exposure surface of cement stabilized soil is gradually neutralized by

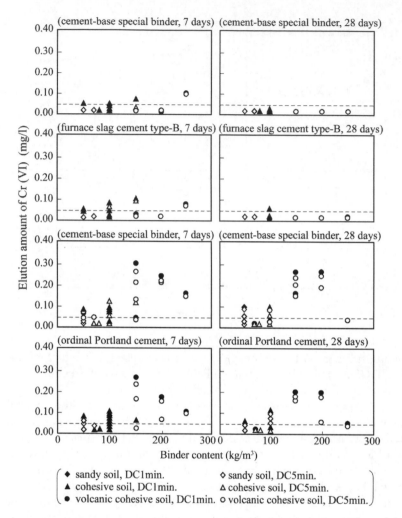

Figure 3.47 Leaching test results of hexavalent chromium from laboratory stabilized soils (Hosoya, 2002).

carbonation due to carbon dioxide in the air and dissolution of alkali components due to rainfall. The alkali components dissolved isn't diffused widely in the surrounding soil due to its buffer action.

Figure 3.48 shows the potential Hydrogen, pH of the cement stabilized soil, the surface water (water run off the surface of stabilized soil without permeation) and the permeated water with time (Japan Cement Association, 2007). The stabilized soil and the permeated water through the stabilized soil show a high pH value for three months, but the permeated water in the unstabilized soil shows neutral in pH. The surface water shows a high pH value at first but gradually decreases in pH and almost neutral after three months.

Table 3.4 pH values of stabilized soils (Japan Cement Association, 2007).

Binder content Soil	75 kg/m³			150 kg/m³		
	3 days	7 days	28 days	3 days	7 days	28 days
A (pH = 8.3)	12.0	11.6	11.4	12.5	12.0	11.7
B (pH = 8.8)	11.7	11.3	11.2	12.0	11.7	11.6

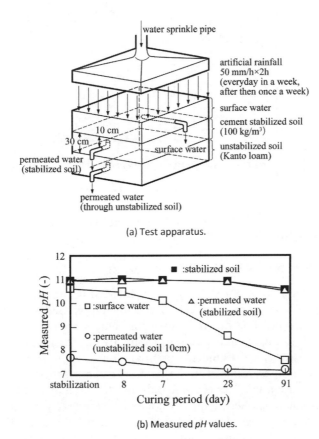

(a) Test apparatus.

(b) Measured pH values.

Figure 3.48 Resolution of alkali from stabilized soil (Japan Cement Association, 2007).

Figure 3.49 shows the pH value distribution in the cement stabilized soil and unstabilized soil in a field, which was measured at 33 months after the stabilization (Japan Cement Association, 2007). The stabilized soil still shows a high pH value of the order of 10 to 12, but a comparatively low pH value at the shallow depth probably due to the dissolution. In the unstabilized soil, the pH value rapidly decreases with depth to about 7 at about 100 mm far from the boundary of the stabilized soil.

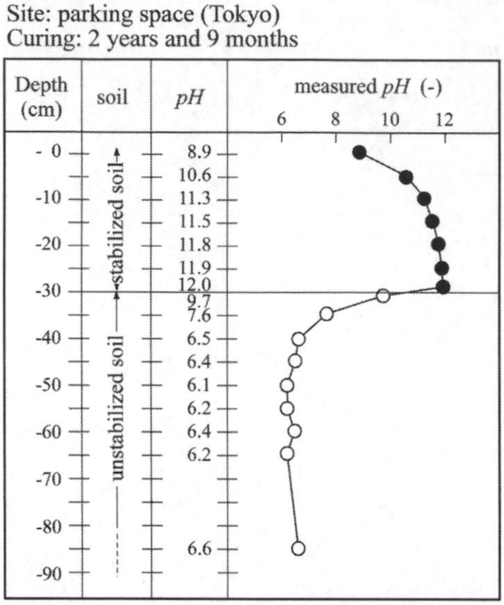

Site: parking space (Tokyo)
Curing: 2 years and 9 months

Figure 3.49 pH distribution in cement stabilized soil and unstabilized soil (Japan Cement Association, 2007).

6   ENGINEERING PROPERTIES OF CEMENT STABILIZED SOIL MANUFACTURED IN SITU

6.1   Mixing degree of in-situ stabilized soils

The engineering properties of stabilized soil mentioned in the previous sections were obtained mostly on laboratory stabilized soil specimens prepared with sufficient mixing degree. In actual production, the original soil and binder are mixed by a deep mixing machine in situ with a lower mixing degree in comparison with laboratory preparation. If the mixing degree and/or the binder content are low, the uniform mixing of original soil and binder cannot be attained in the field. The characteristics of field stabilized soil are, therefore, highly influenced not only by the amount of binder but also by the type of execution machine and quality control during execution. In Japan, various execution machines have been developed and improved incorporating field experiences and experiments as described in Chapter 5 for on-land and marine constructions. The careful quality control program during execution has also been developed and practiced as a routine. In this section, the characteristics of in-situ stabilized soil manufactured by the Japanese machine with careful quality control are briefly introduced.

6.2   Water content distribution

The water content profiles before and after cement stabilization are plotted along the depth in Figure 3.50 (Kawasaki et al., 1978). In the field tests, the Tokyo Port clay (w_L of 93.1% and w_p of 35.8%) was stabilized with ordinary Portland cement with binder content, α of 100 and 135 kg/m^3 and with a water to cement ratio, W/C of

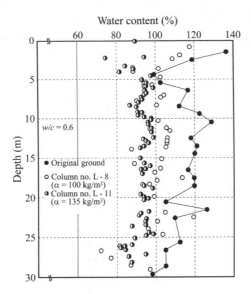

Figure 3.50 Change of water content by in-situ cement stabilization (Kawasaki *et al.*, 1978).

60%. The decrease in water content due to the stabilization is estimated 15 to 30% for α of 100 kg/m^3 and 20 to 30% for α of 135 kg/m^3 by Equation (3.3). It can be seen that the water content after the stabilization decreases about 20% from the original.

Figure 3.51 shows the water content distribution along the depth, where the water contents of the original soil (before stabilization) and the cement stabilized soil were measured in 1981 and 2001 respectively (Ikegami *et al.*, 2002a, 2002b). The ground condition at the site in Yokohama Port consists of three layers: an alluvial clay layer up to the depth of −24 m, an alluvial sand layer from −24 to −37 m, and an alluvial clay layer from the depth of −37 to −49 m. The original ground was stabilized by the wet method of deep mixing to a depth of −49 m. The binder was ordinary Portland cement and the binder content adopted in production was 180 kg/m^3 with a water to binder ratio, W/C of 60% throughout the improvement depth irrespective of the three layers mentioned above. Comparing the water content distributions before and after improvement, it is interesting to see that the stabilized soil clearly remembers the original soil stratification. This is because the mixing tool of the Japanese wet method consists of several vertical rotary shafts and mixing blades attached to each shaft and in-situ mixing is carried out mostly on the horizontal plane. The water contents of the stabilized soils in the upper and lower layers decrease about 10 to 15% due to the stabilization. In the intermediate layer, the alluvial sand layer, the water content of the stabilized soil is almost the same as that of the original soil, because the water content of the original soil was almost the same order of water to binder ratio of the binder slurry.

6.3 Unit weight distribution

The unit weight profiles before and after cement stabilizations are plotted along the depth in Figure 3.52 (Kawasaki *et al.*, 1978). In the field tests, the Tokyo Port clay (w_L of 93.1% and w_p of 35.8%) was stabilized with ordinary Portland cement with

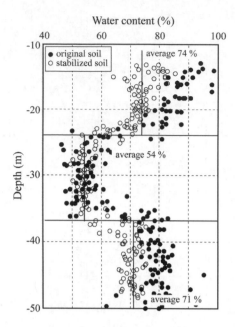

Figure 3.51 Change of water content at Yokohama Port (Ikegami *et al.*, 2002a, 2002b).

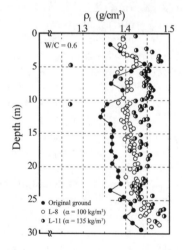

Figure 3.52 Change of wet density by in-situ cement stabilization (Kawasaki *et al.*, 1978).

cement content, α of 100 and 135 kg/m^3 and with a water to cement ratio, W/C of 60%. Although there is scatter in the measured data, it can be seen that the unit weight after the stabilization increases about 4 to 7% from the original.

6.4 Variability of field strength

Major factors which cause strength variability are the variability of the original soil and the degree of mixing. The soil stratification is an important factor in discussing variability if the same mixing process is employed. There are two approaches in the stabilization of stratified soil. One is to select the appropriate type and amount of binder for each layer to achieve relatively uniform strength profile along the depth. The other is to select the appropriate type and amount of binder for the most difficult layer and apply the same mix design to all the layers, which inevitably results in a non-uniform strength profile along the depth but guarantees the required strength even in the most difficult layer. Even in the case of an apparently uniform layer, the natural water content may decrease with the depth in many cases due to the effect of overburden pressure, which can be typically found in a normally consolidated clay layer. As introduced in Chapter 2, the less water content causes a higher strength of stabilized soil in general.

Figures 3.53 shows five examples of field strength profiles, which cover the on-land constructions by the dry method and the wet method and in-water construction by the wet method. In Figure 3.53(a) (Public Works Research Center, 2004), a silty clay and a clay layers having natural water contents of about 70 to 100% were stabilized by the dry method with ordinary Portland cement. The binder content was 120 kg/m^3 to achieve the field strength of 0.4 MN/m^2. The strength of the stabilized soil is ranging from 0.5 to 1.6 MN/m^2, which is higher than the design strength.

Another example of the dry method is shown in Figure 3.53(b), where an organic soil, organic clay, silt and fine sand layers were stabilized with a cement-based special binder (Public Works Research Center, 2004). The binder factor was 400 kg/m^3 for the organic soil layer and 100 kg/m^3 for the silt and sand layers for achieving the design strength of 600 kN/m^2. The field strength varies in a wide range from 1 to 6 MN/m^2 along the full depth of improvement. When looking the different layers independently, the range of field strength is 1 to 3 MN/m^2 for organic soils and 2.5 to 6 MN/m^2 in the sand layer, both of them satisfying the design strength.

In Figure 3.53(c) (Coastal Development Institute of Technology, 2008), a quite uniform clay layer having a natural water content of 110 to 140% was stabilized by the wet method with ordinary Portland cement, α of 74 kg/m^3. The field strength varies in the range of 100 to 600 kN/m^2 with an average of 230 kN/m^2.

Another example of the wet method is shown in Figure 3.53(d), where stratified layers consisted of organic soil, silt with organic soil, silt with sand and sandy silt, were stabilized with a cement-based special binder (Coastal Development Institute of Technology, 2008). The binder content was 200 kg/m^3 for all the layers. The field strength varies to some extent ranging from 0.3 to 0.7 MN/m^2.

Figure 3.53(e) shows an example of the field strength profile in the wet method for marine construction, where a clay layer having a natural water content of 55 to 110% was stabilized with blast furnace slag cement type B, α of 140 kg/m^3 for the depth up to -36 m and 180 kg/m^3 for the further depth. The average strength and the coefficient of variation were 3.76 MN/m^2 and 44.0% for the upper layer, and 6.08 MN/m^2 and 27.0% for the bottom layer, respectively.

According to the Japanese accumulated data, the coefficient of variation in the field strength varies from 50 to 68% for the on-land dry method, and 15 to 50% for the on-land wet method. The reason for the larger coefficient of variation in the dry method may be due to the fact the stratified layers with different soil types are often encountered. For the marine construction by the wet method, the coefficient of variation varies from 20 to 48% (Coastal Development Institute of Technology, 2008).

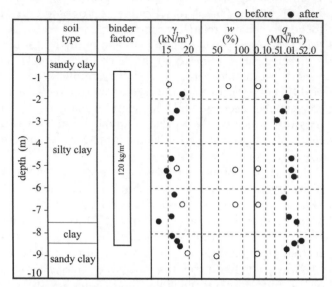

(a) On-land DJM (Public Works Research Center, 2004).

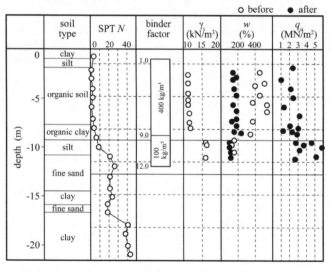

(b) On-land DJM (Public Works Research Center, 2004).

Figure 3.53 Strength distribution along the depth.

6.5 Difference in strength of field produced stabilized soil and laboratory prepared stabilized soil

As explained in the previous chapter, the strength of stabilized soil is influenced by many factors. In comparing the strength of field produced soil and laboratory prepared soil with the same amount of binder, the mixing degree and the curing temperature are

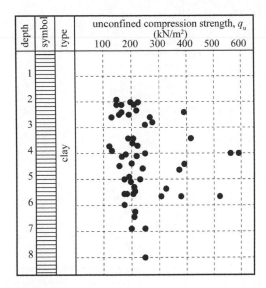

(c) On-land CDM (Coastal Development Institute of Technology, 2008).

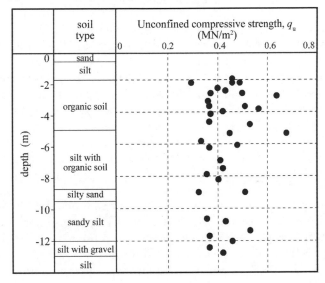

(d) On-land CDM (Coastal Development Institute of Technology, 2008).

Figure 3.53 Continued.

the dominant factors. The mixing degree is generally lower for the field production than that in laboratory preparation. Curing temperature, at least in a moderate climate, is often higher for the field curing than in laboratory curing. Further difference may be caused by the timing of sampling and sample disturbance of the field produced specimens.

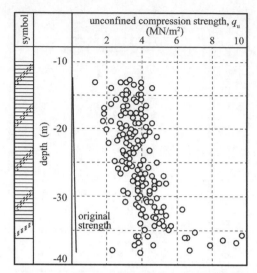

(e) Marine CDM (Coastal Development Institute of Technology, 2008).

Figure 3.53 Continued.

In this subsection, "the strength of field produce stabilized soil" is the unconfined compressive strength of core sample retrieved from the stabilized soil, which is produced by the Japanese wet and dry mechanical deep mixing and cured in situ. "The strength of laboratory prepared stabilized soil" is the unconfined compressive strength of the soil samples prepared and cured following the Japanese standard test procedure. For simplicity the former is often referred to as "field strength", q_{uf} and the latter "laboratory strength", q_{ul}. It is well known in Japan that the strength of field produced stabilized soil, q_{uf} is usually smaller than the strength of laboratory prepared stabilized soil q_{ul}. Figure 3.54 shows the relationship between q_{uf} and q_{ul} (Public Works Research Center, 2004). In the case of on-land constructions (Figures 3.54), the q_{uf} value is as small as 1/2–1/5 of the q_{ul} for clay, but for sand a relatively high field strength is obtained and a ratio larger than unity is often found. In the case of marine constructions (Figure 3.55), on the other hand, the q_{uf} value is almost the same order with the laboratory strength, q_{ul}. The reason why the ratio of q_{uf}/q_{ul} is quite different in on-land constructions and marine constructions is attributed to a relatively large amount of stabilized soil and relatively good mixing degree in marine constructions (see Section 5 in Chapter 5).

6.6 Size effect on unconfined compressive strength

In Japan, the unconfined compression tests are often conducted on a small specimen of 50 mm in diameter and 100 mm in height. The Building Center of Japan conducted a series of compression tests on cement stabilized soils excavated at 26 sites in order to investigate the size effect on the strength. The original soils are classified into five types as shown in Figure 3.56 (The Building Center of Japan, 1997). In the fields, the stabilized soils were manufactured by wet method of deep mixing, in which the

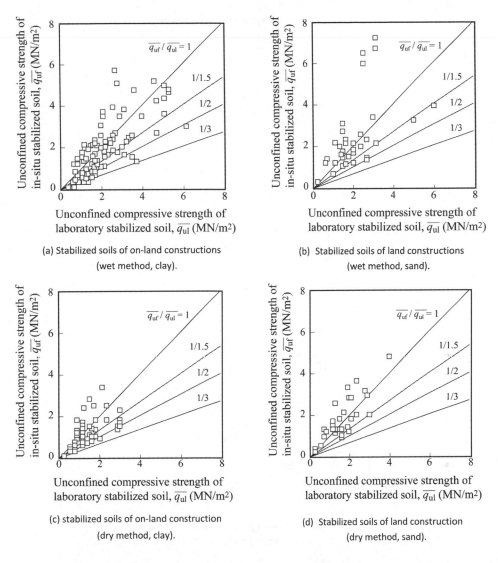

Figure 3.54 Relationship between unconfined compressive strength of laboratory stabilized soil and in-situ stabilized soil (Public Works Research Center, 2004).

amount of cement was 210 to 220 kg/m^3 and the water to cement ratio, W/C was either 60 or 100%. In the tests, the in-situ stabilized soil columns excavated from the fields and trimmed to about 1.0 to 1.2 m in diameter and about 1.5 to 2.4 m in height to determine the unconfined compressive strength of full scale columns. Small specimens with 67 mm in diameter and 130 mm in height were also sampled by coring the in-situ stabilized soil columns and tested in unconfined compression. The coefficient of variation of core samples at each site ranges from 12.4 to 57.3%, and 38.0% in average. Figure 3.56 compares the averages of unconfined compressive strength on

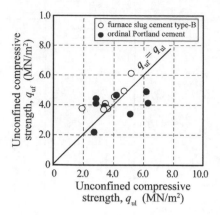

Figure 3.55 Relationship between unconfined compressive strength of laboratory stabilized soil and in-situ stabilized soil in in-water construction (wet method) (Coastal Development Institute of Technology, 2008).

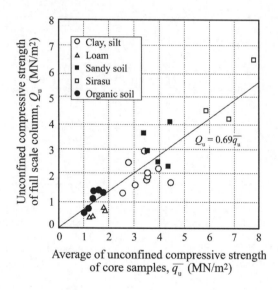

Figure 3.56 Size effect on unconfined compressive strength (The Building Center of Japan, 1997).

the core samples and the overall strength of the full scale column. It can be concluded that the unconfined compressive strength on the full scale stabilized soil column is about 69% of the average unconfined compressive strength of the small size specimens. The unconfined compressive strength of the full scale column can be expressed with the average unconfined compressive strength on the core specimen and the standard deviation, σ as Equation (3.7).

$$Qu = \overline{q}_{u} - 1.33\sigma \tag{3.7}$$

where

$\overline{q}_u$: average unconfined compressive strength on core specimen (kN/m^2)

Qu : unconfined compressive strength of full scale column (kN/m^2)

σ : standard deviation (kN/m^2).

6.7 Strength and calcium distributions at overlapped portion

6.7.1 Test conditions

A column of stabilized soil is constructed by a single stroke (penetration and withdrawal) of the deep mixing machine. A stabilized soil mass in wall, grid or block type improvement is produced in a ground by overlapping these columns. Figure 3.57 shows the cross section of overlapped columns for the case of a two shafts machine. As the improved ground is assumed to be uniform in the current design, it is important to evaluate the strength of the overlapped portion.

The strength characteristics of in-situ stabilized soil manufactured in marine construction were investigated with special emphasis on the characteristics of the overlapped portion (Tanaka and Terashi, 1986). Figure 3.58 shows the cross section with four round stabilized soil columns overlapped each other (Tanaka and Terashi, 1986). The execution machine used in the investigation was a double shafts machine with a set of three stacks of mixing blades at different levels of each shaft. The diameter of blades and spacing between the two shafts were 1.15 m and 0.7 m respectively. A stabilized soil element produced by a single installation process is a pair of round columns with 1.15 m diameter and maximum overlap width of 0.45 m. To distinguish two round columns, one column is termed "S" and the other "L". The overlap between dual shafts is called "machine overlap".

Two stabilized soil elements were produced with ordinary Portland cement, α of 130 kg/m^3, and a W/C ratio of 100%. The stabilized soil elements were produced with the maximum overlap width of 100 mm. The first element is termed as "1" and the second element as "2". The interval of construction of the first and second elements was about 2 to 3 hours. The overlap portion produced by two successive installations is called the "construction overlap." After about 140 days curing in situ, the stabilized soil elements, four round columns overlapped each other, were excavated and lifted

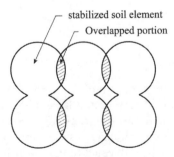

Figure 3.57 Schematic view of overlapped portion.

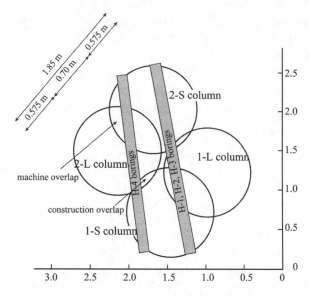

Figure 3.58 In-situ overlapped columns (Tanaka and Terashi, 1986).

up to ground, and four horizontal core borings, named H-1, H-2, H-3 and H-4, were conducted as well as many vertical core borings.

6.7.2 Calcium distribution

Figure 3.59 shows the calcium content distribution of stabilized soil in the horizontal plane along H4 coring that passes through the 1-S column, the construction overlap, the 2-L column, the machine overlap and the 2-S column (Tanaka and Terashi, 1986). The continuous core was sliced into 10 mm thick specimens to determine the detailed calcium distribution. As the size of specimen for the strength test is much larger, the scatter of the calcium content does not directly relate to the scatter of strength. The amount of calcium in the 2-L column was higher than that of the 2-S column, which was caused by controlling the amount of cement slurry as a whole in this particular machine. The amount of calcium in the 2-S column was slightly higher than that of the 1-S column. However, no appreciable difference was found in the calcium distribution between the 1-S column and the construction overlap. The amount of calcium at the machine overlapped portion was almost same as the 2-L column.

6.7.3 Strength distribution

Figure 3.60 shows the strength distribution in the horizontal plane (Tanaka and Terashi, 1986). Three horizontal borings at different levels, H1 to H3 were conducted, which pass through the 1-S column, the machine overlap, the 1-L column, the construction overlap and the 2-S column. The test specimens were cored and trimmed in the horizontal direction and were subjected to unconfined compression tests. Also the vertical borings were conducted to examine the unconfined compressive strength by

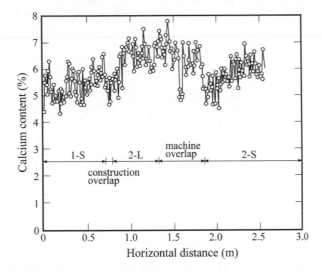

Figure 3.59 Calcium content distribution of stabilized soil columns (Tanaka and Terashi, 1986).

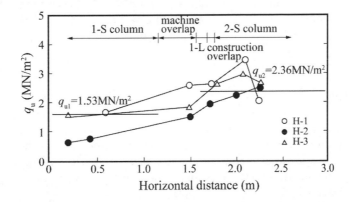

Figure 3.60 Strength distribution of stabilized soil columns (Tanaka and Terashi, 1986).

vertical loadings. As shown in Figure 3.59, the calcium content in the 2-S column was slightly higher than that in the 1-S column. The strength of the 2-S column was higher than that of the 1-S column. The strength at the overlapped portion was between the strengths of the two columns and no appreciable influence of the overlapping operation on the strength was found.

6.7.4 Effect of time interval

Figure 3.61 shows the relationship between the strength of overlapped portion against the time interval of overlapping execution (Yoshida, 1996). The soft silt having a natural water content of 100.6% was stabilized in the field with a cement-based special binder of 200 kg/m³ and W/C ratio of 100%, which was overlapped with various time intervals up to six days. The stabilized soil specimens were sampled within the

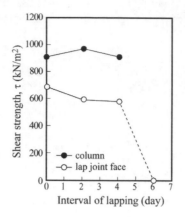

Figure 3.61 Relastionship between interval of overlapping and shear strength (Yoshida, 1996).

column and the overlapped portion, and subjected to the direct shear test to obtain the shear strengths of the column and the overlapped portion. As shown in Figure 3.61, the strength of the overlapped portion is around 66% of that of the core portion as long as the time interval of overlapping execution is within 4 days. However, when the overlapping was carried out after 6 days, the strength of overlapping portion decreases to almost zero.

7 SUMMARY

The current chapter described the engineering characteristics of stabilized soil mainly based on laboratory prepared samples. The general tendency and the correlation of various characteristics and unconfined compressive strength may apply to a variety of admixture stabilization techniques and may help design engineers understand the stabilized soil.

The characteristics of in-situ stabilized soil discussed in section 6, however, are only applicable to the in-situ soil produced by the mechanical mixing process used in Japan, which employs a mixing tool comprising vertical rotary shafts and mixing blades. This is because the quality of in-situ stabilized soil heavily depends upon the mixing process and procedures. It is the responsibility of the deep mixing contractor to collect and accumulate information on the quality of in-situ stabilized soils produced by their own proprietary mixing system.

The knowledge compiled in the present chapter will be summarized in the followings.

7.1 Physical properties

7.1.1 Change of water content and density

When the binder is added to the soil in dry form, the water content of the original soil is decreased due to hydration of the binder. When the binder is added to the soil in the form of binder-water slurry, the change of water content depends upon the initial water content of the soil, the amount of binder and the water to binder ratio of binder slurry. Change of water content can be estimated by Equations (3.2) and (3.3) for quicklime and cement stabilization respectively. The density is increased due to stabilization for dry form binder, but negligibly increased for slurry form binder. The magnitude of increment in density can be estimated by Equations (3.4) and (3.5) for quicklime and cement stabilization.

7.1.2 Change of consistency of soil-binder mixture before hardening

The consistency of the soil-binder mixture changes from that of the original soil due to ion exchange. The liquid limit, w_L decreases with increasing quicklime content, while the plastic limit, w_p increases. As the results, the plasticity index, I_p sharply decreases with increasing quicklime content.

7.2 Mechanical properties (strength characteristics)

7.2.1 Stress–strain behavior

The stress–strain curve of the stabilized soil is characterized by very high strength and small axial strain at failure, while the original soil is characterized by small strength and large axial strain at failure.

   The axial strain at failure of stabilized soil is quite small compared to that of the original soil. In the unconfined compression, the magnitude of axial strain at failure is of the order of a few percent and markedly smaller than that of unstabilized clay. The axial strain at failure decreases with the unconfined compressive strength, q_u.

   The magnitude of E_{50} exponentially increases with the q_u and is 75 to $1000 \times q_u$ depending on the type of soil and type and amount of binder.

   The stress–strain curves and residual strength of stabilized soils are heavily influenced by the confining pressure. The strength ratio of residual strength against peak strength increases with the confining pressure ratio, and the strength ratio is about 50 to 80% for the confining pressure ratio exceeding about 0.1 irrespective of the type of binder.

7.2.2 Poisson's ratio

The Poisson's ratio of stabilized soil is around 0.2 to 0.45, irrespective of the unconfined compressive strength, q_u. The Poisson's ratio on the large size stabilized sands is ranging about 0.2 to 0.3 irrespective of the strength of stabilized soil.

7.2.3 Angle of internal friction

The angle of internal friction, ϕ' of stabilized soil is almost zero as far as the consolidation pressure is lower than the consolidation yield pressure and same as that of the unstabilized soil when the consolidation pressure is higher than the yield pressure. The phenomenon can be seen irrespective of type and amount of binder.

7.2.4 Undrained shear strength

The undrained shear strength, c_u of the stabilized soil is almost constant as long as the consolidation pressure is low. But when the consolidation pressure exceeds the consolidation yield pressure (the pseudo pre-consolidation pressure), the undrain shear strength increases with increasing consolidation pressure. The increasing ratio in the c_u of the stabilized soil is almost the same as that of the unstabilized soil. The phenomenon can be seen irrespective of type and amount of binder.

7.2.5 Dynamic property

The shear moduli measured in the three tests are practically identical for a small strain range between 10^{-6} and 10^{-5}. The G_{max} value is scarcely affected by the confining pressure and the shear stress level in the test condition.

The initial shear modulus, G_0 increases with the unconfined compressive strength and the confining pressure, while the binder content doesn't give a large effect on the G_0 value.

The damping ratio, h_{eq} increases with the shear strain, while the relationship isn't influenced so much by the confining pressure. The h_{eq} slightly decreases with increasing confining pressure, irrespective of the binder content.

7.2.6 Creep and cyclic strengths

Under the sustained load, q_{cr}, the strain rate decreases almost linearly on the double-logarithmic graph with the time duration when the ratio of q_{cr} to the unconfined compressive strength, q_u is smaller than 0.8. The decreasing ratio is almost constant irrespective of the load intensity. The vertical load q_{cr}/q_u higher than about 0.9 exhibits creep failure. When the cyclic loading, whose maximum and minimum load intensity are denoted by σ_{max} and σ_{min}, is applied the axial strains increase gradually with the number of loading cycles, and increases to failure at N_f loading cycles. The linear relation between σ_{max}/q_u and $(\sigma_{max} - \sigma_{min})/q_u$ against $\log N_f$ were found.

7.2.7 Tensile and bending strengths

The tensile strength increases almost linearly with unconfined compressive strength irrespective of the type, amount of binder and initial water content of the soil, but its increment becomes lower with increasing q_u. The tensile strength is about 0.1 to 0.6 of the unconfined compressive strength, which is influenced by the testing procedure.

7.2.8 Long term strength

There are two aspects when the long term strength of stabilized soil is concerned. One is the strength increase with time at the core portion of the stabilized soil column which is negligibly influenced by the surrounding conditions and the other is the possible strength decrease with time due to deterioration in the periphery of the stabilized soil column.

The long term strength of stabilized soil at the core part increases almost linearly with the logarithm of elapsed time, irrespective of laboratory/field manufactured stabilized soil, and the soil type and the type and amount of binder.

The long term strength of stabilized soil at the periphery decreases with elapsed time, and the deterioration portion progresses gradually inward with time especially in the case of exposure to tap water and seawater. The progress of deterioration depth in logarithmic scale is almost linear to logarithmic time, and the slopes in all the test cases are about 1/2 irrespective of the strength of specimens and the exposure conditions.

7.3 Mechanical properties (consolidation characteristics)

7.3.1 Void ratio – consolidation pressure curve

The shape of $e - \log p$ curves of the stabilized soil is similar to ordinary clay samples, which is characterized by a sharp bend at a consolidation yield pressure.

The consolidation yield pressure, p_y of stabilized soil is closely related to its unconfined compressive strength, and the ratio of p_y/q_u of stabilized soil is around 1.3 for the unconfined compressive strength up to 3 MN/m^2, irrespective of the type of original soil, and the type of binder.

7.3.2 Coefficient of consolidation and coefficient of volume compressibility

The ratio of coefficient of consolidation of stabilized soil against original soil is 10 to 100 in a sort of overconsolidated condition, but the ratio approaches to unity in a sort of normally consolidated condition.

The ratio of the coefficient of volume compressibility of the stabilized soil against the original soil is 0.01 to 0.1 in a sort of overconsolidated condition, but the ratio approaches to unity in a sort of normally consolidated condition.

7.3.3 Coefficient of permeability

The coefficient of permeability of stabilized clay is equivalent to or lower than that of the unstabilized clays and whose order is in the 10^{-9} to 10^{-6} cm/sec.

The coefficient of permeability of stabilized sand in logarithm scale decreases almost linearly with the amount of cement slurry irrespective of the soil type. The permeability also decreases with increasing fine grain fraction content irrespective of the amount of cement slurry.

7.4 Environmental properties

7.4.1 Elution of contaminant

The improvement effect on leaching of a hazardous substance by soil stabilization is variable depending upon the type of soil and type of substances. The high improvement effect is achieved for cadmium and lead where the amount of leaching can be reduced lower than the soil elution criteria. For hexavalent chromium, arsenic, mercury, selenium, fluorine and boron, the effect of stabilization is variable depending upon the type of soil and the amount of binder.

The leaching phenomenon of hexavalent chromium is prominent in the case where soil is volcanic soil and in an unsaturated condition, and the binder is ordinary Portland cement.

7.4.2 Resolution of alkali from a stabilized soil

When calcium hydroxide, $Ca(OH)_2$, created by hydration of cement, dissociates in water and the solution shows high alkalinity. The exposure surface of cement stabilized soil is gradually neutralized by carbonation due to carbon dioxide in the air and dissolution of alkali components due to rainfall. The alkali components dissolved isn't diffused widely in the surrounding soil due to its buffer action. The stabilized soil and the permeated water show high pH value for three months, but the permeated water in the unstabilized soil shows neutral in pH. The surface water shows high pH value at first but gradually decreases in pH and almost neutral after three months.

7.5 Engineering properties of cement stabilized soil manufactured in situ

7.5.1 Water content and unit weight by stabilization

The distribution of water content and the unit weight of stabilized soil remember the stratification of original ground when the mixing tool comprising vertical rotary shaft and blades are used. The change of water content depends on the initial water content, binder content and water to cement ratio of the binder water slurry. The change of unit weight in the ordinary conditions is relatively small.

7.5.2 Variability of field strength

According to the Japanese accumulated data, the coefficient of variation in the field strength varies from 50 to 68% for the on-land dry method, and 15 to 50% for the on-land wet method. For the marine construction by the wet method, the coefficient of variation varies from 20 to 48%.

7.5.3 Difference in the strength of field produced stabilized soil and laboratory prepared stabilized soil

The q_{uf} value is as small as 1/2–1/5 of the q_{ul} in the case of on-land constructions. In the case of marine construction, on the other hand, the q_{uf} value is almost the same order with the laboratory strength, q_{ul}.

7.5.4 Size effect on unconfined compressive strength

The unconfined compressive strength on the full-scale column is about 69% of the average unconfined compressive strength of the small size specimens.

7.5.5 Strength distributions at overlapped portion

In the marine construction, the strength at the overlapped portion was between the strengths of the two columns and no appreciable influence of overlapping operation on the strength was found. From the test conducted in on-land construction, it is found that the strength of the overlapped portion is influenced by the time interval of the overlapping operation.

REFERENCES

Coastal Development Institute of Technology (2008) *Technical Manual of Deep Mixing Method for Marine Works*. 289p. (in Japanese).

Enami, A., Yamada, M. & Ishizaki, H. (1993) Dynamic properties of improved sandy soils (Part 2). *Proc. of the 28th Annual Conference of the Japanese Geotechnical Society*. pp. 1065–1066 (in Japanese).

Environment Agency (1975) *Criteria for a Specific Operation of the Ground Storage Tank Outdoors using Deep Mixing Method* (notification) (in Japanese).

Environment Agency (2005) *Criteria for a Specific Operation of the Ground Storage Tank Outdoors using Deep Mixing Method* (notification) (in Japanese).

Hayashi, H., Nishikawa, J., Egawa, T., Terashi, M. & Ohishi, K. (2001) Long-term strength of improved column formed by Deep mixing method. *Proc. of the 56th Annual Conference of the Japan Society of Civil Engineers*. Vol. 3. pp. 378–379 (in Japanese).

Hayashi, H., Nishikawa, J., Ohishi, K. & Terashi, M. (2003) Field observation of long-term strength of cement treated soil. *Proc. of the 3rd International Conference on Grouting and Ground treatment*. Vol. 1. pp. 598–609.

Hayashi, H., Ohishi, K. & Terashi, M. (2004) Possibility of strength reduction of treated soil by Ca leaching. *Proc. of the 39th Annual Conference of the Japanese Geotechnical Society*. pp. 785–786 (in Japanese).

Hirade, T., M. Futaki, K. Nakano & K. Kobayashi (1995) The study on the ground improved with cement as the foundation ground for buildings, part 16. Unconfined compression test of large scale column and sampling core in several fields. *Proc. of the Annual Conference of Architectural Institute of Japan*. pp. 861–862 (in Japanese).

Hosoya, T. (2002) Leaching of hexavalent Chromium from cementitious soil improvement. *Journal of Society of Material Science. Japan*. Vol. 51. No.8. pp. 933–942 (in Japanese).

Ikegami, M., Ichiba, T., Ohnishi, K. & Terashi, M. (2005) Long term property of cement treated soil 20 years after construction. *Proc. of the16th International Conference on Geotechnical Engineering*. pp. 1199–1202.

Ikegami, M., Masuda, K., Ichiba, T., Tsuruya, H. & Ohishi, K. (2002a) Long-term durability of cement-treated marine clay after 20 years. *Proc. of the 57th Annual Conference of the Japan Society of Civil Engineers*. Vol. 3. pp. 121–122 (in Japanese).

Ikegami, M., Masuda, K., Ichiba, T., Tsuruya, H., Satoh, S. & Terashi, M. (2002b) Physical properties and strength of cement-treated marine clay after 20 years, *Proc. of the 57th Annual Conference of the Japan Society of Civil Engineers*. Vol. 3. pp. 123–124 (in Japanese).

Japan Cement Association (2006) *Standard of Measuring Strength of Cement Stabilized Soil, JCAS L-1: 2006*. Japan Cement Association (in Japanese).

Japan Cement Association (2007) *Soil Improvement Manual using Cement Stabilizer (3rd edition)*. Japan Cement Association. 387p. (in Japanese).

Japan Lime Association (2009) *Technical Manual on Ground Improvement using Lime*. Japan Lime Association. 176p. (in Japanese).

Japanese Industrial Standard (2010) *Testing Methods for Industrial Wastewater, JIS K 0102: 2010* (in Japanese).

Kamata, H. & Akutsu, H. (1976) Deep mixing method from site experience. *Journal of the Japanese Society of Soil Mechanics and Foundation Engineering, Tsuchi to Kiso*. Vol. 24. No. 12. pp. 43–50 (in Japanese).

Kaneshiro, T., Moriya, M., Kondou, H. & Takahashi, S. (2006) The leaching behavior of the specific harmful substances from contaminated soil which were stabilized with all-purpose cementious soil stabilizer. *CEMENT & CONCRETE*. No. 714. pp. 12–21 (in Japanese).

Kawasaki, T., Niina, A., Saitoh, S. & Babasaki, R. (1978) Studies on engineering characteristics of cement-base stabilized soil. *Takenaka Technical Research Report*. Vol. 19. pp. 144–165 (in Japanese).

Kitazume, M. & Takahashi, H. (2009) 27 Years' investigation on property of in-situ quicklime treated clay. *Proc. of the 17th International Conference on Soil Mechanics and Geotechnical Engineering*. Vol. 3. pp. 2358–2361.

Kitazume, M., Nakamura, T., Terashi, M. & Ohishi, K. (2003) Laboratory tests on long-term strength of cement treated soil. *Proc. of the 3rd International Conference on Grouting and Ground treatment*. Vol. 1. pp. 586–597.

Kudo, T., Seriu, M., Yoshimoto, K. & Hatakeyama, N. (1993) Deformation and strength properties of cement treated soft clay subjected to cyclic loading. *Proc. of the Symposium on Soil Mechanics and Foundation Engineering*. pp. 275–28 (in Japanese).

Miura, H., Tokunaga, S., Kitazume, M. & Hirota, N. (2004) Laboratory permeability tests on cement treated soils. *Proc. of the International Symposium on Engineering Practice and Performance of Soft Deposits*. pp. 181–186.

Namikawa, T. & Koseki, J. (2007) Evaluation of tensile strength of cement-treated sand based on several types of laboratory tests. *Soils and Foundations*. Vol. 47. No. 4. pp. 657–674.

Niigaki, O., Fukushima, Y., Nodu, M., Yanagawa, Y. & Kasahara, Y. (2001) The property of deep mixing stabilized soil beneath highway embankment after more than 10 years. *Proc. of the 37th Annual Conference of the Japanese Geotechnical Society*. pp. 1117–1118 (in Japanese).

Niina, A., Saitoh, S., Babasaki, R., Miyata, T. & Tanaka, K. (1981) Engineering properties of improved soil obtained by stabilizing alluvial clay from various regions with cement slurry. *Takenaka Technical Research Report*. Vol. 25. pp. 1–21 (in Japanese).

Nishida, T., Terashi, M., Otsuki, N. and Ohishi, K. (2003) Prediction method for Ca leaching and related property change of cement treated soils. *Proc. of the 3rd International Conference on Grouting and Ground treatment*. Vol. 1. pp. 658–669.

Onitsuka, K., Modmoltin, C., Kouno, M. & Negami, T. (2003) Effect of organic matter on lime and cement stabilized Ariake clays. *Journal of Geotechnical Engineering*, Japan Society of Civil Engineers. Vol. 729/III-62. pp. 1–13.

Public Works Research Center (2004) *Technical Manual on Deep Mixing Method for On Land Works*. 334p. (in Japanese).

Saitoh, S. (1988) Experimental study of engineering properties of cement improved ground by the deep mixing method. *Doctoral thesis, Nihon University*. 317p. (in Japanese).

Saitoh, S., Suzuki, Y., Nishioka, S. & Okumura, R. (1996) Required strength of cement improved ground. *Proc. of the 2nd International Conference on Ground Improvement Geosystems*. Vol. 1. pp. 557–562.

Shibuya, S., Tatsuoka, F., Teachavorasinskun, S., Kong, X. J., Abe, F., Kim, Y-S. & Park C-S. (1992) Elastic deformation properties of geomaterials. *Soils and Foundations*. Vol. 32. No. 3. pp. 26–46.

Shimomura, S (2001) New calculation method of water content and density for stabilized soil. *Koei-Forum*. Vol. 9. pp. 163–169 (in Japanese).

Sugiyama, K., Kitawaki, T. & Morimoto, T. (1980) Soil improvement method of marine soft soil by cement stabilizer. *Doboku Sekou*. Vol. 21. No. 5. pp. 65–74 (in Japanese).

Takahashi, H. & Kitazume, M. (2004) Consolidation and permeability characteristics on cement treated clays from laboratory tests. *Proc. of the International Symposium on Engineering Practice and Performance of Soft Deposits*. pp. 187–192.

Tanaka, H. & Terashi, M. (1986) Properties of treated soils formed in situ by deep mixing method. *Report of the Port and Harbour Research Institute*, Vol. 25. No. 2. pp. 89–119 (in Japanese).

Tatsuoka, F. & Kobayashi, A. (1983) Triaxial strength characteristics of cement treated soft clay. *Proc. of the 8th European Regional Conference on Soil Mechanics and Foundation Engineering.* Vol. 1. pp. 421–426.

Terashi, M. & Kitazume, M (1992) An investigation of the long term strength of a lime treated marine clay. *Technical Note of the Port and Harbour Research Institute.* No. 732. 14p. (in Japanese).

Terashi, M., Okumura, T. & Mitsumoto, T. (1977) Fundamental properties of lime-treated soils. *Report of the Port and Harbour Research Institute.* Vol. 16. No. 1. pp. 3–28 (in Japanese).

Terashi, M., Tanaka, H., Mitsumoto, T., Honma, S. & Ohhashi, T. (1983) Fundamental properties of lime and cement treated soils (3rd Report). *Report of the Port and Harbour Research Institute.* Vol. 22. No. 1. pp. 69–96 (in Japanese).

Terashi, M., Tanaka, H., Mitsumoto, T., Niidome, Y. & Honma, S. (1980) Fundamental properties of lime and cement treated soils (2nd Report). *Report of the Port and Harbour Research Institute.* Vol. 19. No. 1. pp. 33–62 (in Japanese).

The Building Center of Japan (1997) *Design and Quality Control Guideline of Improved Ground for Building.* 473p. (in Japanese).

Yoshida, S. (1996) Shear strength of improved soils at lap-joint-face. *Proc. of the 2nd International Conference on Ground Improvement Geosystems.* pp. 461–46.

Chapter 4

Applications

1 INTRODUCTION

When looking at the strength of stabilized soil, the wet method of deep mixing in Japan (CDM) creates stabilized soil with a strength exceeding $1\,\text{MN/m}^2$ in terms of unconfined compressive strength, q_u. The Japanese dry method of deep mixing (DJM) mostly employed in a group column type creates stabilized soil with a strength around $500\,\text{kN/m}^2$. The Swedish lime columns are ordinarily used at a strength less than $150\,\text{kN/m}^2$. The difference in the strength naturally causes differences in the relative stiffness of stabilized and unstabilized soils, which strongly influences the overall behavior of the improved ground as a system. A further difference is that the Japanese stabilized soils are practically impermeable materials, whereas the stabilized soil in the Nordic applications is considered as vertical drainage.

The major purpose of the Nordic applications is the reduction of settlement, and a group of stabilized soil columns is installed underneath a road embankment or around dwellings. In comparison, the Japanese application was initiated to improve the stability of port facilities such as breakwaters and revetments in which the pattern of application was massive stabilization created in-situ by overlapping stabilized columns. The principle of the deep mixing method in Nordic countries and in Japan is the same, but their applications are different.

The current chapter describes the column installation patterns and typical applications in Japan which will help the project owner and geotechnical designer judge the applicability of deep mixing to the project at hand.

2 PATTERNS OF APPLICATIONS

2.1 Size and geometry of the stabilized soil element

Since 1970s, the mechanical deep mixing method (DMM) has frequently been applied to the improvement of soft clays, organic soils and sandy soils for various purposes and in various ground conditions in on-land and marine constructions (Terashi *et al.*, 1979; Terashi and Tanaka, 1981; Kawasaki *et al.*, 1981).

A round column of stabilized soil is produced by a single stroke (penetration and withdrawal) of a one-shaft deep mixing machine. As a deep mixing machine in general has two to eight mixing shafts and blades in Japan, the stabilized soil produced by a single stroke consists of several round columns partially overlapped each other. Such a stabilized soil produced by a multiple shafts machine is also called "column" but sometimes called "element" in this book to avoid confusion with a single round column. The size and geometry of stabilized soil element depend on the diameter of mixing blade and the shaft arrangement as shown in Figure 4.1 (Coastal Development Institute of Technology, 2008; Public Works Research Center, 2004).

The dual shaft machine is most commonly used for the dry method in Japan. The diameter of mixing blade and spacing of mixing shafts are typically 1.0 m and 0.8 m respectively, and the cross sectional area of the stabilized soil element is about 1.50 m². In the wet method for on-land constructions, a stabilized soil element consisted of overlapping two to four round columns has frequently been adopted. The diameter of mixing blade and spacing of mixing shaft are typically 1.0 to 1.3 m and 0.8 to 1.1 m respectively, and the cross sectional area of the stabilized soil element is about 1.50 to 5.00 m². In the wet method for marine constructions, a stabilized soil element consisted of overlapping four or eight round columns has frequently been adopted. The diameter of mixing blade and spacing of mixing shafts are typically 1.0 to 1.6 m and 0.8 to 1.2 m, and the cross sectional area is about 2.2 m² for four columns arrangement and 4.6 to 5.7 m² for eight columns arrangement, respectively.

2.2 Column installation patterns by the mechanical deep mixing method

A stabilized soil mass with any arbitrary shape can be formed in a ground by the installation of stabilized soil columns/elements. Figure 4.2 shows a typical column

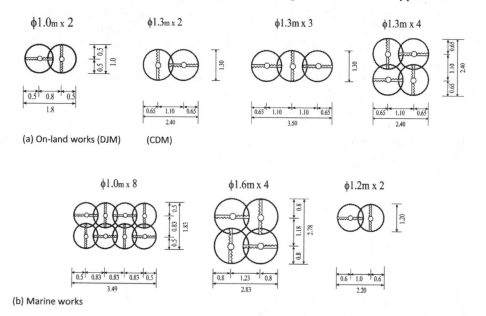

Figure 4.1 Typical arrangements of mixing shafts (Coastal Development Institute of Technology, 2008; Public Works Research Center, 2004).

installation patterns; the group column type, wall type, grid type and block type. To improve the foundation ground for permanent and/or important structures, the block, wall or grid type installation patterns have frequently been applied in Japan. The group column type installation pattern has usually been applied to the foundation of light weight or temporary structures, or embankment in order to improve the stability and/or reduce vertical and horizontal displacements. Careful examination is necessary if the group of individual columns is considered to improve stability of an embankment slope, because progressive failure of individual columns by bending are anticipated due to the low tensile and bending strengths of the stabilized soil (Karastanev *et al.*, 1997; Kitazume *et al.*, 2000; Kitazume and Maruyama, 2007).

A suitable column installation pattern should be chosen considering the type, size and importance of the superstructure, the purpose and function of improvement, the construction cost, and the site condition. The execution of overlapping requires a deep mixing machine with sufficient power and stability, high quality control regarding positioning and verticality of the machine, and tracing the mixing shafts and blades locations during production. The selection of column installation pattern should accompany consideration on the level of execution and quality control techniques available locally.

2.2.1 Group column type improvement

In the group column type improvement, isolated stabilized soil columns or elements are installed in rows with rectangular or triangular arrangements in a ground. The execution requires a relatively short period, and the volume of improvement is small. As the horizontal resistance of the isolated column is not so high, the group column has been widely applied to foundations of relatively low embankments and light weight structures in order to reduce settlement and to increase stability (Figure 4.2(a)).

According to case histories of the Japanese dry method, the improvement area ratio a_s, defined as the ratio of the cross sectional area of stabilized soil columns to the total area of soft ground to be improved by the columns, is typically 0.3 to 0.5 when

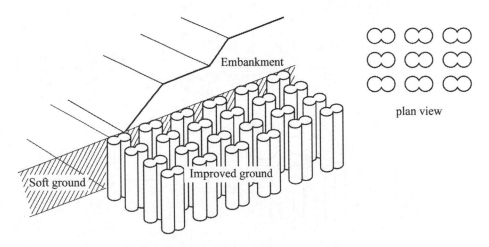

Figure 4.2(a) Group column type improvement.

the settlement reduction of an embankment is the major purpose. When the stability of the embankment side slope is the purpose of improvement, a larger a_s of 0.5 to 0.8 is preferred (Terashi *et al.*, 2009).

With increasing improvement area ratio, the spacing between adjacent columns becomes smaller. In the end the columns touch each other at the periphery which is called tangent columns. When the tangent columns are installed to produce walls that are oriented perpendicular to the embankment centerline, the installation pattern is called "tangent wall" as shown in Figure 4.2(b). Even with the same improvement area ration, the tangent wall is expected to function better than the individual columns when stability of the embankment slope is concerned.

When the columns are in contact with adjacent columns in both directions without overlapping, the installation pattern is called "tangent block" as shown in Figure 4.2(c). As the improvement area ratio exceeds 0.75, a tangent block is expected to show the best function in the group column type improvement and frequently applied to improve foundations of embankment side slope and small buildings.

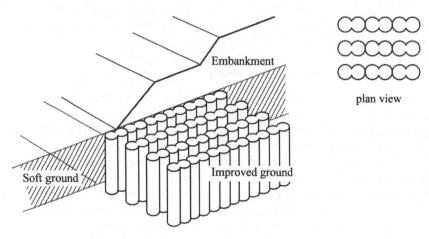

Figure 4.2(b) Group column in tangent wall arrangement.

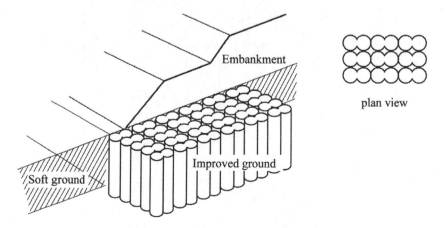

Figure 4.2(c) Group column in tangent block arrangement.

2.2.2 Wall type improvement

In the wall type improvement, the long walls of stabilized soil with or without short walls oriented perpendicular to the centerline of superstructures are produced by overlapping adjacent columns (Figure 4.2(d)). The long wall is expected to function to bear the weight of superstructure and other external loads, and transfer them to the deeper stiff layer. The spacing of the long walls is typically two to three times of their thickness in many cases. The short wall is expected to function to combine the long walls tightly in order to increase the rigidity of the total improved soil mass. The volume of improvement is smaller and is less expensive than the block type improvement. The improvement requires precise execution of overlapping of long and short walls. The column installation pattern is often employed to increase the stability of earth retaining structures such as revetment, to support embankment slopes and to support sheet pile walls.

2.2.3 Grid type improvement

The grid type improvement is an intermediate type between the block type improvement and the wall type improvement. The stabilized soil columns are installed by overlapping execution so that grid shaped improved masses are produced in a ground (Figure 4.2(e)). This pattern is highly stable next to the block type improvement and its cost ranges between the block type and wall type improvements. This improvement has usually been applied for increasing the bearing capacity and stability of ground in marine constructions. The stabilized soil columns function to prevent the shear deformation of original soil within the grid during an earthquake, which can function to prevent the pore water pressure generation there. According to the function, this improvement pattern has also been applied for preventing liquefaction in sandy ground.

As a modified improvement pattern for the grid type, a complicated column installation pattern such as a honeycomb type improvement has sometimes been applied in Europe. However, it should be noted that such installation demands an extremely high construction accuracy in production and three dimensional analysis in the design.

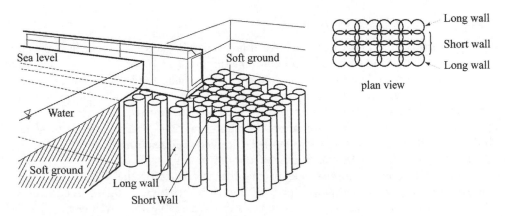

Figure 4.2(d) Wall type improvement.

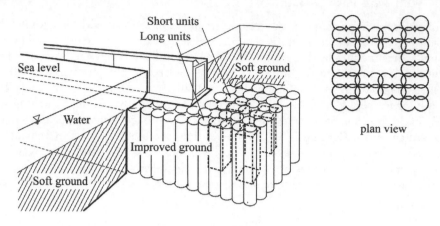

Figure 4.2(e)   Grid type improvement.

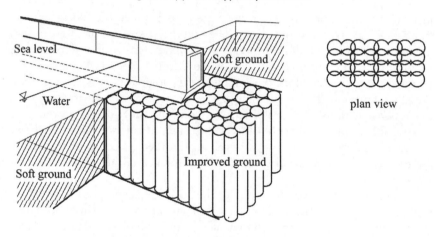

Figure 4.2(f)   Block type improvement.

2.2.4   Block type improvement

In the block type improvement, a huge improved soil mass is formed in a ground by overlapping all the stabilized soil columns (Figure 4.2(f)). This improvement can achieve the most stable improvement, but the cost is higher and the execution period is longer than the other types of improvement. This type of improvement is normally applied to heavy and permanent structures such as breakwater and sea revetment in port and harbor structures. With large width and impermeable characteristics of stabilized soil, this improvement has often been applied to a disposal area for preventing the leaching of waste chemicals to the surroundings.

Table 4.1 shows a comparison of the characteristics of the above mentioned improvements (Coastal Development Institute of Technology, 2002). It is concluded that the block type improvement achieves the most stable improvement, but it is

Table 4.1 Characteristics of improvement types (Coastal Development Institute of Technology, 2002).

Type	Function	Cost	Installation	Design Consideration
Group column type	Effective and efficient for settlement reduction under the full height of embankment.	Installation requires short period, and volume of improvement is small. Low cost.	Overlapping operation is not required.	Requires settlement analysis and bearing capacity of individual columns as a pile foundation.
Group column type by tangent arrangements	Where lateral loads are small, high stability is obtained.	Volume of improvement is larger with increasing improvement area ratio. Cost is lower than wall, grid or block type.	Although overlapoperation is not required, accurate positioning and verticality of columns required for producing tangent arrangement.	Requires design on overall stability and on internal stability of tangent columns.
Wall type	When external loads dominate in one direction, walls function effectively to improve stability.	Volume of improvement is smaller than block type. Lower cost than grid or block type.	Requires precise operation of overlapping of long and short walls.	Requires consideration of unimproved soil between walls. Wall spacing and depth of short wall affected by internal stability.
Grid type	Highly stable next to block Type.	Cost range is between block type and wall type.	Installation sequences are complicated because a grid shape must be formed.	Requires design on three-dimensional internal stress.
Block type	Large solid block resists external loads. Highly stable.	Volume of improvement is greater than other types. High cost.	Takes longer time because all columns are overlapped.	Design of size of block is in the same way as the gravity structures.

expensive. The wall type improvement and the grid type improvement also achieve stable improvement, and are more economical, but both require high quality continuous overlapping executions.

2.3 Column installation pattern by high pressure injection

The improvement pattern of the high pressure injection method is usually either the tangent arrangement or the block type improvement. The block type improvement is desirable for reinforcement and seepage shutoff.

3 IMPROVEMENT PURPOSES AND APPLICATIONS

3.1 Mechanical deep mixing method

The deep mixing by mechanical mixing process has been applied to improvements of soft clays, organic soils and loose sandy soils for various purposes since the middle of the 1970s. The mechanical mixing system employed in Japan consists of vertical rotary shaft(s) with mixing blades at the end of each shaft. Figure 4.3 shows typical improvement purposes of the DM method in Japan for clayey soils and sandy soils (Coastal Development Institute of Technology, 2002). Applications to clayey and organic soils include increasing bearing capacity, reducing settlement, increasing passive earth pressure, reducing active earth pressure and increasing horizontal resistance of pile and sheet wall. Applications to sandy ground, on the other hand, include increasing bearing capacity, reducing settlement and preventing liquefaction.

Figures 4.4(a) and 4.4(b) show typical applications of the deep mixing method to on-land constructions and marine constructions respectively. In on-land constructions,

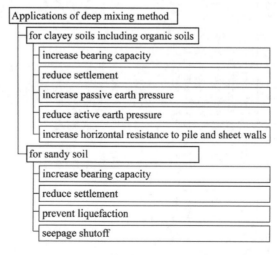

Figure 4.3 Typical improvement purposes of mechanical DM method (Coastal Development Institute of Technology, 2002).

the deep mixing method has been applied to embankments, oil tanks, and building foundations, while the deep mixing method has been applied to breakwaters, sea revetments and piers in marine construction. Other than those exemplified in the figure, the deep mixing is also applied for seepage shutoff, vibration and displacement barrier and immobilization of contaminated soil.

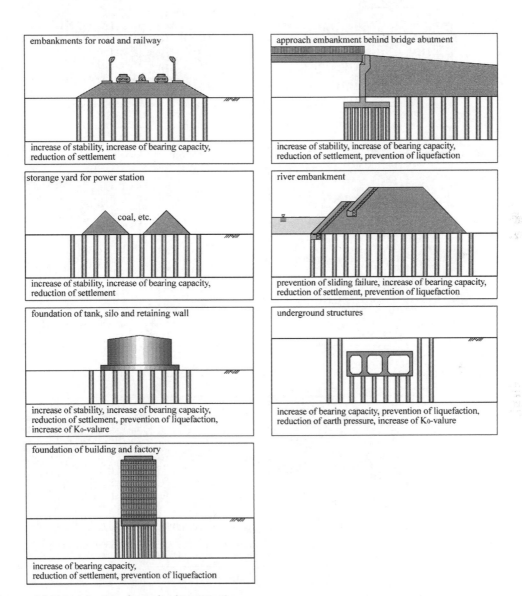

(a) DM applications for on-land construction.

Figure 4.4 Deep mixing applications.

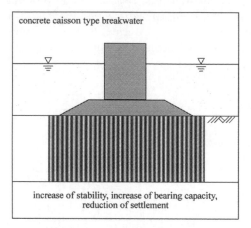

concrete caisson type breakwater

increase of stability, increase of bearing capacity,
reduction of settlement

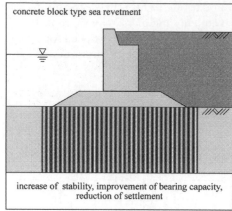

concrete block type sea revetment

increase of stability, improvement of bearing capacity,
reduction of settlement

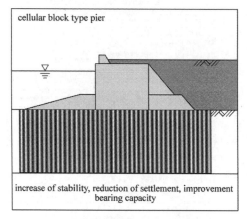

cellular block type pier

increase of stability, reduction of settlement, improvement
bearing capacity

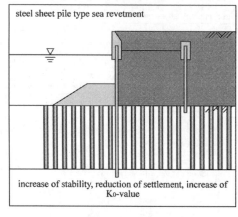

steel sheet pile type sea revetment

increase of stability, reduction of settlement, increase of
K_0-value

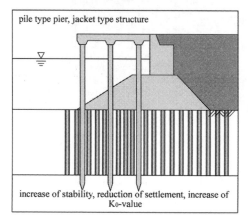

pile type pier, jacket type structure

increase of stability, reduction of settlement, increase of
K_0-value

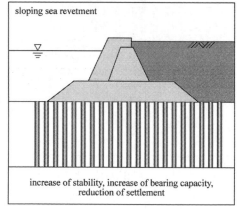

sloping sea revetment

increase of stability, increase of bearing capacity,
reduction of settlement

(b) Deep mixing applications for marine constructions.

Figure 4.4 Continued.

3.2 High pressure injection

High pressure injection has also been applied to improvements of soft clays, organic soils and loose sandy soils for various purposes. Figure 4.5 shows typical improvement purposes of high pressure injection, which include increasing stability of ground for shield machine as well as increasing passive earth pressure, reducing active earth pressure, increasing horizontal resistance of pile and sheet wall and preventing liquefaction (Japan Jet Grouting Association, 2011). The high pressure injection is especially useful when the construction should be carried out in a site with headroom restriction. Such an example is the retrofit of foundation underneath an existing building, which is often carried out from the basement. Figure 4.6 shows typical applications of the high pressure injection technique for support for a shield tunnel (Japan Jet Grouting Association, 2011).

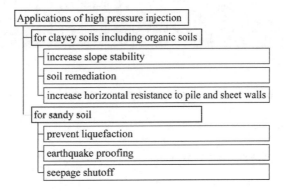

Figure 4.5 Typical improvement purposes of high pressure injection techniques (Japan Jet Grouting Association, 2011).

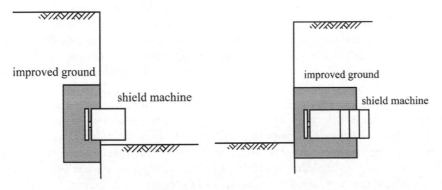

Figure 4.6 High pressure injection applications (Japan Jet Grouting Association, 2011).

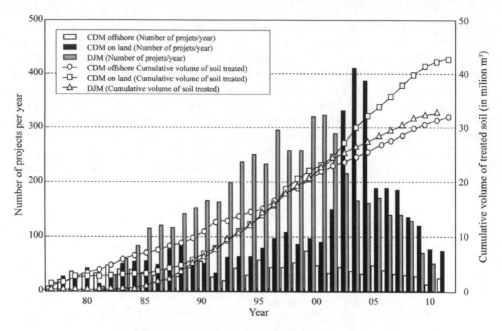

Figure 4.7 Statistics of deep mixing method works in Japan.

4 APPLICATIONS IN JAPAN

4.1 Statistics of applications

4.1.1 Mechanical deep mixing

Figure 4.7 shows the statistics of the number of deep mixing projects and the accumulative volume of stabilized soil in Japan. The total volume of stabilized soil by the deep mixing method from 1977 to 2010 reached more than 100 million m^3; 72.3 million m^3 for the wet method (CDM) and 32.1 million m^3 for the dry method (DJM).

Figure 4.8 shows the purposes of the dry method application found in 4,300 projects on-land (Terashi *et al.*, 2009). The majority are for the issues associated with embankment construction; 37.7% for embankment stability, 26.7% for settlement reduction, and 4.1% for reduction of the impact of embankment construction to nearby structures. Following to embankment are the improvement of foundation for various structures and bridge abutments.

Figure 4.9 shows a comparison of the stabilized soil volumes by the wet method, between on-land applications and marine applications from 1977 to 1999 (Coastal Development Institute of Technology, 2002). For on-land applications, the method has mainly been applied to improve slope stability, to prevent building subsidence and to improve bearing capacity of foundation. In approximately 50% of marine applications, it has been applied to improve foundation of revetment.

Figures 4.10 shows the statistics of the specifications of improved ground in on-land constructions both for the wet and dry methods of deep mixing (Public Works Research Center, 2004).

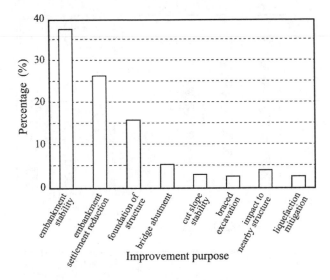

Figure 4.8 Purpose of applications by dry method (Terashi *et al.*, 2009).

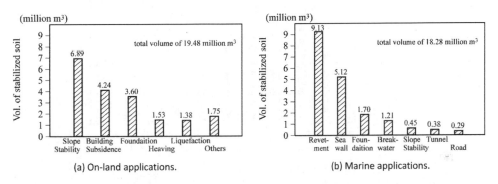

Figure 4.9 Volume of stabilized soil for on-land and marine applications (Coastal Development Institute of Technology, 2002).

Figure 4.10(a) shows the ratio of the width of improvement, *B* to the depth of improvement, *H*. The ratio *B/H* is dependent on the purpose of improvement such as improving stability and bearing capacity, and reducing settlement. About 25% of improved ground has adopted *B/H* smaller than 0.5, most of which are for settlement reduction. Another 25% have a ratio ranging 0.5 to 1.0, and in 50% projects, the *B/H* ratio is larger than 1.0.

Figure 4.10(b) shows the statistics of design strength, where the design strength in the range of 0.2 to 0.6 MN/m² is dominant. The design strength is slightly different depending on the improvement purposes. For the purposes of stability and settlement reduction of an embankment, a design strength smaller than 0.5 MN/m² is preferred. For the purpose of the bearing capacity and horizontal reinforcement of a bridge abutment, a design strength of 0.2 to 0.7 MN/m² is often adopted.

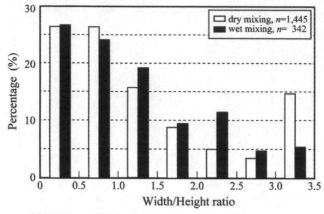

(a) Width and height ratio.

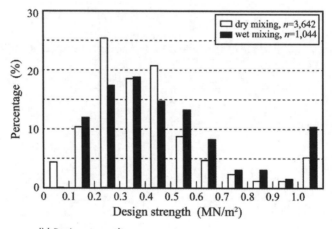

(b) Design strength.

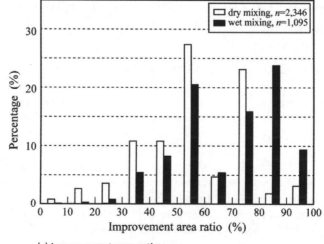

(c) Improvement area ratio.

Figure 4.10 Statistics of deep mixing improved grounds (Public Works Research Center, 2004).

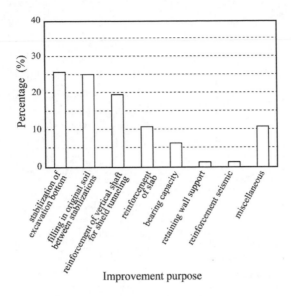

Figure 4.11 Application of high pressure injection (Superjet Association, 2011).

Figure 4.10(c) shows the statistics of the improvement area ratio, a_s. The a_s in the range of 0.5 to 0.6 corresponds to the group of individual columns, and 0.8 to 0.9 corresponds to the tangent block and block type improvements. The improvement area ratio is influenced by the improvement purpose. For the stability and settlement reduction of an embankment, the improvement area ratio is ranging from 0.5 to 0.6. For the settlement reduction of a low embankment, an improvement area ratio smaller than 0.3 is adopted where additional surface stabilization or geotextile is also applied to reduce uneven settlement. For the purpose of the bearing capacity and horizontal reinforcement of a bridge abutment, an improvement area ratio ranging 0.7 to 0.8 are dominant.

4.1.2 Statistics of high pressure injection

Figure 4.11 shows the purposes of the Superjet technique, one of the double fluid high pressure injection techniques (Superjet Association, 2011). The majority are about 32% for stabilization of excavation bottom, 29% for filling in original soil between stabilizations, and 23% for reinforcement of vertical shafts for shield tunneling. "Filling in original soil between stabilizations" means the injection of binder slurry into the unstabilized soil between stabilized soil columns or between the stabilized soil columns and sheet pile wall, with the expectation to improve strength or seepage shut-off.

4.2 Selected case histories

Among many applications of DMM in Japan, 8 examples are selected and briefly introduced in this section: group column type improvements for settlement reduction, tangent group column type improvement for embankment stability, grid type improvement for liquefaction prevention, block type improvements for pile foundation and

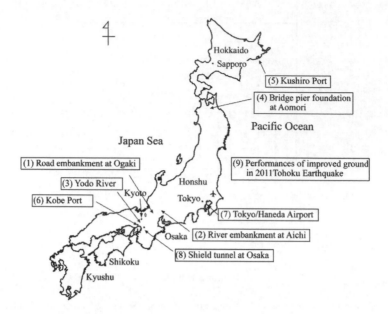

Figure 4.12 Locations of projects introduced.

earthquake disaster mitigation, and block type improvement for foundation in marine construction. The locations of exemplified projects are shown in Figure 4.12.

4.2.1 Group column type – individual columns – for settlement reduction (by courtesy of the Dry Jet Mixing Method Association)

4.2.1.1 Introduction and ground condition

An application of the dry method (DJM) to a road embankment is shown here, where stabilized soil columns were installed in the group column type arrangement to reduce settlement due to the embankment (by courtesy of the Dry Jet Mixing Method Association). The ground condition at the site, Ogaki of Gifu Prefecture, is shown in Figure 4.13(a). The ground consisted of some stratified layers to the depth of -35 m, including an organic clay layer, silty layers, silty clay layers and sand layers. The SPT N-values of the silty clay layer and silty fine sand layer were quite small, and smaller than 10. Especially, the upper silty layer at a depth of -7 to -19 m, was quite soft with the SPT N-value of almost zero.

4.2.1.2 Ground improvement

The stabilized soil columns were constructed under the entire width of embankment as shown in Figure 4.13(b), whose width was 52.5 m. The length of the columns was 30 m. The diameter and spacing of the columns were 1.0 m and 2.5 m respectively, and whose improvement area ratio, a_s was as small as 0.125. The design strength of the stabilized soil column in terms of unconfined compressive strength, q_u was ranging from 670 to 1,050 kN/m². Assuming the field to laboratory strength ratio, q_{uf}/q_{ul} was

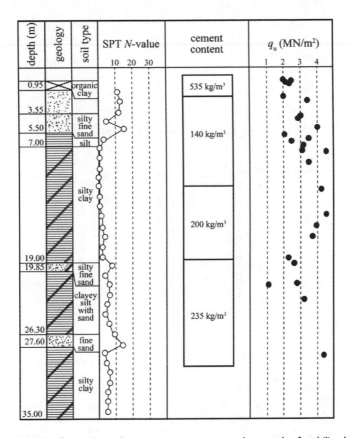

Figure 4.13(a) Ground condition, cement content, and strength of stabilized soil.

0.33 in the design, 140, 200 and 535 kg/m³ of blast furnace slag cement type B were mixed to achieve the design strength.

Figure 4.13(c) shows the dry mixing machines in operation. As the improvement length was large, dual shafts DJM machines were used for the project, in which the spacing of mixing shafts were expanded to 2.5 m. A total of 2,458 stabilized soil columns were constructed, which came up the total volume of 42,972 m³. After the construction, unconfined compression tests were carried out on the core samples for quality assurance. Figure 4.13(a) also shows the strength profile along the depth. The measured unconfined compressive strength ranged from 1 to 5.5 MN/m², which were quite larger than the design strength.

4.2.2 Group column type – tangent block – for embankment stability

4.2.2.1 Introduction and ground condition

An application of the dry method to stability of a river embankment is shown here, where the group column improvement by tangent block was applied to improve the stability of an embankment of 13 m in height (Public Works Research Center, 2004).

Figure 4.13(b) Sectional view of DM improved ground.

Figure 4.13(c) DJM machines in operation.

The ground condition at the site was stratified layers as shown in Figure 4.14(a). The soft layers included an organic silty clay at shallow depth and two silty clay layers down to −26.55 m, with a fine sand layer in between. The SPT N-values of the soft layers were quite small.

4.2.2.2 Ground improvement

The column installation pattern was the tangent block by group columns as shown in Figure 4.14(b), whose width and depth of improvement were 5.2 m and 26.5 m

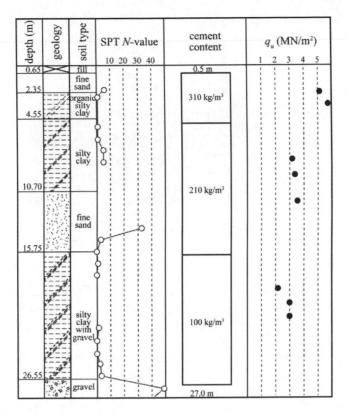

Figure 4.14(a) Ground condition, cement content and strength of stabilized soil.

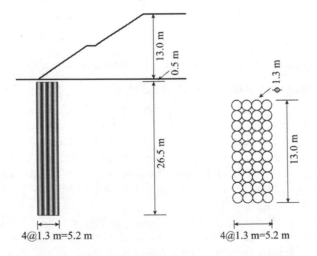

Figure 4.14(b) Sectional view of DM improved ground.

Figure 4.14(c) DJM machines in operation.

respectively, and the improvement area ratio was 0.785. The design strength of the stabilized column, q_u was ranging from 600 to 900 kN/m^2 for the silty clay layers. Assuming the strength ratio, q_{uf}/q_{ul} was 0.25 in the design, 100, 210 and 310 kg/m^3 of blast furnace slag cement type B were mixed to achieve the design strengths.

Figure 4.14(c) shows the dry mixing machines in operation. As the improvement length was large, DJM2090 machines were used for the project. A total of 1,540 stabilized columns were constructed, which came up the total volume of 51,800 m^3. After the execution, unconfined compression tests were carried out on the core samples for quality assurance. Figure 4.14(a) also shows the strength profile along the depth. The unconfined compressive strength of the stabilized soils was ranging from 2 to 6 MN/m^2 depending on the design strength. Different strengths were achieved depending on the soil type and binder factor, but uniform strength was obtained within the same layer. The measured strengths were larger than the design strength.

4.2.3 Grid type improvement for liquefaction prevention

4.2.3.1 Introduction and ground condition

Yodo River flows from Lake Biwa to Osaka Bay through Osaka City. Due to the Hyogoken-Nambu Earthquake in January 1995, the river dike was heavily damaged for a length of 1.8 km because of slope failure due to ground liquefaction (Kamon, 1996). A representative cross section of the damaged dike is shown in Figure 4.15. The top portion of the river dike sank down about 3 m. The damaged dike had to be restored very quickly because there was a risk of flooding during the rainy season which usually commenced in June.

The ground condition at the site is shown in Figure 4.15(b). The ground consisted of a sandy layer and a clay layer. As the SPT N-value of the sandy layer was smaller than 10, the liquefaction might take place again in an earthquake attack in the future.

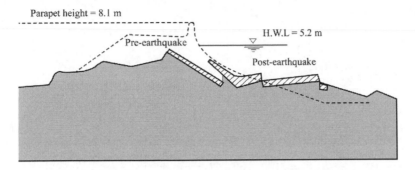

Figure 4.15(a) Cross section of the Yodo River dike after the Hyogoken-Nambu Earthquake.

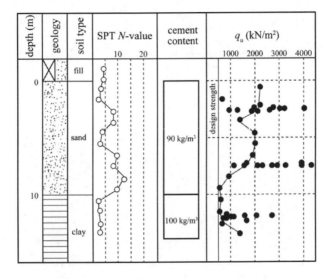

Figure 4.15(b) Ground condition at the site.

4.2.3.2 Ground improvement

Because there were many residential houses in the neighborhood along the river dike, it was necessary to avoid noise and vibratory problems during the construction. This was one of the reasons why the deep mixing method was applied there. The cross section of the improved ground is shown in Figure 4.15(c), where grid type improvement was applied to prevent liquefaction of the ground and to improve the stability of the river embankment. The grid of the stabilized soil columns was about 5 m by 5.4 m. The design strength of the stabilized column, q_u was 500 kN/m^2. Assuming the strength ratio, q_{uf}/q_{ul} was 0.25 in the design, 90 or 100 kg/m^3 of blast furnace slag cement type B were mixed to achieve the design strength for the sandy layer and clay layer respectively.

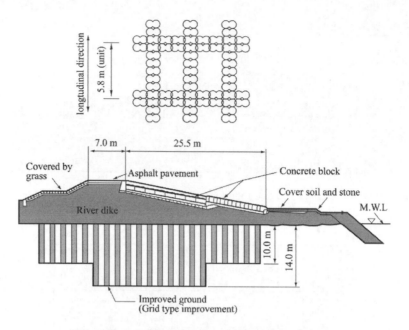

Figure 4.15(c) Sectional view of DM improved ground.

Figure 4.15(d) Deep mixing machines in operation.

Figure 4.15(d) shows the deep mixing machines in operation. The river embankment of more than 7 km long was improved. In the construction period, more than 50 DJM and CDM machines were simultaneously put into operation for rapid restoration. After the construction, unconfined compression tests were carried out on the core samples for quality assurance. Figure 4.15(b) shows the strength profile along the depth. The measured unconfined compressive strength ranged from 0.5 to 4.5 MN/m^2.

Table 4.2 Ground condition.

Soil	depth	SPT N-value	γ (kN/m^3)	ϕ	c (kN/m^2)
Humic soil (Ap)	0		11	0	8
Alluvial clay (Ac)	0		16	0	20
Alluvial sandy soil (As)	8		17	32	0
Diluvial clay (Noc2)	10		16	0	256

4.2.4 Block type improvement to increase bearing capacity of a bridge foundation (Tokutomi et al., 2009)

4.2.4.1 Introduction and ground condition

The dry method was applied to a bridge pier foundation for the rapid Shinkansen train. The construction site was an alluvial flat and marshy area in Shichinohe Edasawa, Aomori Prefecture. The soil properties of the soil layers are tabulated in Table 4.2. A humic soil layer, Ap was sedimented at a depth of 5 to 7 m which was underlain by an alluvium layer, Ac at a depth of 10 to 12 m. The water content of the humic soil layer was considerably high, 150 to 1,000% and its SPT N-value was almost 0. An alluvial sandy layer, As and a diluvial clay layer, Noc2 were stratified underneath. The SPT N-value of the diluvial clay layer was around 10 and the undrained shear strength was 256 kN/m^2.

4.2.4.2 Ground improvement

In order to assure the stability of the pier and reduce its settlement, the deep mixing method was applied. The layout of the deep mixing improvement is shown in Figure 4.16(a), where the rectangular area of 12 m by 12.3 m was stabilized down to about 7 or 8 m from the bottom of the bridge foundation. The humic soil layer was also stabilized by the method. The improvement area ratio, a_s was 0.96 and the design strength of the stabilized soil was 1,050 kN/m^2 for assuring the stability of the pier. A series of laboratory mix tests was carried out to determine the mix condition, in which three types of binder, blast furnace slag cement type B, cement-based special binders for high water content soils and for organic soils were used. A field trial test was also carried out to investigate the strength ratio of field strength to laboratory strength, and to determine the amount of binder for production. Based on the test results, the cement-based special binder for high water content soils was selected and the binder contents for different layers were determined from place to place; 440 to 620 kg/m^3 for the Ap layer, 200 to 280 kg/m^3 for the Ac layer, and 110 to 150 kg/m^3 for the As layer.

After the construction, the unconfined compressive strength of the stabilized soil was measured on the core samples. Figure 4.16(b) shows an example of the strength profile along the depth, where the humic soil layer was not stabilized, and the upper Ac layer was stabilized with the binder content of 530 kg/m^3 and the lower Ac and the As layers were stabilized with α of 200 kg/m^3. The strength of the Ac layer was quite large due to the large binder content. The strength of the lower Ac and the As layers were smaller than that in the upper Ac layer, but quite higher than the design strength.

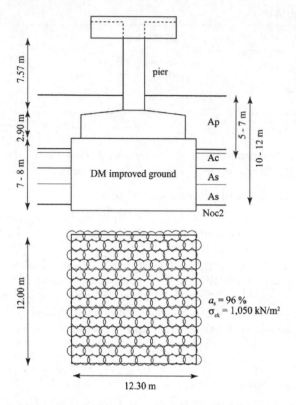

Figure 4.16(a) Sectional view of DJM improved ground.

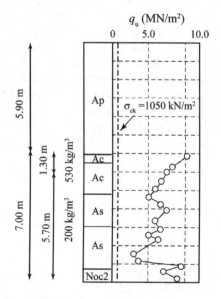

Figure 4.16(b) Ground condition at the site.

After the execution, the vertical loading tests were also carried out on five stabilized soil columns and confirmed the quite small settlement.

4.2.5 Block type improvement for liquefaction mitigation (Yamazaki, 2000)

4.2.5.1 Introduction and ground condition

Kushiro Port, Hokkaido, had been subjected to several huge earthquakes, where liquefaction took place in reclaimed land. Figures 4.17(a) and 4.17(b) show a sectional

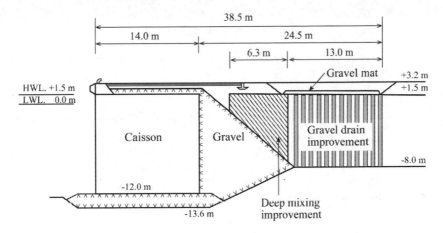

Figure 4.17(a) Application of DMM at Kushiro Port.

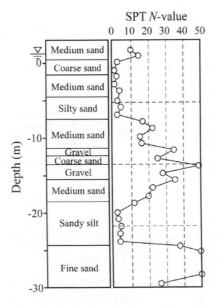

Figure 4.17(b) Ground profile at No. 2 Pier at Kushiro Port.

(c) Unimproved area.   (d) Improved area.

Figure 4.17(c)   Damages of revetment by liquefaction and *(d)* Performance of improved area in Hokkaido Toho-Oki Earthquake (Yamazaki, 2000).

view of the sea revetment and the ground profile at No. 2 Pier at Kushiro Port. The reclaimed layers up to the depth of -8.0 m consisted of several sand layers whose SPT N-value was quite small. In 1993, the pier was subjected to the large earthquake of Magnitude of 7.9 (Kushiro Oki Earthquake) and heavily damaged by liquefaction in the reclaimed layers (Inatomi *et al.*, 1997). Displacements of the concrete caisson of 0.20 to 0.305 m in horizontal and 0.30 to 0.15 m in vertical were reported.

4.2.5.2   Ground improvement

After the earthquake, the reclaimed sand layers were improved by the wet method of deep mixing and the gravel drain method, both for liquefaction mitigation. Figure 4.17(a) shows an application at Kushiro Port for liquefaction mitigation of backfill, where sandy soil was stabilized by the wet method with a block type improvement with 1.0 in improvement area ratio. The design field strength, q_u, was as small as $100\,\mathrm{kN/m^2}$, which was sufficient to increase liquefaction resistance of the sand (Zen *et al.*, 1987).

The revetment was subjected to the Hokkaido Toho-Oki earthquake of Magnitude of 8.5 later in 1994. Figures 4.17(c) and 4.17(d) show the performance of unimproved and improved areas respectively (Yamazaki, 2000). Figure 4.17(c) shows damages of the revetment by liquefaction and cracks at the unimproved area. However, due to the ground improvement, negligible damage took place at the improved area, which has revealed the high applicability of the deep mixing method for liquefaction mitigation and reinforcement of sea revetment.

4.2.6   Grid type improvement for liquefaction prevention

4.2.6.1   Introduction and ground condition

The wet method was applied to the foundation of a building at Kobe Port, where sandy ground was improved by a grid type improvement to prevent excess pore water pressure generation during an earthquake by restraining the shear deformation of a

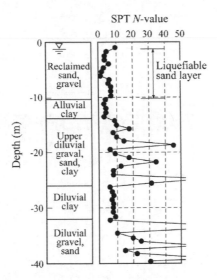

Figure 4.18(a)   Ground condition at Kobe Port (Tokimatsu *et al.*, 1996; Suzuki *et al.*, 1996).

liquefiable sand layer. A 14-story building located on Meriken Wharf in Kobe was experienced the Hyogoken-Nambu earthquake in 1995. Figure 4.18(a) shows the soil profile at the site which consisted of 10 to 12 m of soft reclaimed sand and gravel layers over the seabed (Tokimatsu *et al.*, 1996; Suzuki *et al.*, 1996). The seabed soil consisted of alternating layers of clay, sand and gravel. As small SPT N-value lower than 10, the top layer had been anticipated to liquefy due to earthquake excitation. The building was supported by cast-in-place reinforced concrete piles with a diameter of 2.5 m extending down to dense diluvial sand and gravel at a depth of 33 m.

4.2.6.2   Ground improvement

The section and plan diagrams of the deep mixing improved ground are shown in Figure 4.18(b). A grid type improvement was applied to prevent liquefaction in the upper loose fill. More than 1,000 stabilized soil columns with a diameter of 1.0 m were constructed where 200 kg/m^3 of blast furnace slag cement type B was mixed to obtain 2,400 kN/m^2 in q_u for the sand layer and 3,600 kN/m^2 for the clay layer. The improvement area ratio was approximately 0.2. The unconfined compressive strength of the stabilized soil after about six weeks curing was 4 to 6 MN/m^2 (Suzuki *et al.*, 1996).

The building and the improved ground were subjected to the large earthquake in 1995. As the reclaimed ground around the building was not improved, liquefaction took place during the earthquake. Figure 4.18(c) shows the damage of the quay wall near the building after the earthquake. The concrete caisson type quay walls were subjected to a large excess pore water pressure due to the liquefaction, and they on the

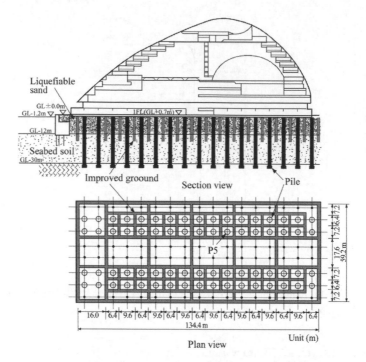

Figure 4.18(b) Grid type improved ground.

Figure 4.18(c) Damage of sea revetment near the building.

west, south and east displaced horizontally towards the sea by 1 m, 2 m, and 0.6 m respectively and the ground behind the quay walls settled by 0.5 m, 0.6 m and 0.3 m. Sand boils and ground cracks were observed at the ground surface outside of the building. In the building, however, there was no crack at the surface of the improved ground as shown in Figure 4.18(d). The head of the cast-in-place piles supporting the building

Figure 4.18(d) Parking area in the building.

was found to be intact. Moreover, negligible differential settlement was observed on the first floor of the building. They have indicated that the grid type improvement to restrain the shear deformation of loose sand could mitigate the liquefaction damage to pile foundation and superstructure.

4.2.7 Block type improvement for the stability of a revetment (Kitazume, 2012; Kawamura et al., 2009)

4.2.7.1 Introduction and ground condition

Tokyo/Haneda International Airport was founded in 1931 as the first primary airport in Japan. The airport had been expanded several times to cope with the rapid increase in air transportation. In order to cope with the recent and expected future increase in air transportation, the construction of a fourth runway was commenced in 2006 and completed in 2010 (Figure 4.19(a)). As soft grounds were stratified at the construction site, various ground improvement techniques including the sand drain method, the sand compaction pile method and the deep mixing method were employed depending on the location and the requirements.

The ground condition and the major soil properties at the site were extensively studied and are summarized in Figure 4.19(b). The ground can be roughly divided into five layers. The most upper layer between -20 m and around -35 m has a high plasticity index ranging from 60 to 100 and a high water content ranging from 100 to 150%. The undrained shear strength and the pre-consolidation pressure increased linearly with the depth, which indicated the clay was lightly over-consolidated condition of OCR of 1.3. The second upper layer from -35 to -60 m was a clay layer underlain locally by a sand layer. The upper two layers should be improved to increase the stability of superstructures and to reduce the residual settlement of the man-made island for the fourth runway.

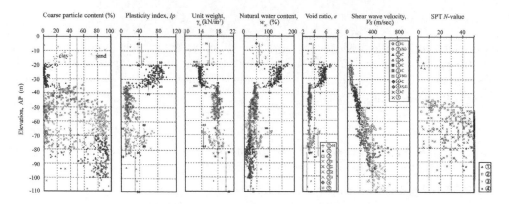

Figure 4.19(a) Sky view of construction site and existing airfield on March 15th, 2009.

Figure 4.19(b) Soil properties at construction site.

4.2.7.2 Ground improvement

Almost all part of the sea revetment was an embankment constructed on the improved ground by the sand compaction pile method. However, caisson type quays were constructed at two corners of the island (CW and CN revetments). A block type improvement of 60 m in width was constructed to a depth of −45 m, as shown in Figure 4.19(c) for the CW revetment. Table 4.3(a) summarizes the properties of the soil layers (Kitazume, 2012). The table shows that the properties of the four soil layers were much different each other so that the mixing condition should be adequately determined for each layer to assure the design strength of stabilized soil. The mixing conditions were designed as tabulated in Table 4.3(a) based on the laboratory mix tests, where blast furnace slag cement type B of 110 to 165 kg/m^3 was mixed with the soil to obtain the average field unconfined compressive strength, q_{uf}, of 3,375 kN/m^2 at 28 days curing. A total of about 620,000 m^3 soft soils was stabilized by four DM vessels within five months. Figure 4.19(d) shows the DMM vessels in operation.

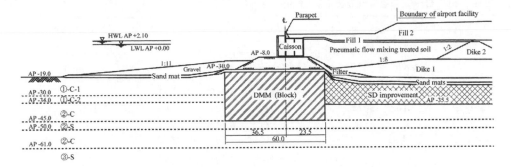

Figure 4.19(c) Cross section of DMM improvement at CW revetment (Kitazume, 2012).

Table 4.3(a) Soil property and cement content.

		Soil property					Binder content	
	Depth	w_n (%)	ρ_t (g/cm^3)	w_l (%)	w_p (%)	I_p	CW rev. (kg/m^3)	CN rev. (kg/m^3)
surface layer	−19 to −21 m	168–177	1.29	–	–	–	165	165
clay 1-C1	−21 to −30 m	132–145	1.34–1.36	132–137	51–54	78–85	140	145
clay 1-C2	−30 to −34 m	42–117	1.38–1.79	41–118	22–47	19–70	130	135
clay 2-C	−34 to −45 m	35–52	1.75–1.84	32–55	18–24	14–31	110	120
sand 2-S	−45 m deeper	37	1.827	–	–	–		

Figure 4.19(d) DMM vessels in operation (by courtesy of the Tokyo/Haneda International Airport Construction Office).

Table 4.3(b) Original cement factor and field strength.

| | | No. of specimen | Field strength, q_{uf28} | | | | Binder content | |
	Depth		Average (kN/m²)	max. (kN/m²)	min. (kN/m²)	COV (%)	CW rev. (kg/m³)	CN rev. (kg/m³)
surface layer	−19 to −21 m	20	3,409	5,608	2,391	27.1	165	165
clay 1-C1	−21 to −30 m	36	4,009	7,981	2,568	28.9	140	145
clay 2-C2	−30 to −34 m	16	3,929	6,116	2,257	21.3	130	135
sand 2-C	−34 to −45 m	44	4,534	7,595	2,617	26.4	110	120
total		116	4,094	7,981	2,257	28.3		

Table 4.3(c) Modified cement factor and field strength.

| | | No. of specimen | Field strength, q_{uf91} | | | | Binder content | |
	Depth		Average (kN/m²)	max. (kN/m²)	min. (kN/m²)	COV (%)	CW rev. (kg/m³)	CN rev. (kg/m³)
surface layer	−19 to −21 m	30	3,568	6,923	2,027	35.8	160	160
clay 1-C1	−21 to −30 m	16	4,010	6,052	2,009	31.7	160	160
clay 1-C2		72	4,410	7,313	2,013	29.8	120	125
clay 2-C2	−30 to −34 m	32	4,561	7,726	2,092	33.9	110	120
sand 2-C	−34 to −45 m	88	3,871	6,076	2,038	26.2	80	85
total		238	4,066	7,313	2,009	31.4		

At 24 to 26 days after the construction, soil sampling was carried out at several points for quality assurance. The stabilized soils sampled were subjected to unconfined compression test to investigate the strength of in-situ stabilized soil. The summary of the test results is shown in Table 4.3(b). The table revealed that the average of q_{uf} was 4,094 kN/m², 20% higher than the target value, and the coefficient of variation (COV) was 28.3%, lower than the design value, 35%. According to that, the mix design was modified for the subsequent construction in order to reduce the amount of ground heaving and the cost, where the amount of cement was decreased by 3 to 27%. Table 4.3(c) shows the q_{uf} values of the stabilized soils after the modification. The table clearly shows that the average strength of the stabilized soils was 4,066 kN/m² and the stabilized soil constructed in-situ satisfied the acceptance criteria.

4.2.8 Jet grouting application to shield tunnel (Noda et al., 1996)

4.2.8.1 Introduction and ground condition

The jet grouting method was applied to reinforcement of the starting point of a rail way shield tunnel with a diameter of 10.8 m in Osaka, where the sand and silty layers

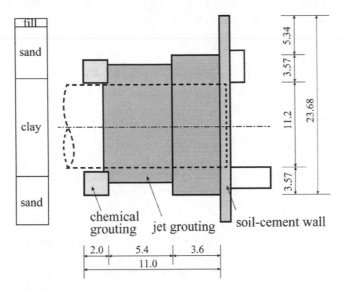

Figure 4.20 Sectional view of high pressure injection improved ground.

Table 4.4 Execution specifications.

Item		Value
high pressure water	pressure	40 MN/m^2
	flow rate	70 to 90 litters/min.
binder	pressure	2 to 3 MN/m^2
	flow rate	180 liters/min.
compressed air	pressure	0.6 to 0.7 MN/m^2
	flow rate	1000 to 2000 liters/min.
rotation speed		5 rpm
withdrawal speed		5 cm/min.

were stratified to a depth of about -25 m. The triple fluid type jet grouting was applied for reinforcing the vertical shaft.

4.2.8.2 Ground improvement

Figure 4.20 shows the sectional view of the site, where a stabilized soil with 9.0 m in width and about 18.3 m in height was constructed by the triple fluid type of jet grouting. The diameter and design strength of the stabilized soil, q_u, were 1.8 m and 1,000 kN/m^2 respectively. The execution specifications are tabulated in Table 4.4.

After the construction, unconfined compression tests on the core samples of the stabilized sand layers and clay layers were carried out. The tests revealed that the q_u values were higher than the design strength, ranging from 4,670 to 8,720 kN/m^2 and

$6,170 \, kN/m^2$ in average for the sand layer, and ranging from 1,090 to $4,700 \, kN/m^2$ and $1,950 \, kN/m^2$ in average for the clay layer.

5 PERFORMANCE OF IMPROVED GROUND IN THE 2011 TOHOKU EARTHQUAKE

5.1 Introduction

The 2011 earthquake off the Pacific coast of Tohoku was a magnitude 9.0 (Mw) undersea mega thrust earthquake that occurred on 11 March 2011. It was the most powerful known earthquake ever to have hit Japan, and one of the five most powerful earthquakes in the world since modern record-keeping began in 1900. The earthquake resulted in a major tsunami that brought destruction along the Pacific coastline of Japan and resulted in the loss of thousands of lives and devastated entire towns. The degree and extent of damage caused by the earthquake and resulting tsunami were enormous, with most of the damage being caused by the tsunami. The aftermath of the 2011 Tohoku earthquake and tsunami included both a humanitarian crisis and massive economic impacts. The tsunami created over 300,000 refugees in the Tohoku region.

The Cement Deep Mixing Association, the Dry Jet Mixing Association and Chemical Grouting Co. Ltd. conducted field surveys in the Tohoku and Kanto areas to investigate the performance of the improved grounds by deep mixing. Table 4.5 summarizes the number of survey for wet and dry methods of deep mixing. Though a few slight deformations were found in some improved grounds, as a whole no serious deformation and damage was found in the improved ground and superstructures even they were subjected to quite a large seismic force. The results of the field surveys are briefly introduced here.

5.2 Improved ground by the wet method of deep mixing

5.2.1 Outline of survey

The field surveys on the improved grounds by the wet method were carried out in Tohoku and Kanto areas after the earthquake. A total of 815 projects were recorded in the CDM Association, while 400 sites of them were surveyed (Table 4.5). No ground deformation and damages were found in the survey.

Table 4.5 Summary of the survey.

	Aomori	Iwate	Akita	Yamagata	Miyagi	Fukushima	Ibaragi	Chiba	Saitama	Tokyo	Kanagawa	Total
Wet method												
no. of projects	28	17	23	21	38	10	77	74	73	302	152	815
no. of surveys	15	9	8	9	27	2	27	28	37	153	85	400
Dry method												
no. of projects	12	4	–	–	19	3	33	49	3	–	–	123
no. of surveys	8	2	–	–	14	1	21	28	3	–	–	77

5.2.2 Performance of improved grounds

5.2.2.1 River embankment in Saitama Prefecture

A part of the river embankment at the Naka River, Saitama Prefecture, was improved by the DM method. The steel pipe sheet pile wall installed at the front of the river embankment was improved by the DM for increasing the horizontal resistance of the wall and stability of the embankment. Jet grouting was also applied between the pipe and the DM ground to increase the lateral resistance. The DM improved ground had about 7.0 m in width and 8.9 m in height, and where the design strength, q_{uck} and the improvement area ratio, a_s were 1.0 MN/m^2 and 0.97 for the upper part, 0.6 M N/m^2 and 0.58 for the lower part, as shown in Figure 4.21(a). The improvement execution was carried out in 2005 by the on-land type machine installed on the small barge, as shown in Figure 4.21(b). No damage was found in the embankment and the improved ground, even they were subjected to the large ground motion of the seismic force of 5.0 upper in Japanese Magnitude-Shindo (seismic intensity scale) as shown in Figure 4.21(c). In contrast to the improved ground, damage at river embankment without any ground improvement was found as shown in Figure 4.21(d).

5.2.3 River embankment in Ibaraki Prefecture

A part of the river embankment at the Oshitsuke Nitta River, Ibaraki Prefecture, was improved by the block type. The original ground beneath the embankment consisted of stratified layers to a depth of −30 m which included sand layers with fine particle. The ground was anticipated to liquefy during an earthquake. The ground was improved by the wet method for preventing liquefaction, which was 16.2 m in width and 35 m in depth, and the design strength of 100 kN/m^2 as shown in Figure 4.22(a). No damage was found in the embankment and the improved ground as shown in Figure 4.22(b), even they were subjected to the seismic force of 6.0 lower in Japanese Magnitude-Shindo (seismic intensity scale). In contrast to the improved ground, damage at the river embankment without any ground improvement was found as shown in Figure 4.22(c).

5.2.4 Road embankment in Chiba Prefecture

A road embankment at Soga, Chiba Prefecture, was improved by the grid type CDM method for liquefaction prevention. The original ground beneath the embankment is a fine sand layer to a depth of −7 m, which was anticipated to be highly liquefiable during an earthquake. The improved ground had about 5.8 m in width, 6.0 m in height, and an improvement area ratio of 0.5, and whose strength, q_u was 200 kN/m^2 (Figure 4.23(a)). No damage was found in the embankment and the improved ground as shown in Figure 4.23(b), even they were subjected to a seismic force of 5.0 upper in Japanese Magnitude-Shindo (seismic intensity scale). In contrast, Figure 4.23(c) shows heavy damage of a road embankment due to liquefaction which was located in the neighborhood and wasn't improved.

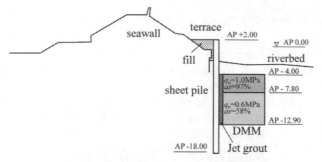

(a) Cross section of the Naka River embankment.

(b) DM machine at site in 2005.

(c) River embankment at the Naka River.

(d) River embankment without improvement.

Figure 4.21 Comparison of CDM improved ground and unimproved ground.

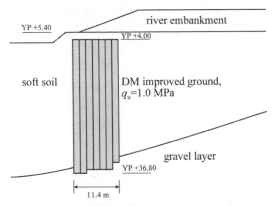

(a) CDM improved ground at the Oshitsuke Nitta River, Ibaragi Prefecture.

(b) River embankment at the Oshitsuke
Nitta River.

(c) Damage at River embankment without
improvement.

Figure 4.22 Comparison of CDM improved ground and unimproved ground.

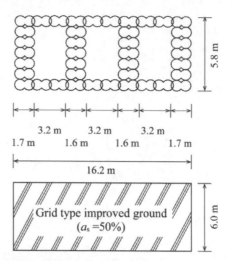

(a) CDM improved ground at Soga, Chiba Prefecture.

(b) Road at Soga. (c) Road without improvement.

Figure 4.23 Comparison of CDM improved ground and unimproved ground.

5.3 Improved ground by the dry method of deep mixing

5.3.1 Outline of survey

The field surveys on the improved grounds by the dry method were carried out in Tohoku and Kanto areas on July and August, which covered improved grounds exceeding a volume of $5,000\,\text{m}^3$ and constructed after 2000. A total of 123 projects were recorded in the DJM Association, while 77 sites of them were surveyed (Table 4.5). No ground deformation and damages were found in the survey.

5.3.2   Performance of improved ground

5.3.2.1   River embankment in Chiba Prefecture

The foundation for a river embankment in Chiba Prefecture was improved by the grid type as shown in Figure 4.24(a), where the width and height of the improved ground were 21.0 m and 21.0 m respectively. The improvement area ratio and the design strength were 0.506 and q_{uck} of 600 kN/m² respectively. No damage was found in the embankment and the improved ground, as shown in Figure 4.24(b).

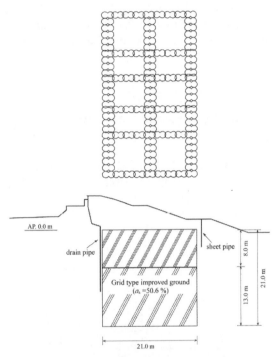

(a) Layout of DJM improved ground.

(b) River embankment after earthquake.

Figure 4.24   DJM improved ground for the river embankment in Chiba Prefecture.

5.3.2.2 Road embankment in Chiba Prefecture

The ground beneath a road embankment with a height of about 9 m in Chiba Prefecture was improved by the grid type (Figure 4.25(a)). The improvement area ratio and the design strength were 0.545 and q_{uck} of 300 kN/m^2 respectively. No damage was found in the embankment and the improved ground as shown in Figure 4.25(b).

5.3.2.3 Box culvert in Chiba Prefecture

The foundation of a box culvert in Chiba Prefecture was improved by the group column type with the improvement area ratio of 0.63 and the design strength, q_{uck} of 400 kN/m^2. The width and height of the improved ground were 14.2 m and 29 m respectively. No damage was found in the embankment and the improved ground as shown in Figure 4.26.

5.4 Improved ground by Grouting method

5.4.1 Outline of survey

The field surveys on the Jet Grout and Chemical Grout improved grounds were carried out in Tohoku and Kanto areas by Chemical Grouting Co. Ltd. No ground deformation and damages were found in the survey.

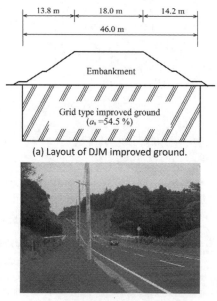

(a) Layout of DJM improved ground.

(b) Road embankment after earthquake.

Figure 4.25 DJM improved ground for the road embankment in Chiba Prefecture.

Figure 4.26 DJM improved ground for the box culvert in Chiba Prefecture.

5.4.2 Performance of improved ground

5.4.2.1 River embankment at Tokyo

A river embankment in Tokyo was improved by the grid type of jet grouting method with a design strength, q_{uck} of $1800\,kN/m^2$. No damage was found in the embankment as shown in Figure 4.27(a). In contrast, Figure 4.27(b) shows heavy damage of embankment without improvement in the neighborhood.

(a) Jet grouting improved ground.

(b) Embankment without improvement.

Figure 4.27 Comparison of jet grouting improved ground and unimproved ground.

(a) Jet grouting improved ground.

(b) Road without improvement.

Figure 4.28 Comparison of chemical grouting improved ground and unimproved ground.

5.4.2.2 Approach road to immerse tunnel in Kanagawa Prefecture

An approach road to the immerse tunnel in Kanagawa Prefecture was improved by the jet grouting method with a design strength, q_{uck} of 85 kN/m². No damage was found in the road as shown in Figure 4.28(a), while Figure 4.28(b) shows heavy damage which was located in the neighborhood and wasn't improved.

5.5 Summary

The field surveys on the improved grounds are briefly introduced. No serious damage was found in the improved grounds by the deep mixing and the superstructures even they were subjected to quite large seismic force. It can be concluded that the soil stabilization by deep mixing guarantees a high performance and high applicability for mitigating damages due to earthquake.

REFERENCES

Coastal Development Institute of Technology (2002) *The Deep Mixing Method – Principle, Design and Construction.* A.A. Balkema Publishers, 123p.
Coastal Development Institute of Technology (2008) *Technical Manual of Deep Mixing Method for Marine Works.* 289p. (in Japanese).

Inatomi, T., Uwabe, T., Iai, S., Tanaka, S., Yamazaki, H., Miyai, S., Nozu, A., Miyata, M. & Fujimoto, Y. (1997) Damage to port structure by the 1994 East Off Hokkaido Earthquake. *Technical Note of the Port and Harbour Research Institute*. No. 856. 583p. (in Japanese).

Japan Jet Grouting Association (2011) *Technical Manual of Jet Grouting Method, Ver. 19*. 82p. (in Japanese).

Kamon, M. (1996) Effect of grouting and DMM on big construction projects in Japan and the 1995 Hyogoken-Nammbu earthquake. *Proc. of the 2nd International Conference on Ground Improvement Geosystems*. Vol. 2. pp. 807–823.

Karastanev, D., Kitazume, M., Miyajima, S. & Ikeda, T. (1997) Bearing capacity of shallow foundation on column type DMM improved ground. *Proc. of the 14th International Conference on Soil Mechanics and Foundation Engineering*. Vol. 3. pp. 1621–1624.

Kawamura, K., Noguchi, T., Kurumada, Y., Junde, S., Watanabe, M. & Nakanishi, M. (2009) Application of cement deep mixing method in Tokyo International Airport D-Runway Project – Vol. 1 Construction outline. *Proc. of the International Symposium on Deep Mixing and Admixture Stabilization*. pp. 247–252.

Kawasaki, T., Niina, A., Saitoh, S., Suzuki, Y. & Honjyo, Y. (1981) Deep mixing method using cement hardening agent. *Proc. of the 10th International Conference on Soil Mechanics and Foundation Engineering*. Vol. 3. pp. 721–724.

Kitazume, M. (2012) Ground improvement in Tokyo Haneda Airport Expansion project. *Ground Improvement*. Vol. 165, Issue GI2. pp. 77–86.

Kitazume, M. & Maruyama, K. (2007) Internal stability of group column type deep mixing improved ground under embankment loading. *Soils and Foundations*. Vol. 47. No. 3. pp. 437–455.

Kitazume, M., Okano, K. & Miyajima, S. (2000) Centrifuge model tests on failure envelope of column type DMM improved ground. *Soils and Foundations*. Vol. 40. No. 4. pp. 43–55.

Noda, H., Noguchi, Y., Hara, M. & Kai, K. (1996) Case of jet grouting for 10.8 m diameter shield. *Proc. of the 2nd International Conference on Ground Improvement Geosystems*. Vol. 1. pp. 295–298.

Public Works Research Center (2004) *Technical Manual on Deep Mixing Method for On Land Works*. 334p. (in Japanese).

Superjet Association (2011) *Technical Manual of Superjet Method*. 44p. (in Japanese).

Suzuki, Y., Saitoh, S., Onimaru, S., Kimura, T., Uchida, A. & Okumura, R. (1996) Grid-shaped stabilized ground improvement by deep cement mixing method against liquefaction for a building foundation. *Journal of the Japanese Society of Soil Mechanics and Foundation Engineering, Tsuchi to Kiso*. pp. 46–48 (in Japanese).

Terashi, M. & Tanaka, H. (1981) Ground improved by deep mixing method. *Proc. of the 10th International Conference on Soil Mechanics and Foundation Engineering*. Vol. 3. pp. 777–780.

Terashi, M., Ooya, T., Fujita, T., Okami, T., Yokoi, K. & Shinkawa, N. (2009) Specifications of Japanese dry method of deep mixing deduced from 4300 projects, *Proc. of the International Symposium on Deep Mixing and Admixture Stabilization*. pp. 647–652.

Terashi, M., Tanaka, H. & Okumura, T. (1979) Engineering properties of lime-treated marine soils and D.M.M. method. *Proc. of the 6th Asian Regional Conference on Soil Mechanics and Foundation Engineering*. Vol. 1. pp. 191–194.

Tokimatsu, K., Mizuno, H. & Kakurai, M. (1996) Building damage associated with geotechnical problems. *Soils and Foundations*. pp. 219–234.

Tokutomi, Y., Kurokawa, T. and Shimano, A. (2009) Improvement of bridge pier foundation by DJM method. *Proc. of the International Symposium on Deep Mixing and Admixture Stabilization*. pp. 275–278.

Yamazaki, H. (2000) Current and trend of ground improvement techniques for liquefaction prevention for port facilities (2) – Effectiveness and trend of techniques. *Japan Society of Civil Engineers Magazine, Civil Engineering*. Vol. 85. pp. 60–62 (in Japanese).

Zen, K., Yamazaki, H., Watanabe, A., Yoshizawa, H. & Tamai, A. (1987) Study on a reclamation method with cement-mixed sandy soils – Fundamental characteristics of treated soils and model tests on the mixing and reclamation. *Technical Note of the Port and Harbour Research Institute*. No. 579. 41p. (in Japanese).

Chapter 5

Execution – equipment, procedures and control

1 INTRODUCTION

In this chapter, deep mixing equipment, construction procedure and quality control methods will be introduced for the representative deep mixing techniques in Japan, which are the dry method of deep mixing, DJM and the wet method of deep mixing, CDM, the high pressure injection mixing, Jet Grouting, and the hybrid of mechanical mixing and high pressure injection mixing. The descriptions in this chapter are based on the latest information as of 2012. As described in Chapter 1, a variety of ground improvement techniques has evolved to cope with changing needs since the deep mixing was developed in the middle of 1970s. The diversified applications of the method (Chapter 4) and the pursuit of cost effectiveness have continuously promoted the improvement of existing execution systems and the development of new systems. Project owners and design engineers are encouraged to update the information periodically.

The purpose of construction is to install stabilized soil columns or elements so that the improved ground, a composite system comprising stabilized soil and unstabilized soil, may meet the function required by geotechnical design. The responsibility for achieving the requirements are shared by owner, designer, general contractor and deep mixing contractor, depending on the adopted contractual scheme. It is necessary for the owner and designer to have sufficient knowledge on the capability and limitation of locally available execution systems and experience of local contractor, and for the contractors to understand the design intent behind the given specifications (Chapter 6).

1.1 Deep mixing methods by mechanical mixing process

Regardless the contractual scheme, the construction of mechanical deep mixing is carried out in the following steps.

1. Examination of specifications
2. Examination of necessary information
3. Selection of appropriate execution system
4. Laboratory mix test for the process design
5. Field trial test
6. Process design

7   Establish quality control and quality assurance plan
8   Establish verification test procedure and the measures in case of non-compliance
9   Production with quality control (QC)
10   Post construction quality assurance (QA)

Requirements for stabilized soil columns/elements are given in the specification by 1) required engineering characteristics of stabilized soil (often in terms of unconfined compressive strength) and acceptable variability, 2) geometric layout (plan location, verticality and depth) of stabilized soil columns/elements, and 3) acceptance criteria. Geometric layout includes the needs of overlapping operation and the end-bearing condition to the underlying stiff layer,

Necessary information to establish a construction plan includes; soil condition at the construction site and other site conditions which affect or limit the construction. Soil condition include stratification, strength profile, physical and chemical properties of the soil such as grain size distribution, natural water content, liquid limit, plastic limits, organic matter content and pH. Other conditions include geometry and topography of the site, obstacles, environmental restrictions such as noise and vibration, characteristics of nearby structures, and relevant local regulations.

Selection of an appropriate execution system is possible only when the clear specifications and necessary information are provided by the owner. A variety of deep mixing techniques are available as shown earlier in Table 1.4 of Chapter 1. These execution systems are developed to effectively accomplish the locally preferred column installation patterns and to meet the local soil conditions. Capability of equipment such as maximum depth of improvement, maximum capacity of binder delivery, ease of overlapping operation and end bearing differs from one system to another. It is emphasized that the owner/designer should consider the capability of locally available techniques before the geotechnical design and writing the specifications.

A laboratory mix test and field trial test conducted before the production is as important as quality control during production and post-construction quality assurance. The details of quality control/quality assurance (QC/QA) related activities will be described in details in Chapter 7.

1.2   Deep mixing methods by high pressure injection mixing process

On the contrary to mechanical mixing, the concept of QC/QA is different in the high pressure injection mixing, especially in the horizontal high pressure injection. This is perhaps caused by the nature of horizontal high pressure injection. The extent of improvement in the radial direction cannot be controlled by the injection process but heavily governed by the strength of the original soil. Therefore, the process design of high pressure injection is carried out based solely on the accumulated experience of the contractor. As will be seen in the corresponding sections in this chapter, standard operational parameters appropriate for soil type are proposed by the contractor and the contractor guarantees the minimum size and minimum strength of stabilized soil columns/elements. Hence the pre-production QA measures are rarely undertaken. Quality control is focused on keeping the standard operational parameters during construction. The construction of high pressure injection deep mixing is carried out in the following steps.

1   Examination of specifications and necessary information
2   Selection of appropriate execution system
3   Production according to standard operational parameters
4   Post construction QA.

2   CLASSIFICATION OF DEEP MIXING TECHNIQUES IN JAPAN

As summarized earlier in Chapter 1, admixture stabilization techniques to improve soft soils by binder are diversified and include in-situ and ex-situ mixings (Table 1.3). Deep mixing techniques are further sub-divided into five groups based on mixing process as shown in Table 1.4. The techniques most commonly employed for in-situ deep mixing in Japan can be divided into three groups: mechanical mixing by vertical rotary shafts with mixing blades at the bottom end of each mixing shaft, high pressure injection mixing, and combination of the mechanical mixing and high pressure injection mixing. The various methods in these groups are classified in Figure 5.1. In the mechanical mixing techniques, binder is injected into a ground with relatively low pressure and forcibly mixed with the soil by mixing blades equipped to vertical mixing shaft(s). The binder is used either with powder form (dry method) or slurry form (wet method). The Dry Jet Mixing (DJM) method is the most common dry method of deep mixing and has usually been applied for on-land works (Dry Jet Mixing Association, 2010). The Cement Deep Mixing (CDM) method, the most common wet method of deep mixing, has frequently been applied for both in-water and on-land works (Cement Deep Mixing Method Association, 1999). In the high pressure injection technique, on the other hand, ground is disturbed by a high pressure jet of water and/or air, while at the same time binder slurry is injected and mixed with the soil. The combination of mechanical mixing and high pressure injection mixing exploits the features of both basic techniques (Endo, 1995).

3   DRY METHOD OF DEEP MIXING FOR ON-LAND WORKS

3.1   Dry jet mixing method

3.1.1   Equipment

3.1.1.1   System and specifications

The Dry Jet Mixing (DJM) method is a dry method of deep mixing, which was put into practice in 1980 and has been frequently applied to on-land works in Japan (Dry Jet Mixing Association, 2010). The system of the method consists of a DM machine and the binder plant. The binder plant consists of a generator, air compressor(s), an air tank, a binder silo, binder feeder(s), and a control room, as shown in Figure 5.2. The DJM machine consists of a mixing tool and a crawler crane with a leader as a base carrier as shown in Figure 5.3. The DJM machines are manufactured by a single company, Kobelco Cranes Co., Ltd., and the specifications of the system are listed in

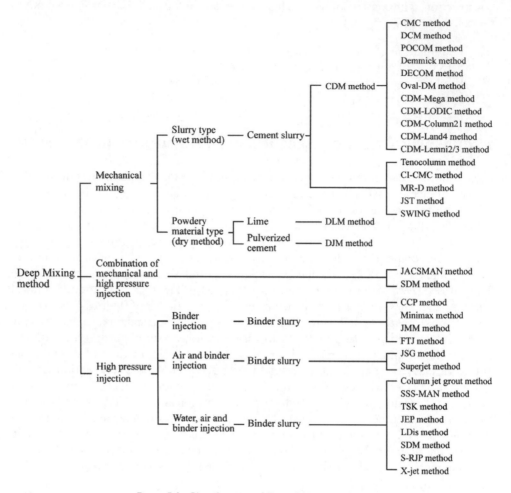

Figure 5.1 Classification of Deep Mixing methods.

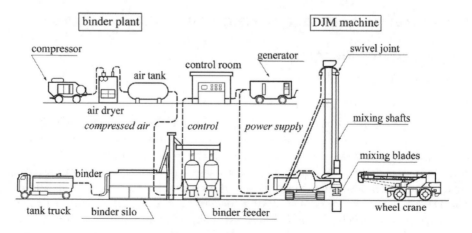

Figure 5.2 Equipment for DJM.

Figure 5.3 DJM machine in operation (by the courtesy of Dry Jet Mixing Association).

Table 5.1 Specifications of DJM machines (Dry Jet Mixing Association, 2010).

Type	DJM1070	DJM2070	DJM2090	DJM2110
Max. depth	20 m	26 m	33 m	33 m
Base carrier				
Leader	26 m	34 m	42 m	42 m
Capacity	70 kW (hydraulic)	55 kW × 2 (electric)	90 kW × 2 (electric)	110 kW × 2 (electric)
Mixing tool				
Number of shafts	1	2	2	2
Spacing of shafts	–	0.8, 1.0, 1.2, 1.5 m	0.8, 1.0, 1.2, 1.5 m	0.8, 1.0, 1.2, 1.5 m
Diameter of blade	1.0 m	1.0 m	1.0 m	1.0 m to 1.3 m
Applicable soils				
Clay	$N < 3$ (max. $N = 6$)	$N < 3$ (max. $N = 6$)	$N < 4$ (max. $N = 7$)	$N < 6$ (max. $N = 9$)
Sand	$N < 10$ (max. $N = 18$)	$N < 10$ (max. $N = 18$)	$N < 14$ (max. $N = 23$)	$N < 20$ (max. $N = 32$)
Binder delivery rate for one shaft	25–120 kg/min.	25–120 kg/min.	25–120 kg/min.	25–120 kg/min.
Binder silo	300 kN	300 kN	300 kN	300 kN

Table 5.1. The crawler cranes with a lifting capacity of 240 to 930 kN are used as a base carrier. The DJM machine can be classified into four types depending on their size of base carrier and the maximum stabilization depth. The DJM1070 machine, the smallest type, has a single mixing shaft and is capable of stabilizing soil up to 20 m in depth. The other types, the DJM2070, DJM2090 and DJM2110, have two mixing

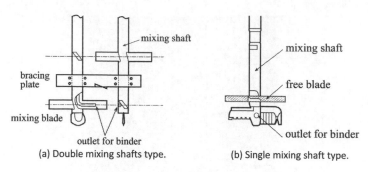

(a) Double mixing shafts type.   (b) Single mixing shaft type.

Figure 5.4 Mixing shafts and mixing blades for DJM machine.

shafts. The DJM2110, the largest type, is capable of stabilizing soils up to 33 m in depth.

3.1.1.2 Mixing tool

As shown in Table 5.1, the DJM machines have one or two mixing shafts. The set of mixing shaft(s) is suspended along the leader and laterally clamped to the leader at two levels: one at the top of the mixing shafts and another at the gear box installed to the bottom of the leader. As the locations of the heavy motor and gear box to drive the mixing shaft(s) are always at the bottom of the leader, the DJM machine is superior in stability. Binder is supplied to each shaft by independent binder feeder to enable even delivery of binder to each shaft. A swivel joint is installed at the top of each mixing shaft for binder supply. The motor(s) for driving the mixing shaft(s) are; one 70 kW motor for the DJM1070, two 55 kW motors for the DJM2070, two 90 kW motors for the DJM2090 and two 110 kW motors for the DJM2110. A shroud covering the mixing tool is placed at the ground surface to minimize the surface spoil.

In the double shafts machine, the spacing of mixing shafts is adjustable either to 0.8, 1.0, 1.2 or 1.5 m. When the spacing of shafts is smaller than the diameter of the mixing blade, the stabilized soil element is partially overlapped two round columns. When the spacing and diameter are the same, two round column tangent each other is constructed, and when the spacing is larger than the diameter two isolated round columns are constructed by a single installation process. Thus the cross sectional area of a stabilized soil element constructed by a single installation process ranges from 0.8 to 1.5 m^2. A double shafts machine has a bracing plate to keep the distance of two mixing shafts (see Figures 5.4(a) and 5.5). The plate is also expected to function to increase mixing degree by preventing the "entrained rotation phenomenon," a condition in which disturbed soil adheres to and rotates with the mixing blades without efficient mixing of soil and binder. For a single shaft machine, a "free blade," an extra blade about 100 mm longer than the diameter of the mixing blade (see Figure 5.4(b)), is installed when necessary, close to one of the mixing blades to prevent the "entrained rotation phenomenon." Two shafts of the double shaft machine rotate in the opposite direction, which increase the degree of mixing and also improve the stability of the machine.

Figure 5.5 Mixing shafts and mixing blades for DJM machine (by the courtesy of Dry Jet Mixing Association).

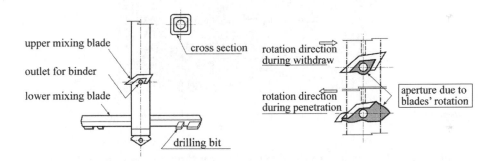

Figure 5.6 Mixing shaft and blades of DJM machine.

The mixing shaft is about 200 mm by 200 mm square shape so that the injected air into a ground can be exhausted through the openings at the four sides of the shaft during rotation, as shown in Figure 5.6. A duct with 50 mm in diameter is installed in the mixing shaft, through which the binder is delivered to the mixing blades with the aid of air pressure. The mixing blades are installed at the bottom end of the shaft ordinarily at two different levels, which are intersected at right angles each other. A small cone attached at the very end of the mixing shaft helps the machine penetrate vertically down into a soil efficiently. The diameter of the mixing blades is typically 1.0 m. The upper mixing blade is a "C" shape and the lower is an "L" shape in general, as shown in Figures 5.6 and 5.7. Several drilling bits are installed on both blades. Two outlets of binder are installed on the shaft close to the mixing blades behind the

Figure 5.7 Mixing shaft and blades of DJM machine (by the courtesy of Dry Jet Mixing Association).

Figure 5.8 Binder Plant for DJM method (by the courtesy of Dry Jet Mixing Association).

rotation direction, so that they are not blocked by the soil. The upper outlet is used for the withdrawal injection and the lower one is for the penetration injection.

3.1.1.3 Binder plant

A binder silo and binder feeders are placed at the site for supplying binder to the DJM machine, as shown in Figure 5.8. A silo of maximum capacity of 300 kN in general is placed at the site for storage of binder as shown in Figure 5.9. An air tank is installed to store compressed air whose maximum air pressure and capacity are $700 \, kN/m^2$ and $2 \, m^3$ respectively, which can supply $4 \, m^3$ air of $450 \, kN/m^2$ per one minute. The binder is added to the air flow at the binder feeder through the feed wheel, where the flow rate of binder is controlled by the rotation speed of the wheel. The binder is supplied through the swivel joints at the tops of the mixing shafts to the mixing blades with the aid of compressed air. Two binder feeders are installed for a double shaft machine and the binder feed rate for each shaft is independently controlled, monitored and recorded. The binder feed rate is adjustable and ranges from 25 to 120 kg/min. for each shaft.

Figure 5.9 Binder Plant for DJM method (by the courtesy of Dry Jet Mixing Association).

Figure 5.10 Control unit for DJM method (by the courtesy of Dry Jet Mixing Association).

3.1.1.4 Control unit

A control room is placed close to the binder feeder, where all the measured data during production of stabilized soil columns/elements are continuously monitored, controlled and recorded by the control unit as shown in Figure 5.10. The data for the dry method include the air pressure, flow rate of air, the amount of binder, the rotation speed of mixing blades, the depth of mixing tool, the penetration and withdrawal speeds of mixing shafts, power consumption, *etc.* As the rig operator on the DJM machine is responsible for controlling the geometric layout, verticality, rate of vertical shaft movement and depth of improvement, the relevant data are fed back to the cab for display. The plant operator is responsible for the other mixing process including rotational speed of the mixing blade and the amount of binder, which are in most cases preset based on the process design and computer controlled. The plant operator and rig operator keep communication such by wireless device and modify the production process to some extent when adjustment is necessary.

3.1.2   Construction procedure

3.1.2.1   Preparation of site

Field preparation is carried out in accordance with the site specific conditions, which includes suitable access for the plant and machinery, leveling of working platform, and inspection of obstacles at and below ground level at the construction site. The binder plant usually requires about 150 to 200 m² in total. Before actual operation, execution circumstances should be prepared to assure smooth execution and prevention of environmental impact. A sand blanket with about 0.5 to 1.0 m in thickness is usually spread on the ground as a working platform. Several steel plates with about 1.5 m by 4.0 m are preferably placed on the sand mat so as to assure the bearing capacity of the DJM machine.

3.1.2.2   Field trial test

It is recommended to conduct a field trial test in advance in, or adjacent to the construction site, in order to confirm the smooth execution. In the test, all the equipment monitoring the amount of binder, rotation speed of mixing blades and the penetration and withdrawal speeds of mixing shafts are calibrated. In the case where the stabilized soil columns/elements should reach and have firm contact with the stiff bearing layer (fixed type improvement), a field trial installation should be carried out to measure the change in the electric or hydraulic power required for driving the mixing shafts and the penetration speed of the mixing shafts at the stiff layer so that they can help detect if the mixing blades have reached the stiff layer in the actual production. Such a trial installation is often conducted without delivering the binder, but should be conducted in the vicinity of the existing boring to compare with the known soil stratification.

When there is less experience in similar soil conditions, it is recommended to carry out a field mixing trial and to confirm that the strength and integrity of the trial column/element meet the design requirement.

3.1.2.3   Construction work

The DJM machine is usually placed along the columns' alignment where the tower and mixing tools face a direction perpendicular to the moving direction of the base carrier, as shown in Figure 5.11, so that the steel plates placed on the sand mat can be moved forward efficiently during the successive execution.

After setting the machine at the prescribed position, the mixing tool is penetrated into the ground while rotating the mixing shafts. There are two basic execution procedures depending on the injection sequence of the binder (Figure 5.12): (a) injecting binder during the penetration of mixing shafts and (b) injecting binder during the withdrawal of the mixing shafts. Each injection sequence has its respective advantages and disadvantages. The penetration injection is beneficial for the homogeneity of strength of a stabilized soil column in which the soil binder mixture is subjected to mixing twice. However, it is possible to deadlock or cause serious damage to the machine if any trouble occurs in the mixing machine during the penetration. The withdrawal injection method has the opposite benefits and disadvantages to the penetration injection method. The location of the injection outlets should be different for each injection method. For the penetration injection method, the injection outlets should locate at

Figure 5.11   DJM machine in operation.

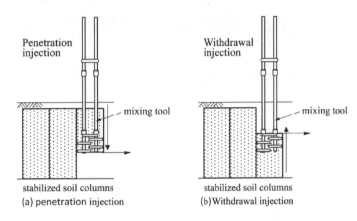

Figure 5.12   Basic execution procedures of the deep mixing method.

the lower mixing blades, but they should be at the upper mixing blades for the withdrawal injection as shown earlier in Figure 5.6. Basically, the withdrawal injection is applied to the DJM method. However, the penetration injection is applied in limited cases when the soft soil is sensitive and causes difficulty in exhausting air or when the extra mixing is thought necessary.

The ordinary execution process of the DJM method is shown in Figure 5.13, where binder is injected during the withdrawal stage. During the penetration, the mixing blades are rotating to cut, disaggregate and disturb the soil to reduce the strength of ground so as to make the mixing tools penetrate by their self-weight. The DJM2110 machine can penetrate a local stiff layer where SPT N-value is less than 9 for a mud stone layer, and SPT N-value is less than 32 for a sandy layer. While penetrating into a stiff layer, the shaft movement and rotation of blade may be reduced and should be considered in estimating the cycle time. In a particular case where a considerably hard layer exists in the ground, pre-drilling of the layer may be necessary in advance of the mixing work.

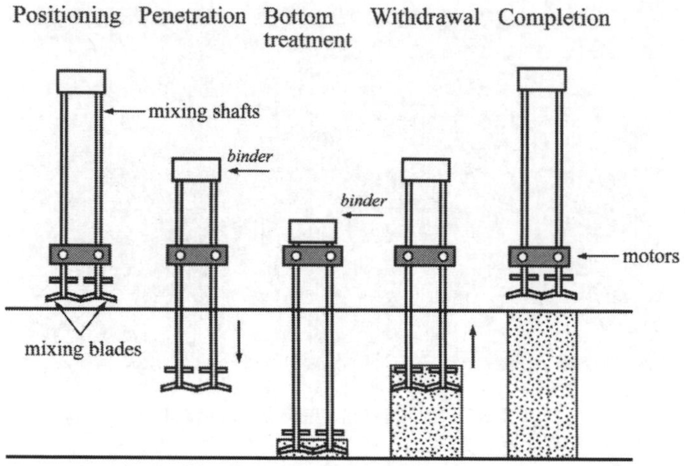

Positioning Penetration Bottom Withdrawal Completion
treatment

Figure 5.13 Execution process of DJM method.

Table 5.2 Typical operational parameters of DJM method.

Type	DJM1070	DJM2070	DJM2090	DJM2110
Mixing shaft				
Penetration speed (m/min.)	1.0–2.0	1.0–2.0	1.0–2.0	1.0–2.0
Withdrawal speed (m/min.)	0.7	0.7	0.9	0.9
Mixing blades rotation speed				
Penetration (rpm)	24	24	32	32
Withdrawal (rpm)	48	48	64	64
Blade rotation number (N/m)	274	274	284	284

In the withdrawal stage, the direction of the mixing blade rotation is reversed and binder is injected and mixed with the soil. During the withdrawal, the flow rate of binder, the rotation speed and the withdrawal speed is controlled to the values predetermined by process design. When the withdrawal speed changes from the pre-determined value, the binder rate is adjusted accordingly.

The typical operational parameters for withdrawal injection are summarized in Table 5.2, which are somewhat different depending on the type of machine. The "blade rotation number" as defined by Equation (7.2) in Chapter 7 has been introduced to assess the degree of mixing. A blade rotation number of around 300 is necessary to assure sufficient homogeneity of the stabilized soil column according to experience and research efforts in Japan. Based on the typical operational parameters of withdrawal injection, the blade rotation number becomes 274 for the DJM1070 and DJM 2070 machines and 284 for the DJM2090 and DJM2110 machines.

The volume of injected air is in general 4 m^3/min. which is controlled by the air pressure of 450 kN/m^2 at its maximum. The required air pressure is equal to or slightly

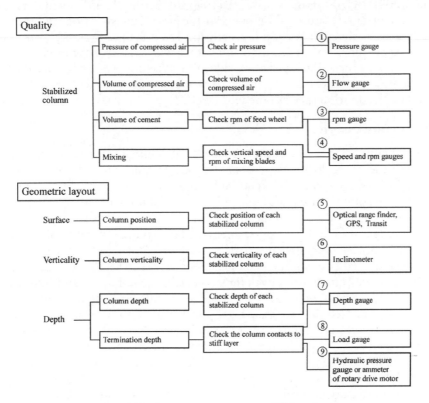

Figure 5.14 Operation monitoring in DJM method.

larger than the sum of the water pressure in the ground at the outlet and the injection pressure. As the former increases with the depth, it is desirable to increase the air pressure with the depth in the penetration stage and decrease in the withdrawal stage. If an excessive amount of air is injected, the soil close to the mixing shaft is blown out and results in a doughnut like stabilized soil column.

During the execution, the spilled out soil is excavated and removed by a backhoe to prevent any adverse influence to the execution. As the spilled out soil contains binder, the soil should be handled with care according to the local regulation. The amount of spoil in the dry method is generally smaller than the wet method. In some cases, the stabilized soil volume becomes smaller than that of the original soil and it becomes necessary to fill the depression at the column top with imported materials.

3.1.2.4 Quality control during production

To produce stabilized soil columns/elements that meet the design requirements on quality and dimension, it is essential to control and monitor the quality of binder, geometric layout, and operational parameters such as amount of binder, rotation speed of mixing blades, shaft speed, *etc*. Figure 5.14 shows the operational parameters for the DJM method and items for geometric layout. The verticality of the mixing tool is usually evaluated by the measurement of the verticality of the leader, and is controlled within 1/200 to 1/100 in many cases. During production the measured data are fed back to the plant operator and rig operator for precise construction. In practice, the rotation speed of the mixing shafts is usually fixed as shown in Table 5.2, depending on the type of machine. The penetration and withdrawal speeds of the mixing shafts are controlled to the prescribed speed by sending out the wire which suspends the mixing tool. The flow rate of binder is adjusted to the penetration and withdrawal speeds by controlling the rotation speed of the feed wheel.

3.1.3 Quality assurance

After the construction work, in-situ stabilized soil columns/elements should be investigated in order to verify the design quality, such as continuity, uniformity, strength, permeability and dimension. In Japan, full depth coring and unconfined compression test on the core samples are most frequently conducted for verification. The number of core borings is dependent upon the number of stabilized soil columns/elements in the project. In the case of on-land works, three core borings are generally conducted when the total number of columns/elements is less than 500. When the total number exceeds 500, one additional core boring is conducted for every further 250 columns/elements.

The continuity and uniformity of the stabilized soil column are confirmed by visual observation of the continuous core. Determination of the engineering properties of the stabilized soil is based on unconfined compressive strength on samples selected from the continuous core. The number of test depends upon the construction's condition and the soil properties. In general three core barrels are selected from three levels and three specimens are taken from each core barrel and subjected to the unconfined compression test for each core boring. Properties other than unconfined compressive strength can be correlated with unconfined compressive strength as discussed in Chapter 3.

The quality of the core sample primarily depends on the uniformity of stabilized soil. However, it further relies on the quality of boring machine, coring tool and skill of the workmen. If the coring is not properly conducted, a low quality sample with some cracks can be obtained. A double tube core sampler or triple tube core sampler has been used for core sampling of stabilized soil. It is recommended to use sampler of relatively large diameter such as 86 or 116 mm in order to take good quality samples. The quality of core sample is usually evaluated by visual inspection and/or the Rock Quality Designation (RQD) index, defined by Equation (7.3) in Chapter 7. The RQD index measures the percentage of "good rock" within a borehole and provides the rock quality as shown in Table 7.3 in Chapter 7.

Recently, in-situ tests have also been applied for quality verification together with the unconfined compression test, which is briefly introduced in Chapter 7.

4 WET METHOD OF DEEP MIXING FOR ON-LAND WORKS

For the wet method of deep mixing a variety of deep mixing machines are developed by deep mixing contractors to meet the purpose of improvement and applications and their specifications are quite variable. Due to the limitation of pages, ordinary types of wet mixing machine for on-land and in-water constructions, and two types of special machines for on-land works are briefly introduced in this section.

4.1 Ordinary cement deep mixing method

4.1.1 Equipment

4.1.1.1 System and specifications

The Cement Deep Mixing (CDM) method is one of the wet methods of deep mixing, which was originally developed for in-water works in the 1970s but has also been frequently applied for on-land works (Cement Deep Mixing Method Association, 1999). A system of the method consists of a DM machine and a binder plant as shown in Figure 5.15. The binder plant consists of a binder silo, water tank, binder-water mixer, agitator tank, pumping unit and control room. The CDM machine consists of a mixing tool and a crawler crane with a leader, as shown in Figure 5.16. The crawler cranes with a lifting capacity of 250 to 550 kN are often used as a base carrier. The CDM machine can be classified into four groups depending on their size of base carrier and the maximum stabilization depth, and the major specifications of the system are tabulated in Table 5.3 (Cement Deep Mixing Method Association, 1999).

4.1.1.2 Mixing tool

The ordinary CDM machines for on-land works have two mixing shafts. The set of mixing shafts are suspended along the leader and laterally clamped at the top of the mixing tool and the bottom of the leader. The motor and gear box are installed on the top of the shafts. Binder slurry is supplied to each shaft by an independent pumping unit to enable even delivery of binder slurry to each shaft. A swivel joint is installed at the top of each mixing shaft for binder slurry supply. The motor for driving mixing shafts is different for each group, two 45 kW motors for the 10 m class, two 50 to 60 kW motors for the 20 m class, two 75 to 90 kW motors for the 30 m class and two 90 kW motors for the 40 m class. The spacing of the mixing shafts are either 0.8, 1.0 or 1.1 m for the diameter of mixing blades of 1.0, 1.2 and 1.3 m respectively, to produce a stabilized soil element consisting of two partially overlapped round columns. The cross sectional area of a stabilized soil element ranges from 1.5 to 2.6 m^2.

A double shafts machine usually has a bracing plate to keep the distance of the two mixing shafts (see Figure 5.18). The plate is also expected to function to increase

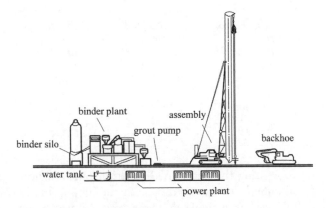

Figure 5.15 Typical CDM machine for on-land work.

Figure 5.16 Typical CDM machine for on-land work in operation.

Table 5.3 Specifications of CDM machines (Cement Deep Mixing Method Association, 1999).

Type	10 m class	20 m class	30 m class	40 m class
Max. depth	10 m	20 m	30 m	40 m
Base carrier				
Leader	20 m	30 m	40 m	50 m
Motor capacity	45 kW × 2	50–60 kW × 2	75–90 kW × 2	90 kW × 2
	250 kVA	300 kVA	400 kVA	450 kVA
	(electric)	(electric)	(electric)	(electric)
Mixing tool				
Number of shafts	2	2	2	2
Spacing of shafts	0.8, 1.0, 1.1 m	0.8, 1.0, 1.1 m	0.8, 1.0, 1.1 m	0.8, 1.0 m
Diameter of blade	1.0, 1.2, 1.3 m	1.0, 1.2, 1.3 m	1.0, 1.2, 1.3 m	1.0, 1.2 m
Applicable soils				
Clay	$N < 4$	$N < 4$	$N < 4$	$N < 4$
	(max. $N = 8$)	(max. $N = 8$)	(max. $N = 8$)	(max. $N = 8$)
Sand	$N < 6$	$N < 6$	$N < 6$	$N < 6$
	(max. $N = 15$)	(max. $N = 15$)	(max. $N = 15$)	(max. $N = 15$)
Binder mixer	$2\,m^3 \times 1$	$2\,m^3 \times 2$	$2\,m^3 \times 2$	$3.5\,m^3 \times 2$
Binder agitator	$3.5\,m^3$	$3.5\,m^3$	$3.5\,m^3$	$3.5\,m^3$
Binder slurry				
delivery rate for one shaft	$20\,m^3/hr.$	$20\,m^3/hr.$	$20\,m^3/hr.$	$20\,m^3/hr.$
Binder silo	300 kN	300 kN	300 kN	300 kN

mixing degree by preventing the "entrained rotation phenomenon," a condition in which disturbed soil adheres to and rotates with the mixing blade without efficient mixing of soil and binder. For a single shaft machine, a "free blade," an extra blade about 100 mm longer than the diameter of mixing blade (see Figure 5.17), is installed

Figure 5.17 Mixing blades and free blade for CDM method for on-land works.

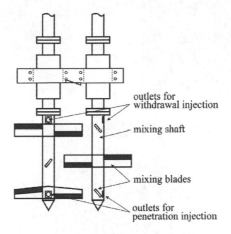

Figure 5.18 Typical mixing blades for CDM method for on-land work.

when necessary, close to one of the mixing blades to prevent the "entrained rotation phenomenon." Two shafts of the double shaft machine rotate in the opposite direction, which increase the degree of mixing and also improve the stability of the machine. The mixing shaft is a 267 mm circular shape. A duct with 50 mm in diameter is installed in the mixing shaft, through which binder slurry is supplied to the mixing blades. A stack of blades is installed at the bottom end of the mixing shaft, which consists of excavation blade and mixing blades, as shown in Figure 5.18. The excavation blade is installed at the very end of the mixing shaft, on which forks made by hard metal are fixed so that the machine can excavate and screw in a soil efficiently. The mixing blades at different levels are intersected at right angles each other. Two outlets of binder slurry are installed on the shafts at different levels close to the mixing blades, so that the outlets are not blocked by the soil. The upper outlet is used for withdrawal injection and the lower one is for penetration injection. The shape and the number of mixing blades have been developed to assure a high mixing degree as much as possible, and now have various variations depending upon the contractors, as shown in Figure 5.19.

(a) CDM method.

(b) CDM method.

(c) CDM Mega method.

(d) CDM Column method.

(e) CDM Land4 method.

Figure 5.19 Various types of mixing shaft and blades for CDM method (by the courtesy of Cement Deep Mixing Method Association).

Figure 5.20 Binder silo and pumping unit (by the courtesy of Cement Deep Mixing Method Association).

4.1.1.3 Binder plant

A binder plant is prepared for producing and supplying binder slurry to the CDM machine. A silo of maximum capacity of 300 kN in general is prepared for storage of binder, as shown in Figure 5.20. Binder slurry is usually manufactured by every $1\,m^3$ in a mixer of 2.0 to $3.5\,m^3$ capacity, and temporarily stored in an agitator of $3.5\,m^3$ in capacity. The water to binder ratio (W/C) of binder slurry is usually 60 to 100%. The binder slurry thus manufactured is supplied to each mixing shaft of the CDM machine by the independent pump, where a total of about 100 to 350 l/min. in volume is supplied to the machine by the help of a pumping pressure of about 2.5 MN/m². The pumping pressure is controlled to assure a constant flow of binder slurry during production.

4.1.1.4 Control unit

A control unit is installed in a control room in many cases, but in some cases on the CDM machine, where the binder condition, the amount of each material, the rotation speed of mixing blades, the penetration and withdrawal speeds of mixing shafts, *etc.* are continuously monitored, controlled, and recorded as shown in Figure 5.21.

In the case where the control unit is installed in the control room (Figure 5.21(a)), as the rig operator on the CDM machine is responsible for controlling the geometric layout, verticality, rate of vertical shaft movement and depth of improvement, the relevant data are fed back to the cab for display. The plant operator is responsible for the other mixing process including rotational speed of mixing blade and the amount of binder, which are in most cases preset based on the process design and computer controlled. The plant operator and rig operator keep communication such by wireless device and modify the production process to some extent when an adjustment is necessary.

In the case where the control unit is installed on the CDM machine (Figure 5.21(b)), the rig operator on the CDM machine is responsible for controlling not only the geometric layout, verticality, rate of vertical shaft movement, depth of improvement

(a) Controle unit in control room.

(b) Controle unit on execution machine.

Figure 5.21 Control unit for CDM method (by the courtesy of Cement Deep Mixing Method Association).

but also the other mixing process including rotational speed of mixing blade and the amount of binder.

4.1.2 Construction procedure

4.1.2.1 Preparation of site

Similarly to the dry method, field preparation is carried out in accordance with the specific site conditions, which includes suitable access for plant and machinery, leveling of the working platform. The binder plant usually requires about $200\,m^2$ in total. Before actual operation, execution circumstances should be prepared to assure smooth execution and prevention of environmental impact. A sand blanket with about 0.5 to 1.0 m in thickness is usually spread on the ground as a working platform. Several steel plates with about 1.5 m by 4.0 m are preferably placed on the sand mat so as to assure the bearing capacity of the CDM machine.

4.1.2.2 Field trial test

It is recommended to conduct a field trial test in advance in, or adjacent to the construction site, in order to confirm the smooth execution. In the test, all the equipment monitoring the amount of binder, rotation speed of mixing blades and penetration and withdrawal speeds of mixing shafts are calibrated. In the case where the stabilized soil columns should reach and have firm contact with the stiff bearing layer (fixed type improvement), a field trial installation should be carried out to measure the change in the electric or hydraulic power required for driving the mixing shafts and the penetration speed of the mixing shafts at the stiff layer so that they can help detect if the mixing blades have reached the stiff layer in the actual production. Such a trial installation is often conducted without delivering the binder, but should be conducted in the vicinity of existing boring to compare with the known soil stratification.

When there is less experience in similar soil conditions, it is recommended to carry out a field mixing trial and to confirm that the strength and integrity of the trial column/element meet the design requirement.

Figure 5.22 CDM machine in operation.

4.1.2.3 Construction work

The CDM machine is usually placed along the columns' alignment where the tower and mixing tools face a direction perpendicular to the moving direction of the base carrier, as shown in Figure 5.22, so that the steel plates placed on the sand mat can be moved forward efficiently during the successive installation.

After setting the machine at the prescribed position, the mixing tool is penetrated into a ground while rotating the mixing shafts. There are two basic execution procedures depending on the injection sequence of binder (Figure 5.12): (a) injecting binder slurry during penetration of the mixing shafts and (b) injecting binder slurry during withdrawal of the mixing shafts. The location of the injection outlet is different for each injection method. For the penetration injection method, the injection outlets should locate at the lowest mixing blades, but they should be at the uppermost mixing blades for the withdrawal injection. The penetration injection is frequently applied to the CDM method for on-land works.

The ordinary execution process of the CDM method is shown in Figure 5.23, where binder slurry is injected during the penetration stage. During the penetration, the mixing blades are rotating to disaggregate and disturb the soil to reduce the strength of ground so as to make the mixing tools penetrate by their self-weight. The binder slurry is injected during penetration and mixed with the disaggregated soil. The mixing also continues in the withdrawal stage. The flow rate of binder slurry is kept constant while the penetration speed is controlled constant so as to assure the design amount of binder should be mixed. The CDM machines can penetrate a local stiff layer where SPT N-value is less than 8 for a clay layer and SPT N-value is less than 15 and the thickness is less than 3 m for a sandy layer. While penetrating a stiff layer, the shaft movement and rotation of mixing blades may be reduced and should be considered in estimating the cycle time. In a particular case where a considerably hard layer exists in the ground, pre-drilling of the layer may be necessary in advance of the mixing work. In the withdrawal stage, the direction of the mixing blade rotation is reversed and the binder is mixed with the soil again.

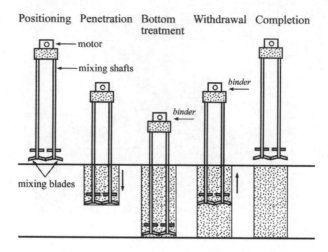

Positioning Penetration Bottom Withdrawal Completion
treatment

Figure 5.23 Execution process of CDM method.

Table 5.4 Typical execution specification of CDM method (Cement Deep Mixing Method Association, 1999).

Type	Injection during penetration	Injection during withdrawal
Mixing shaft		
Penetration speed (m/min.)	1.0	1.0
Withdrawal speed (m/min.)	1.0	0.7
Mixing blades rotation speed		
Penetration (rpm)	20	20
Withdrawal (rpm)	40	40
Blade rotation number (N/m)	360	350

The stabilized soil columns should reach a stiff layer sufficiently in the case of the fixed type improvement. In practical execution, a rapid change in the penetration speed of the mixing shaft, the required torque and rotation speed of the mixing blades are useful to detect whether the mixing blades have reached the stiff layer. When the mixing tool reached the stiff layer, the machine stays there for several minutes or goes up and down about one meter with continuing injection of binder slurry and mixing to assure sufficient contact of the column with the stiff layer.

The typical operational parameters for the wet method are summarized in Table 5.4. Different operational parameters are used for the penetration injection and withdrawal injection in order to achieve the same level of mixing degree. The "blade rotation number" as defined by Equation (7.2) in Chapter 7 of about 350 is attained by the set of typical operational parameters both for penetration and withdrawal injection. This number is proposed to assure sufficient homogeneity of the stabilized soil column according to experience and research efforts.

During the execution, the spilled out soil are excavated and removed by a backhoe to prevent any adverse influence to the execution. As the spilled out soil contains binder, the soil should be handled with care according to the local regulation.

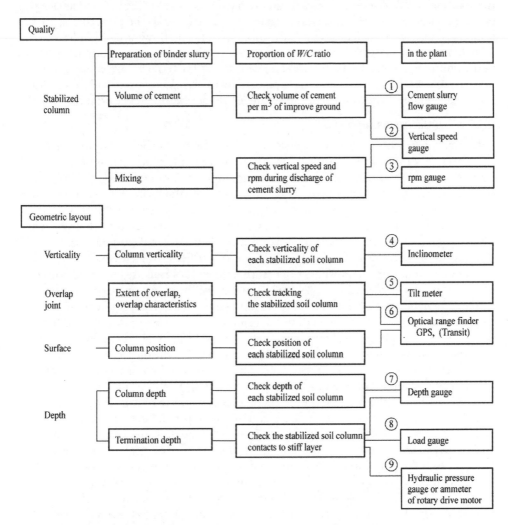

Figure 5.24 Operation monitoring for CDM method on-land works (after Cement Deep Mixing Method Association, 1999).

4.1.2.4 Quality control during production

To produce stabilized soil columns/elements that meet the design requirements on quality and dimension, it is essential to control and monitor the quality of binder, geometric layout, and operational parameters such as amount of binder, rotation speed of mixing blades, shaft speed, *etc.* Figure 5.24 shows the operational parameters for the CDM method and items for geometric layout (after Cement Deep Mixing Method Association, 1999). The verticality of the mixing tool is usually evaluated by the measurement of the verticality of the leader, and is controlled within 1/200 to 1/100 in many cases. During production the monitoring data are fed back to the plant operator in the control room or the rig operator in the cab on the machine for precise construction. In practice, the rotation speed of mixing shafts is usually fixed as shown in Table 5.4. The penetration and withdrawal speeds of mixing shafts are controlled to the prescribed speed by sending out the wire which suspends the mixing tool. The flow rate of binder slurry is adjusted to the penetration or withdrawal speed by controlling the pumping pressure at the pumping units. The W/C ratio and density of binder slurry are controlled to the design value in the binder plant. The binder slurry should be used within about one hour after preparation to prevent the setting of binder before injection into the soil.

4.1.2.5 Quality assurance

After the construction work, in-situ stabilized soil elements should be investigated in order to verify the design quality, such as continuity, uniformity, strength, permeability and dimension. In Japan, full depth coring and unconfined compression test on the core samples are most frequently conducted for verification. The number of core borings is dependent upon the number of stabilized soil elements in the project. In the case of on-land works, three core borings are generally conducted in the case where the total number of elements is less than 500. When the total number exceeds 500, one additional core boring is conducted for every further 250 elements.

The continuity and uniformity of the stabilized soil column are confirmed by visual observation of the continuous core. Determination of the engineering properties of the stabilized soil is based on the unconfined compressive strength on samples selected from the continuous core. The number of test depends upon the construction's condition and the soil properties. In general three core barrels are selected from three levels and three specimens are taken from each core barrel and subjected to the unconfined compression test for each core boring. Properties other than unconfined compressive strength can be correlated with unconfined compressive strength as discussed in Chapter 3.

The quality of the core sample primarily depends on the uniformity of stabilized soil. However, it further relies on the quality of boring machine, coring tool and the skill of workmen. If the coring is not properly conducted, a low quality sample with some cracks can be obtained. A double tube core sampler or triple tube core sampler has been used for core sampling of stabilized soil. It is recommended to use samplers of relatively large diameter such as 86 or 116 mm in order to take good quality samples. The quality of core sample is usually evaluated by visual inspection and/or the Rock Quality Designation (RQD) index, defined by Equation (7.3) in Chapter 7. The RQD index measures the percentage of "good rock" within a borehole and provides the rock quality as shown in Table 7.3 in Chapter 7.

Recently, in-situ tests have also been applied for quality verification together with the unconfined compression test, which is briefly introduced in Chapter 7.

4.2 CDM-LODIC method

4.2.1 Equipment

4.2.1.1 System and specifications

During the execution of deep mixing, injected binder slurry causes heaving and horizontal displacement of soft ground to some extent as described later in Section 6.3. The CDM-LODIC method (low displacement and control method), a variation of the CDM methods for on-land work, was developed in 1985 for minimizing the heaving and horizontal displacement during the execution (Horikiri *et al.*, 1996; Cement Deep

Figure 5.25 CDM-LODIC machine (by the courtesy of Cement Deep Mixing Method Association).

Table 5.5 Specifications of CDM-LODIC machine (Cement Deep Mixing Method Association, 2006).

	φ 1.0 m type	*φ 1.2 m type*	*φ 1.3 m type*
Max. depth	40 m	40 m	30 m
Driving motor	75 to 90 kW × 2	90 to 110 kW × 2	90 to 110 kW × 2
Mixing tool			
Spacing of shafts	0.8 m	1.0 m	1.1 m
Diameter of blade	1.0 m	1.2 m	1.3 m
Sectional area	1.50 m^2	2.17 m^2	2.56 m^2
Applicable soils			
Clay	$c < 100$ kN/m^2, $N < 10$	$c < 80$ kN/m^2, $N < 8$	$c < 60$ kN/m^2, $N < 6$
Sand	$N < 30$	$N < 25$	$N < 20$

Mixing Method Association, 2006). If a soil equivalent to the volume of the mixing tool and amount of binder slurry can be removed before/during the binder slurry injection, it is basically possible to substantially reduce the displacement of surrounding ground and influence on nearby structures.

Similarly to the ordinary CDM machine, a system for the CDM-LODIC method consists of a DM machine and a binder plant as shown in Figure 5.25. The binder plant consists of a binder silo, water tank, binder-water mixer, agitator tank, pumping unit and a control room. The CDM-LODIC machine consists of a mixing tool and a crawler crane with a leader. Crawler cranes with a lifting capacity of 250 to 400 kN are often used as a base carrier. The CDM-LODIC machine can be classified into three groups depending on their size of base carrier and the maximum stabilization depth, and the major specifications of the system are tabulated in Table 5.5 (Cement Deep Mixing Method Association, 2006).

Figure 5.26 Mixing shafts and blades of CDM-LODIC machine (by the courtesy of Cement Deep Mixing Method Association).

4.2.1.2 Mixing tool

The CDM-LODIC machine has two mixing shafts (Figure 5.26). The set of mixing shafts are suspended along the leader through an electric motor. Binder slurry that is manufactured in the plant is supplied from the top of the shafts though the rotary joints. The machine can be categorized into three groups depending on the diameter of mixing blades, 1.0, 1.2 and 1.3 m, as shown in Table 5.5. The spacing of mixing shafts is designed 0.2 m smaller than the diameter of the mixing blade in order to produce two round columns partially overlapped each other with 0.2 m overlap. The diameter of mixing blade of 1, 1.2 and 1.3 m results in a cross-sectional area of 1.50, 2.17 and 2.56 m^2 respectively.

The mixing shaft is 267 mm in diameter and has a duct of 50 mm in diameter for supplying binder slurry. A stack of mixing blades consists of excavating blade and mixing blades. The excavating blade is usually installed at the bottom on which forks made by hard metal are fixed so that the mixing tool can disaggregate the soil efficiently. The shape and stack of mixing blades have been developed to assure a high mixing degree as much as possible. The distance of the shafts are kept constant by a bracing plate at the bottom of the shafts. The plate is expected to function to increase the mixing degree by preventing the "entrained rotation phenomenon," a condition in which disturbed soft soil adheres to and rotates with the mixing blade without efficient mixing of the binder slurry and soil. The rotation speed of the mixing shaft is controlled in general to either 20 or 40 rpm. In the CDM-LODIC method, helical screw blade with about 50 mm in width are installed along the mixing shaft to excavate and bring the soil to the ground surface, as shown in Figure 5.27.

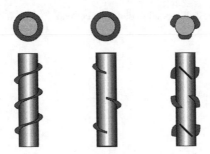

Figure 5.27 Typical shapes of helical screw blade for CDM-LODIC method.

4.2.1.3   Binder plant

A plant and pumping unit is for producing and supplying binder slurry to the CDM-LODIC machine, which is the same as that of the ordinary CDM method.

4.2.1.4   Control unit

A control unit is placed in a control room, where the binder condition, the amount of each material, the rotation speed of mixing blades, the speed of shaft movement, *etc.* are continuously monitored, controlled and recorded. These monitoring data can be fed back to the operator on the CDM-LODIC machine for precise construction of stabilized soil columns.

4.2.2   Construction procedure

4.2.2.1   Preparation of site

Similarly to the DJM and the ordinary CDM methods, field preparation is carried out to assure smooth execution and prevention of environmental impact.

4.2.2.2   Field trial test

It is recommended to conduct a field trial test in advance in, or adjacent to, the construction site, with the same purposes as the DJM and the ordinary CDM.

4.2.2.3   Construction work

The construction process for the CDM-LODIC method is shown in Figure 5.28, which employs withdrawal injection (Cement Deep Mixing Method Association, 2006). The characteristic of the CDM-LODIC method is the soil removal during both penetration and withdrawal stages. During the penetration stage, mixing blades disaggregate the soil and the helical screw conveys a certain amount of original soil to the ground surface. In the last couple of meters to the design depth, binder slurry is injected from the lower injection outlet to produce the lower part of the stabilized soil column which roughly corresponds to the distance between the excavation blade to the uppermost mixing blade. After the bottom treatment, the injection port is switched to the upper outlet to continue the production of the column by withdrawal injection. During

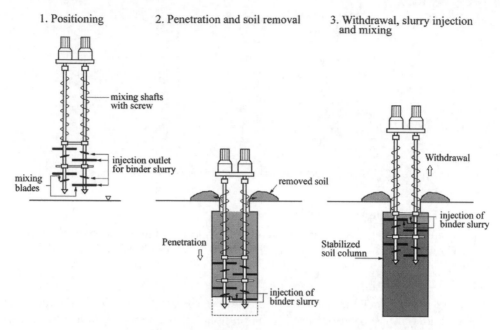

Figure 5.28 Execution process of CDM-LODIC method (Cement Deep Mixing Method Association, 2006).

the withdrawal stage, the unstabilized soil above the mixing blades is excavated and conveyed to the ground surface by the helical screws.

Ideally, amount of soil to be removed during penetration is the volume of the mixing tool and that during the withdrawal stage is the amount of injected binder slurry. The spilled out soil are excavated and removed by backhoe to prevent adverse influence to the operation. As the spilled out soil may contain a small amount of cement, the soil should be handled with care by the local regulations.

The amount of soil removed is influenced by the shape, sectional area and pitch of the helical screw, rotation speed of mixing blade and the time, which is formulated by Equation (5.1). The coefficient of efficiency of soil removal, K depends upon the soil property and execution condition, the mean value of which is around 0.1, ranging from 0.02 to 0.2 (Kamimura *et al.*, 2009a).

$$V = K \times \alpha \times N$$
$$= K \times P \times S \times N \tag{5.1}$$

where
K : coefficient of efficiency of soil removal
N : total number of rotation of helical screw during production
P : pitch of helical screw (m)
S : sectional area of helical screw (m^2)
V : amount of soil removed (m^3)
α : characteristic of helical screw (m^3).

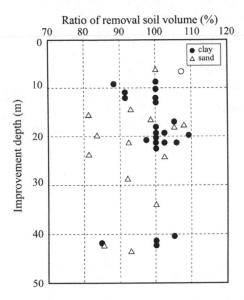

Figure 5.29 Rate of removal soil volume (Kaminura *et al.*, 2009).

4.2.3 Quality control during production

During production, the quality and dimension of stabilized soil columns and their geo-metric layout are monitored, controlled and recorded in the same way as the ordinary wet method of deep mixing (see Figure 5.24). For the CDM-LODIC method, the volume of extracted soil is measured and reported for each stabilized soil column.

4.2.4 Quality assurance

After the construction work, the quality of the in-situ stabilized soil columns should be verified in advance of construction of the superstructure in order to confirm the design quality, such as uniformity, strength, permeability or dimension. Full depth coring, observation of core and testing of selected specimens are conducted for quality assurance in the same way as that for the ordinary CDM.

4.2.5 Effect of method – horizontal displacement during execution

Figure 5.29 compares the ratio of removal soil volume and the column length measured in the previous projects (Kaminura *et al.*, 2009a), in which the ratio is defined as the ratio of soil volume removed to the binder slurry volume injected. Soil removal was found successful and the ratio ranges from 80 to 110%.

Figure 5.30 shows the horizontal displacement measured by inclinometer during the improvement operation (Horikiri *et al.*, 1996). In the figure, two case records by the ordinary CDM machine and the CDM-LODIC method are plotted. It is obvious that the CDM-LODIC method can reduce the horizontal displacement considerably.

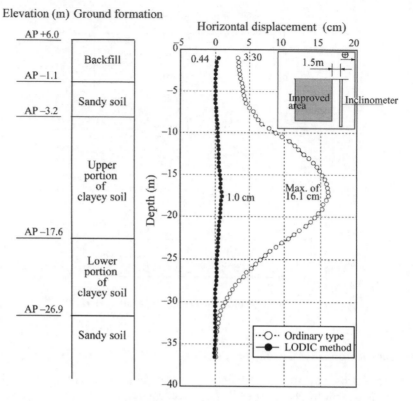

Figure 5.30 Measured horizontal displacement during improvement operation (Horikiri *et al.*, 1996).

Figure 5.31 shows the accumulated data on the relationship between the horizontal distance from the machine normalized by the depth of improvement and the horizontal displacement of the ground (Horikiri *et al.*, 1996). The figure clearly shows that the horizontal displacements caused by the CDM-LOIC method are less than 20 mm and are quite smaller than those by the ordinary CDM method.

4.3 CDM-Lemni 2/3 method

4.3.1 Equipment

4.3.1.1 System and specifications

The ordinary CDM machine for on-land works has in most cases two mixing shafts as shown in Table 5.3. In order to reduce the construction cost by increasing the construction speed, there have been various attempts of expanding the diameter of the mixing blade or increasing the number of mixing shafts (Terashi, 2003). In the same direction, a new deep mixing technique called "CDM-Lemni 2/3 method" was developed. The CDM-Lemni 2/3 machine is a triple shafts machine as shown in Figure 5.32 to produce three round columns partially overlapped each other. However, in this method, binder

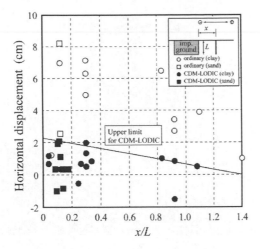

Figure 5.31   Lateral displacement of surrounding ground during improvement operation (Horikiri *et al.*, 1996).

Figure 5.32   CDM-Lemni 2/3 machine (by the courtesy of Cement Deep Mixing Method Association).

slurry is injected only from two outer shafts, which enables the use of ordinary double shafts CDM equipment without substantial modification (Cement Deep Mixing Method Association, 2005, Kamimura *et al.*, 2009b).

Similarly to the ordinary on-land CDM, a system of the method consists of a DM machine and a binder plant. The CDM machine consists of a crawler crane with a leader and mixing tool. Crawler cranes with a lifting capacity of 500 to 750 kN are

Figure 5.33 Mixing shafts and blades of CDM-Lemni 2/3 machine (by the courtesy of Cement Deep Mixing Method Association).

Table 5.6 Specifications of CDM-Lemni 2/3 machine (Cement Deep Mixing Method Association, 2005).

	ϕ 1.0 m type	ϕ 1.2 m type	ϕ 1.3 m type
Max. depth	30 m	30 m	30 m
Driving motor	75 to 90 kW × 3	90 to 110 kW × 3	90 to 110 kW × 3
Mixing tool			
Spacing of shafts	0.8 m	1.0 m	1.1 m
Diameter of blade	1.0 m	1.2 m	1.3 m
Sectional area	2.19 m²	3.21 m²	3.79 m²
Applicable soils			
Clay	$c < 100$ kN/m², $N < 10$	$c < 80$ kN/m², $N < 8$	$c < 60$ kN/m², $N < 6$
Sand	$N < 30$	$N < 25$	$N < 20$

often used as a base carrier. The mixing tool has three mixing shafts. The mixing tool is suspended along the leader. Binder slurry that is manufactured in the plant is supplied from the top of two outer shafts through the rotary joints.

The CDM-Lemni 2/3 machine can be categorized into three groups depending on their diameter of mixing blades, 1.0, 1.2 and 1.3 m, as tabulated in Table 5.6 (Cement Deep Mixing Method Association, 2005). The motor power for driving the mixing blades are different for each group, three sets of 75 to 90 kW motors, 90 to 110 kW motors and 90 to 110 kW motors are installed on the top of mixing shafts respectively.

4.3.1.2   Mixing tool

The CDM-Lemni 2/3 machine has three mixing shafts (Figure 5.33). The set of mixing shafts are suspended along the leader through an electric motor. Binder slurry that is manufactured in the plant is supplied from the top of the shafts though the rotary joints. The machine can be categorized into three groups depending on the diameter of mixing blades, 1.0, 1.2 and 1.3 m, as shown in Table 5.6. The spacing of mixing shafts

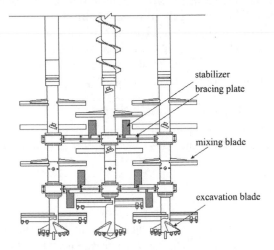

stabilizer

bracing plate

mixing blade

excavation blade

Figure 5.34 Mixing shafts and blades of CDM-Lemni 2/3 machine (by the courtesy of Cement Deep Mixing Method Association).

is designed 0.2 m smaller than the diameter of the mixing blade in order to produce three round columns partially overlapped each other with 0.2 m overlap. The diameter of mixing blade of 1, 1.2 and 1.3 m results in a cross-sectional area of 2.19, 3.21 and 3.49 m^2 respectively.

The mixing shaft is 267 mm in diameter and has a duct of 50 mm in diameter for supplying binder slurry. A stack of mixing blades consists of excavating blade and mixing blades. The excavating blade is usually installed at the bottom on which forks made by hard metal are fixed so that the mixing tool can disaggregate the soil efficiently. The shape and stack of mixing blades have been developed to assure a high mixing degree as much as possible. The distance of the shafts are kept constant by bracing plates at the bottom of the shafts. The plates are expected to function to increase the mixing degree by preventing the "entrained rotation phenomenon," a condition in which disturbed soft soil adheres to and rotates with the mixing blade without efficient mixing of the binder slurry and soil. The rotation speed of the mixing shaft is controlled in general to either 20 or 40 rpm.

As the binder slurry injection outlets are equipped only to the two outer shafts, soil binder mixture produced at outer mixing shafts should flow into the central shaft to produce three round columns. To enable this flow, the outer right and left shafts rotate in the same direction while the central one rotates in the opposite direction. This rotation pattern causes soil binder mixture to flow between the three axes in the form of two lemniscates (i.e. in the form of a ∞, the symbol for infinity).

Two types of auxiliary devices, helical screw and stabilizers, were developed to ensure the smooth lemniscate motion of soil binder mixture and efficient mixing. The helical screw is installed along the central shaft above the mixing blades as shown in Figure 5.34. The CDM-Lemni 2/3 method employs the penetration injection. During penetration, the screw cuts and conveys the original soil around the central shaft upward, reduces the pressure around the central shaft, and makes it easier for the soil-binder mixture produced in the outer columns to flow into the central column. The stabilizers, meanwhile, guide the flow smoothly from the outer columns to the central column as shown in Figure 5.34. During withdrawal, the mixing shafts rotate in the reverse direction. The screw along the central shaft pushes the soil black, resulting in a vertical mixing of the soil binder mixture, and the stabilizers promote the smooth flow and mixing, further improving the mixing degree.

4.3.1.3   Binder plant

The plant and pumping unit is for producing and supplying binder slurry to the outer two shafts of the CDM-Lemni 2/3 machine, which can produce binder slurry of $60 \, m^3/h$.

4.3.1.4   Control unit

A control room is placed in the binder plant, where the admixture condition, quantity of each material, rotation speed of mixing blade, speed of shaft movement, *etc.* are continuously monitored, controlled, and recorded. These monitoring data can be fed back to the operator for precise construction of the column.

4.3.2   Construction procedure

4.3.2.1   Preparation of site

Similarly to the DJM and the ordinary CDM method, field preparation is carried out to assure smooth execution and prevention of environmental impact.

4.3.2.2   Field trial test

It is recommended to conduct a field trial test in advance in, or adjacent to, the construction site, with the same purpose as the DJM and the ordinary CDM. in order to confirm the smooth execution.

4.3.2.3   Construction work

The construction procedure for the CDM-Lemni 2/3 method is almost the same as that of the ordinary CDM method. As the penetration injection is adopted for the CDM-Lemni 2/3, binder slurry is injected from the bottom tip of the outer two mixing shafts. In the withdrawal stage, the mixing blades rotate reversibly in the horizontal plane and the soil – binder mixture is mixed again. Typical penetration and withdrawal speeds of the shafts are 0.6 to 1.0 m/min. and 1.0 m/min. respectively, and the rotation speed of the mixing blades is about 20 and 40 rpm during the penetration and the withdrawal stages respectively. This corresponds to the "blade rotation number" of more than 350.

4.3.3   Quality control during execution

To assure the quality and the dimension of stabilized soil elements, it is essential to keep the designed condition by monitoring the binder condition, quantity of each material, rotation speed of mixing blades, mixing shafts speed, *etc*. As same as the ordinary CDM machine, the operation monitoring in the CDM-Lemni 2/3 method covers quality and quantity control monitoring, as shown in Figure 5.24. These monitoring data can be fed back to the operator for precise construction of stabilized soil elements.

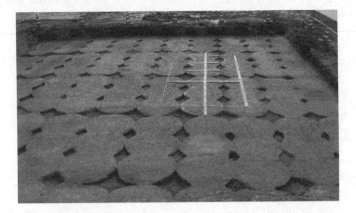

Figure 5.35 Top view of stabilized soil columns by CDM-Lemni 2/3 method (Cement Deep Mixing Method Association, 2005).

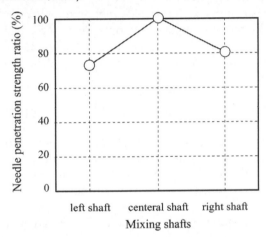

Figure 5.36 Strength profile of stabilized soil columns by CDM Lemini 2/3 method (Cement Deep Mixing Method Association, 2005).

4.3.3.1   Quality assurance

After the construction work, the quality of the in-situ stabilized soil elements should be verified in advance of construction of superstructure in order to confirm the design quality, such as uniformity, strength, permeability or dimension. Full depth coring, observation of core and testing of selected specimens are conducted for quality assurance in the same way as that for the ordinary CDM.

4.3.3.2   Effect of method

Figure 5.35 is the top view of the stabilized soil elements by the CDM-Lemni 2/3 method, which shows quite uniform stabilized soils were constructed (Cement Deep Mixing Method Association, 2005). Figure 5.36 shows the comparison of the strength distribution of stabilized soil column that were measured at its top by the needle penetration tests and normalized by that at the central column (Cement Deep Mixing

Method Association, 2005). The strengths at the outer two stabilized soil columns are slightly smaller than that at the central column.

5 WET METHOD OF DEEP MIXING FOR IN-WATER WORKS

5.1 Cement deep mixing method

5.1.1 Equipment

5.1.1.1 System and specifications

For near shore construction works such as port and harbor facilities or man-made island constructions, a variety of deep mixing barges specially designed for improving sea-bottom sediment are available in Japan. In-water construction works at rivers or lakes where the special barges are not easily accessible, deep mixing equipment for on-land works may be used by mounting them on a flat bottom and shallow draft barge (Cement Deep Mixing Method Association, 1999).

A special barge is equipped with leaders, mixing tools, binder silos, binder-slurry mixers, agitator tanks, pumping units and an operation room, as shown in Figure 5.37. The special barges are classified into three categories, small, medium and large size based on the cross sectional area of the stabilized soil element installed by a single stroke of the deep mixing tool. The cross sectional area of three categories is about 2.2, 4.6 and 5.7 m^2. The size of barge, the maximum improvement depth from the water surface, major specifications of mixing tool and binder slurry plant are summarized in Table 5.7 (Coastal Development Institute of Technology, 2008).

There are two types in the position of the leader; center and front end of the barge, as shown in Figure 5.38. Although the barge is anchored during production, wind and wave forces influence the stability of the barge by causing motion in the pitching and rolling directions. In addition, during the penetration stage, the leader is subjected to an upward force due to the reaction of ground and buoyancy force of mixing tool, while a downward force in the withdrawal stage. The barge with the mixing tool at its

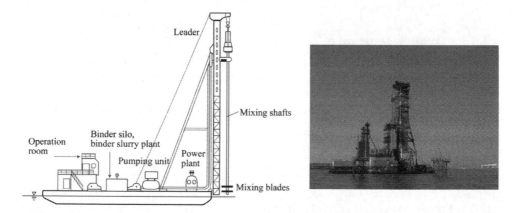

Figure 5.37 CDM barge for in-water works (Cement Deep Mixing Method Association, 1999).

Table 5.7 Specifications of CDM barges for marine works (Coastal Development Institute of Technology, 2008).

Class, Name	Barge				Improvement machine				Plant		
	length (m)	breadth (m)	depth (m)	draft (m)	area (m²)	depth (m)	blades	position	binder silo (kN)	mixer (m³)	pump (l/min.)
5.7 m² class											
Decom 7	63.0	30.0	4.5	3.2	5.74	−70	1.0 m × 8	C	4000 × 4	3.5 × 2	250 × 8
Pocom 2	48.0	30.5	4.1	3.3	5.75	−67	1.01 m × 8	C	2500 × 4	3.5 × 1	150 × 12
DCM 3	47.5	28.0	4.5	3.0	5.74	−70.5	1.0 m × 8	C	2000 × 2	2.0 × 2	250 × 8
Decom 5	60.0	27.0	4.0	2.7	6.91	−60	1.6 m × 4	F	3000 × 4	3.5 × 1	250 × 8
Kokaku	70.0	32.0	4.5	2.65	5.47	−52.0	1.4 m × 4	F	3000 × 4	3.0 × 2	440 × 8
4.6 m² class											
DCM 6	56.0	26.0	4.2	2.2	4.64	−60	0.95 m × 8	F	1500 × 2	2.0 × 2	250 × 8
DCM 8	48.0	22.5	3.5	1.5	4.05	−41	1.4 m × 4	F	1500 × 2	2.0 × 1	350 × 4
CMC 7	67.0	30.0	4.0	2.0	4.63	−55	1.3 m × 4	F	1500 × 2	2.0 × 2	600 × 4
CMC 8	53.0	24.0	4.0	2.3	4.63	−45	1.3 m × 4	F	1500 × 2	2.0 × 2	220 × 4
Pocom 10	52.0	22.8	4.0	2.9	4.65	−49	1.31 m × 4	F	1500 × 2	2.0 × 1	200 × 4
Pocom 11	50.0	26.4	3.6	2.5	4.65	−40	1.31 m × 4	F	1500 × 2	2.0 × 1	200 × 4
Pocom 12	60.0	30.0	4.0	2.5	4.65	−52	1.31 m × 4	F	2000 × 2	2.5 × 1	350 × 8
Decom 8	55.6	24.0	4.3	2.85	4.68	−52	1.4 m × 4	F	1500 × 1	2.0 × 1	250 × 8
2.2 m² class											
Pocom 8	38.0	16.8	2.3	1.4	2.23	−29	1.22 m × 2	F	1200 × 1	2.2 × 1	250 × 4
CMC 3	40.0	18.0	3.5	2.3	2.20	−40	1.2 m × 2	F	500 × 1	1.5 × 1	600 × 2
CMC 5	40.0	18.0	3.5	2.3	2.20	−40	1.2 m × 2	F	500 × 2	1.5 × 1	600 × 2
Decom S-3	30.0	15.0	3.0	1.5	2.23	−27	1.22 m × 2	F	650 × 1	2.0 × 1	445 × 2
Decom S-5	35.0	12.0	2.2	1.3	2.23	−30	1.22 m × 2	F	500 × 1	2.2 × 2	445 × 2
Decom S-7	36.0	15.0	2.5	1.4	2.23	−30	1.22 m × 2	F	1000 × 1	1.8 × 1	440 × 2

(a) Center leader type (Pocom 2).

(b) Center leader type (DCM 3).

(c) Front leader type (Kokaku).

(d) Front leader type (Pocom 12).

(e) Front leader type (CMC 7).

Figure 5.38 CDM barge for in-water works in operation (by the courtesy of Cement Deep Mixing Method Association).

front end is fluctuated in the pitching direction, which requires sophisticated control to penetrate and withdraw the mixing shafts vertically. The barge with the mixing tool at its center is preferred from the view point of the stability of the barge and control of the deep mixing work. However, the barge with the mixing tool at the front end

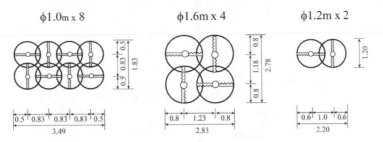

Figure 5.39 Typical arrangements of mixing shafts of CDM method for in-water work (Cement Deep Mixing Method Association, 1999).

(a) Four mixing shafts type (Pocom 12).

(b) Eight mixing shafts type (Decom 7).

Figure 5.40 Mixing blades of CDM machine for in-water work.

is far superior when the improvement in the close vicinity of the existing structures is required.

5.1.1.2 Mixing tool

The mixing tool for in-water works usually has two to eight mixing shafts (Figure 5.39). Figure 5.40 shows examples of the bottom end of mixing tools that have four mixing shafts and eight mixing shafts (Cement Deep Mixing Method Association, 1999). The distance of mixing shafts is smaller than the diameter of the mixing blade so that the stabilized soil element consisting of round stabilized soil columns partially overlapped each other is produced by a single operation, as shown in Figure 5.39. Depending on the diameter of mixing blade and the number of shafts, the cross sectional area of the stabilized soil element ranges from 2.20 m² to 6.91 m² (Table 5.7). The machine usually has a bracing plate to keep the distance of the two mixing shafts (see Figure 5.41). The plate is also expected to function to increase the mixing degree by preventing the "entrained rotation phenomenon," a condition in which disturbed soil adheres to and rotates with the mixing blade without efficient mixing of soil and binder. A mixing shaft is adjusted to rotate in an opposite direction to the adjacent shaft in order to increase the degree of mixing and also improve the stability of the mixing tool.

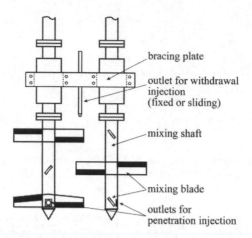

Figure 5.41 Sliding pipe for binder injection and mixing blades (Cement Deep Mixing Method Association, 1999).

The mixing tool with multiple shafts is suspended along the leader and clamped laterally at the top gearbox and at the bottom of the leader. The binder slurry prepared in the plant on the barge is supplied from the top of each shaft through the swivel joint. Several outlets are installed on each shaft at two different levels; at the upper most mixing blades or at the lowest mixing blades as shown in Figure 5.41. The lower outlet is used for the penetration injection and the upper outlet is used for the withdrawal injection. The location of the outlet in the radial direction is either on the shaft surface or at some distance on the mixing blade. Some mixing machines installed a sliding injection pipe between the mixing shafts, which is used for the bottom treatment and/or the withdrawal injection processes.

A stack of mixing blades consists of an excavation blade and mixing blades. The excavation blade is installed at the very end of the mixing shaft on which forks made by hard metal are fixed so that the machine can excavate and screw in a soil efficiently. The maximum depth of stabilization by the available machines is 70 m from the water surface. The mixing tool can penetrate local stiff layers to reach the desired depth. A machine with a relatively large capacity can penetrate a layer whose SPT N-value and thickness are 8 and 4 m for clayey soil, and 15 and 4 m for sandy soil, respectively. The shape and stack of mixing blades have been developed to assure a high mixing degree as much as possible.

5.1.1.3 Plant and pumping unit

One to four silos with a capacity of 500 to 4000 kN each are installed on the barge for storage of binder. The mixers, agitator tanks and pumping units are installed on the barge for producing and supplying binder slurry to the mixing tool. Binder slurry is manufactured by every 1.5 to 3.5 m³ in a mixer. The water and cement ratio (W/C) of the slurry is usually 60 to 100%. The binder slurry thus manufactured is temporarily stored in one or two agitator tanks with a capacity of 2 to 20 m³ each, then supplied

(a) Control desk for operating pumping units.   (b) Control desk for operating mixing tools.

Figure 5.42 Control desks on CDM barge, Pocom 12.

to the mixing tool by the pumping unit, where about 150 to 600 l/min. in volume is supplied to each mixing shaft by the help of a pumping pressure of 300 to 500 kN/m^2. An air entraining (AE) agent or water reducing agent is often used together with the binder slurry to improve the fluidity of the slurry and/or to prevent the setting of binder before injection into the soil.

5.1.1.4 Control room

A control room is installed on the barge, as shown in Figure 5.42, where the positioning of barge and operation of the machine are conducted, and the binder condition, the amount of each material, the rotation speed of mixing blades, the penetration and withdrawal speeds of mixing shafts, *etc.* are continuously monitored, controlled and recorded.

5.1.2 Construction procedure

The construction procedure for in-water works is similar to that for on-land works, which includes preliminary survey, positioning, field trial test and construction work.

5.1.2.1 Site exploration and examination of execution circumstances

Before production, execution circumstances should be examined to ensure smooth operation and prevent environmental impact. The execution schedule can be delayed by weather and wave conditions. Thus weather and wave conditions should be examined in advance when planning the execution schedule; wave height, wind direction, wind velocity and tides should be carefully considered. According to experience, in-water work is difficult to conduct in conditions where the maximum wind velocity exceeds 10 m/sec., the maximum significant wave height exceeds 0.5 m, or the minimum visibility is less than 1,000 m. Environmental impacts such as water contamination, noise, vibration *etc.* which can occur during the execution should obviously be kept to a minimum.

Any obstacles on or below the seabed in the construction site can delay the operation schedule, or cause damage to the mixing blades. Before operations, the seabed should be surveyed carefully and any obstacles should be removed. This process is particularly important with regard to blind shells that can cause human damage. This soil survey can usually be carried out by means of a magnetic prospecting probe.

5.1.2.2　Positioning

In order to position the CDM barge at the prescribed position, several anchors are extended at first. The positioning methods have four alternatives; collimation of two transit apparatuses, collimation using a transit and an optical range finder, an automatic positioning system with three optical finders, and a positioning system with GPS. Recently, the GPS system has been frequently used for positioning. The barge is positioned by controlling the extended anchors.

5.1.2.3　Field trial test

It is recommended to conduct a field trial test in advance at a ground in or adjacent to the construction site, in order to ensure smooth execution at the construction site. In the test, all the monitoring equipment, such as amount of binder, rotation speed of mixing blades and penetration and withdrawal speeds of the mixing shafts are calibrated. In the case where the stabilized soil columns should reach and have firm contact with a stiff bearing layer (fixed type improvement), a field trial test should be carried out to measure the change in the electric or hydraulic power required for driving the mixing shafts and the penetration speed of the mixing shaft at the stiff layer so that they can help to detect if the mixing blades have reached the stiff layer in the actual construction.

When there is less experience in similar soil conditions, it is recommended to carry out a field mixing trial and to confirm that the strength and integrity of the trial column/element meet the design requirement.

5.1.2.4　Construction work

After setting the CDM barge at the prescribed position, the mixing tool is penetrated into a ground while rotating the mixing shafts. There are two basic execution procedures depending on the injection sequence of binder (Figure 5.12): (a) injecting binder during the penetration of mixing shafts and (b) injecting binder during the withdrawal of mixing shafts. The penetration injection is applied to the $2.2\,m^2$ class, while the withdrawal injection is applied to the 4.6 and $5.7\,m^2$ classes. The location of the injection outlet is different for each injection method. For the penetration injection method, the injection outlets should locate at the lowest mixing blades, but they should be at the uppermost mixing blades for the withdrawal injection.

For the withdrawal injection, the construction procedure is shown in Figures 5.43. The mixing shafts are penetrated into a ground by sending out the wires. During the penetration, the mixing blades at the bottom end of the mixing shafts cut and disturb the soil to reduce the strength of the original soil to make the mixing shafts penetrate by their self-weight. Table 5.8 summarizes the typical execution specifications of CDM method (Cement Deep Mixing Method Association, 1999). The $5.7\,m^2$ class CDM machines can penetrate a local stiff layer where SPT N-value is less than 8 and the thickness is less than $4\,m$ for a clay layer, and SPT N-value is less than 15 and the thickness is less than $4\,m$ for a sandy layer. The smaller class machines have

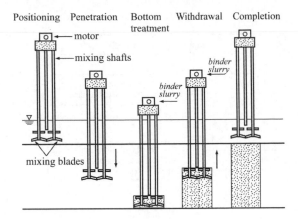

Figure 5.43 Execution process of CDM method for in-water work (Cement Deep Mixing Method Association, 1999).

Table 5.8 Typical execution specification of CDM method for in-water work (Cement Deep Mixing Method Association, 1999).

Class	clay		sand	
	SPT N-value	thickness	SPT N-value	thickness
2.2 m²	<6	<2.0 m	<10	<2.0 m
	<8	<1.0 m	<15	<1.0 m
4.6 m²	<8	<3.0 m	<15	<3.0 m
5.7 m²	<8	<4.0 m	<15	<4.0 m

lower capacity for penetration as shown in the table. For a soil layer exceeding these conditions, water jetting may be required for the penetration. In some cases where the soil is considerably hard, pre-drilling may be necessary in advance of the mixing work.

The stabilized soil elements should reach the stiff layer sufficiently in the case of the fixed type improvement. Rapid change in the penetration speed of the mixing shafts, the required torque and rotation speed of the mixing blades are useful to detect whether the shafts have reached the stiff layer. For the withdrawal injection, the bottom treatment is an inevitable process and carried out by one of the following procedures. When the mixing tool reaches the stiff layer, the mixing tool is lifted up and down several times while continuing to inject binder slurry from the lower binder injection outlets in order to assure sufficient mixing at the lower portion of the column and to attain the reliable contact of the column with the stiff layer. Instead of re-stroking, some machines have an injection pipe between the mixing shafts, as shown in Figure 5.41 in order to assure sufficient contact of the column with the stiff layer. When the mixing tool reach the stiff layer, the injection pipe temporarily extends down to the bottom level of the blades to inject the binder slurry, and the binder is mixed with the soil after the pipe returned to the original position.

In the withdrawal stage, the vertical speed of the shafts is kept constant. At the same time, the binder slurry is injected from the upper binder injection outlets into the

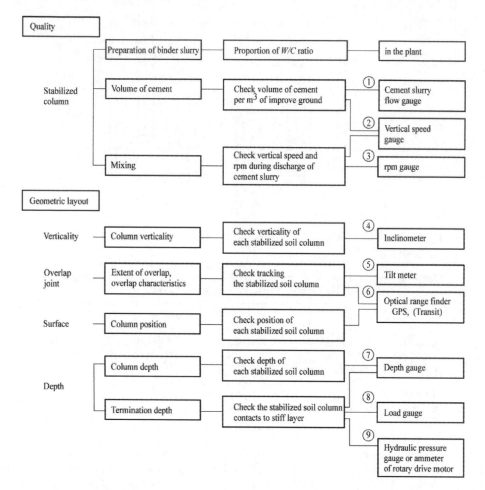

Figure 5.44 Operation monitoring for CDM method for in-water works (after Cement Deep Mixing Method Association, 1999).

ground at a constant flow rate. The mixing blades rotate in the horizontal plane and mix the soil and binder. Thus a stabilized soil element having a cross section as shown in Figure 5.39 is manufactured in situ. The speed of shaft movement and the rotation speed are determined so as to satisfy the rotation number of about 360 (see Equation (7.2)). Typical penetration and withdrawal speed is about 0.3 to 1.0 m/min. and the rotation speed of the mixing blade is 20 to 40 rpm.

5.1.3 Quality control during production

To produce stabilized soil columns/elements that meet the design requirements on the quality and dimension, it is essential to control and monitor the quality of binder, geometric layout, and operational parameters such as amount of binder, rotation speed of mixing blades, shaft speed, *etc.* Figure 5.44 shows the operational parameters for the CDM method and items for geometric layout (after Cement Deep Mixing Method

Association, 1999). The verticality of the mixing tool is usually evaluated by measurement of the verticality of the leader, and is controlled within 1/200 to 1/100 in many cases. During the execution the monitoring data are fed back to the operators in the control room for precise construction. In practice, the rotation speed of the mixing shafts is usually fixed at 20 or 40 rpm. The penetration and withdrawal speeds are controlled to the prescribed speed by sending out the wire which suspends the mixing tool. The amount of binder slurry is adjusted to the penetration and withdrawal speeds by controlling the pumping pressure at the pumping units. The W/C ratio and density of binder slurry are controlled to the design value. The binder slurry is manufactured every 1.5 to 3.5 m^3 and used up within about one hour to prevent the setting of binder before injection into the soil.

5.1.3.1 Quality assurance

After the construction work, in-situ stabilized soil elements should be investigated in order to verify the design quality, such as continuity, uniformity, strength, permeability or dimension. In Japan, full depth coring and unconfined compression test on the core samples are most frequently conducted for verification. The number of core borings is dependent upon the volume of the stabilized soil. In the case of in-water works, one core boring is generally conducted for every 10,000 m^3. When the total volume exceeds 100,000 m^3, one additional core boring is conducted for every further 50,000 m^3.

The continuity and uniformity of the stabilized soil elements are confirmed by visual observation of the continuous core. Determination of the engineering properties of the stabilized soils is based on unconfined compression tests on samples selected from the continuous core. The number of test depends upon the construction's condition and the soil properties. In general three core barrels are selected from three levels and three specimens are taken from each core barrel and subjected to the unconfined compression test for each core boring. Properties other than unconfined compressive strength can be correlated with unconfined compressive strength as discussed in Chapter 3.

The quality of the core sample primarily depends on the uniformity of stabilized soil. However, it further relies on the quality of the boring machine, the coring tool and the skill of the workmen. If the coring is not properly conducted, a low quality sample with some cracks can be obtained. A double tube core sampler or triple tube core sampler has been used for core sampling of stabilized soil. It is recommended to use samplers of relatively large diameter such as 86 or 116 mm in order to take good quality samples. The quality of the core sample is usually evaluated by visual inspection and/or the Rock Quality Designation (RQD) index, defined by Equation (7.3) in Chapter 7. The RQD index measures the percentage of "good rock" within a borehole and provides the rock quality as shown in Table 7.3 in Chapter 7.

6 ADDITIONAL ISSUES TO BE CONSIDERED IN THE MECHANICAL MIXING METHOD

6.1 Soil improvement method for locally hard ground

Where soil stratifications are complicated by past geological history, it is not unusual to encounter a local stiff layer before reaching the designed depth. The DM machine can penetrate a relatively hard layer as shown in Tables 5.1, 5.3, 5.5, 5.6 and 5.8.

However, in some cases, the mixing blades and shafts of the DM machine may be damaged and/or stuck in the ground. When penetrating such a hard layer, therefore, it is necessary to carry out pre-boring with an auger machine, or use a machine with larger capacity.

6.2 Noise and vibration during operation

Figure 5.45 shows the relationship between the noise and vibration levels and the distance from the source, in which the field values caused by piling and various ground improvement techniques are plotted for comparison (Japanese Society of Soil Mechanics and Foundation Engineering, 1985). The figure indicates that the noise and vibration levels caused by the deep mixing method are relatively small among the different soil improvement techniques and they satisfy the Japanese regulation values except at a close vicinity of the source.

6.3 Lateral displacement and heave of ground by deep mixing work

6.3.1 On-land work

As a result of injecting binder into a ground, the surrounding soil may be displaced horizontally and the ground surface may heave to some extent. Figure 5.46 shows the measured lateral displacement at the ground surface in on-land works in reference to the local topography (Mizuno *et al.*, 1988). Although the amount of ground movement is relatively small compared with the in-water works as shown later, the ground moves horizontally 0.1 to 0.4 m near the excavation or cut slope. The amount of displacement is dependent upon the improvement area ratio, amount of binder injected per column and the installation sequence in the improved area. It is important to estimate the amounts of lateral displacement and ground heave and their influence on nearby structures.

In order to reduce the influence, the CDM-LODIC method was developed and successfully employed in a number of construction sites (see Figures 5.30 and 5.31).

6.3.2 In-water work

Figure 5.47 shows a typical case record on upheaved ground at Yokohama Port in which $160 \, kg/m^3$ of binder slurry (W/C ratio of 60%) was injected into a ground (Cement Deep Mixing Method Association, 1999). The extent of the upheaved ground is not uniform and depends on many factors such as the soil profile, the thickness of improved layer, the improvement area ratio, and the installation sequence. According to accumulated field experiences, the total volume of upheaved soil is almost equivalent to that of the binder slurry injected, and the upheaved volume within the improved ground area is approximately 70% of the volume of the binder slurry injected. Since the upheaved soil is disturbed and softened, it is usually handled by one of the following procedures: 1) dredge and dispose of the soil up to the determined depth; 2) improve to a level close to the surface, then dredge and dispose of the upper surface layer; 3) improve the soil to the surface of the upheaved soil. The first procedure has been used in most cases in Japan to obtain the required water depth for quay structures.

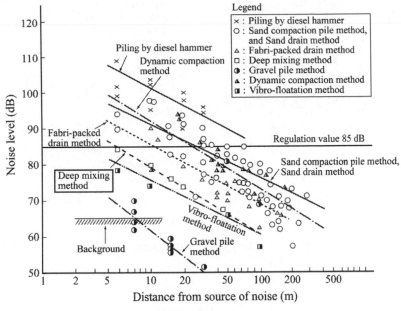

(a) Noise level and distance from the source.

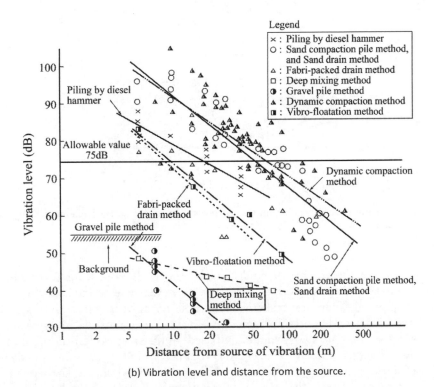

(b) Vibration level and distance from the source.

Figure 5.45 Noise and vibration during operation (Japanese Society of Soil Mechanics and Foundation Engineering, 1985).

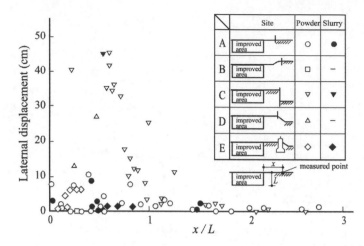

Figure 5.46 Lateral displacement of surrounding ground during improvement operation (Mizuno et al., 1988).

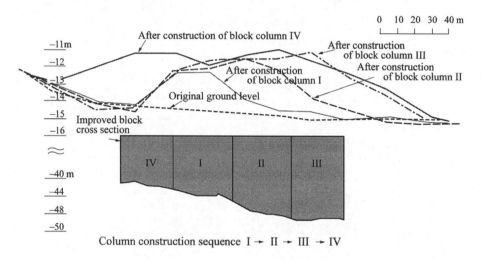

Figure 5.47 A typical case record on up heaved ground at Yokohama Port (Cement Deep Mixing Method Association, 1999).

Reduction of the volume of binder and/or the W/C ratio leads to a reduction of the volume of upheaved soil. Contractors, however, tend to increase the volume of binder to avoid failure to acceptance criteria in terms of the strength of stabilized soil. If the owner and contractor agree in advance, it is possible to 1) reduce the binder volume to the necessary minimum based on the strength test of initial several production columns, and 2) to reduce the volume of binder slurry by reducing the W/C ratio after confirming the constructability in the initial phase. The reduction of the W/C ratio can provide the increase of the strength of stabilized soil, which in turn can reduce the amount of the volume of binder.

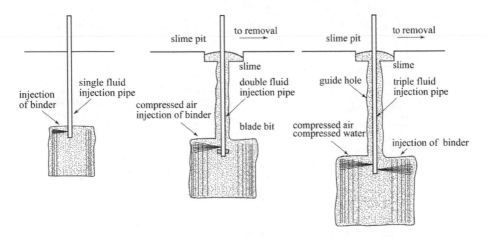

Figure 5.48 Classification of high pressure injection techniques.

7   HIGH PRESSURE INJECTION METHOD

In the high pressure injection technique, binder slurry is injected at a high pressure of 10 to 60 MN/m² through the nozzle into a soil. The binder slurry mixes with the surrounding soil as the injection pipe is slowly rotated and withdrawn from the drilled hole. The high pressure injection tool is designed to withstand high injection pressures using proper materials as well as specialized seals between the rod joints.

Basically, there are three types of the method: single fluid technique, double fluid technique and triple fluid technique, as shown in Figure 5.48. In the single fluid technique, neat binder slurry is injected into a ground. This technique produces a stabilized soil column with a small amount of slime, mixture of soil and binder slurry during execution. While the volume of binder slurry that was not released as slime may cause adverse influence such as ground heaving and horizontal displacement. Because of this, the technique is now used in a limited number of cases in Japan.

In order to increase the diameter of the stabilized soil column and to prevent ground heaving and horizontal movement during execution, compressed air is injected together with binder slurry in the double and triple fluid techniques, where a large amount of slime is uplifted by the help of buoyancy effect of air bubbles. In the double fluid technique, binder slurry and compressed air are injected, while binder slurry, compressed air and pressurized water are injected in the triple fluid technique. As large amount of binder slurry is injected into a ground and a large amount of slime is removed from the ground in the double fluid and triple fluid techniques, it can be said that they are a sort of soil mix/replace technique instead of soil mix technique.

In the high pressure injection techniques, a large number of techniques have been developed and available in Japan (Figures 5.1 and 5.49). Among them, the Chemical Churning Pile or Chemical Churning Pattern (CCP) method, a typical single fluid technique, creates a stabilized soil column of about 1 to 2 m in diameter. In the double fluid, the JSG method and the Superjet method create a stabilized soil column of about 2 m and 5 m in diameter respectively. In the triple fluid technique, the CJG method, the

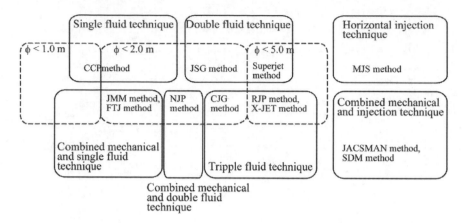

Figure 5.49 Classification of high pressure injection techniques.

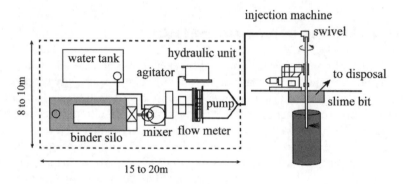

Figure 5.50 Layout of column jet grout system (Japan Jet Grouting Association, 2011).

RJP method and the X-JET method are typical techniques, which create a column of about 2 m, 5 m and 2.5 m in diameter respectively. The Superjet method and the X-jet method are frequently applied in recent years in Japan.

In this section, the CCP method, the JSG method and the Superjet method, and the Column Jet Grout Method and the X-jet method are briefly introduced as single fluid, double fluid and triple fluid techniques, respectively.

7.1 Single fluid technique (CCP method)

7.1.1 Equipment

The system of the single fluid technique is the most simple, in which neat binder slurry is injected through a small nozzle at high pressure and mixes with in-situ soil. The equipment for the technique consists of an injection machine, binder silo, water tank, batching plant, mixer and agitator, and hydraulic unit, as shown in Figure 5.50 (Japan Jet Grouting Association, 2011).

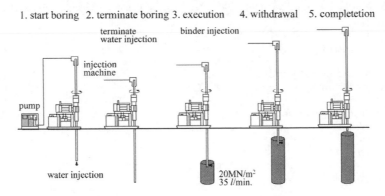

1. start boring 2. terminate boring 3. execution 4. withdrawal 5. completetion

terminate water injection

binder injection

injection machine

pump

water injection

20MN/m² 35 l/min.

Figure 5.51 Execution process of CCP method (Japan Jet Grouting Association, 2011).

The injection machine consists of a boring machine and an injection pipe. Binder slurry made in the mixer is transferred to the injection machine by the hydraulic pump, and injected into a ground. The injection pipe for the technique is a hollow cylinder with 40.5 mm in outer diameter. An injection nozzle is installed on the side surface of the pipe near its bottom so that the binder slurry is injected in the horizontal direction. Several binders specially designed for the technique are available as shown later in Table 5.10, which are a composite of cement as a mother material and chemical additives for achieving various target strengths of stabilized soil column.

7.1.2 Construction procedure

7.1.2.1 Preparation of site

The field preparation is carried out in accordance with the specific site conditions, which includes suitable access for plant and machinery. The basic layout of the equipment is illustrated in Figure 5.50. The plant usually requires about 120 to 200 m². Before actual operation, execution circumstances should be prepared to assure smooth operation and prevention of environmental impact.

7.1.2.2 Construction work

The construction work of the technique is illustrated in Figure 5.51 (Japan Jet Grouting Association, 2011). After locating the injection machine at the prescribed position, the injection pipe is penetrated into a ground by the help of flushing water of pressure of about 3 MN/m². The injection pipe can penetrate a local stiff layer whose SPT N-value and cohesion are lower than about 15 and 50 kN/m² respectively. When reaching the design depth, the flushing port at the tip of the pipe is closed and the binder slurry is injected at high pressure of 20 MN/m² through the nozzle into the soil. Simultaneously, the injection pipe is rotated and withdrawn stepwise by about 25 to 50 mm interval. Table 5.9 summarizes standard process control values for sand and clay grounds, by which the production of a stabilized soil column having at least a

Table 5.9 Standard operational parameters for CCP method (Japan Jet Grouting Association, 2011).
(a) For sand layer.

SPT N-value of original soil	$N < 5$	$5 < N < 10$	$10 < N < 15$
Binder injection pressure (MN/m²)	20.0	20.0	20.0
Withdrawal speed (m/min.)	0.25	0.25	0.25
Rotation speed (rpm)	20	20	20
Flow rate of binder slurry (m³/min.)	0.035	0.035	0.035
Diameter of column (m)	0.40	0.35	0.30

(b) For clay layer.

Cohesion of original soil (kN/m²)	$C < 10$	$10 < C < 30$	$30 < C < 50$
Binder injection pressure (MN/m²)	20.0	20.0	20.0
Withdrawal speed (m/min.)	0.25	0.25	0.25
Rotation speed (rpm)	20	20	20
Flow rate of binder slurry (m³/min.)	0.035	0.035	0.035
Diameter of column (m)	0.50	0.45	0.30

diameter as shown in Table 5.9 is guaranteed empirically (Japan Jet Grouting Association, 2011). Some amount of slime, mixture of soil and binder slurry, is lifted up along the injection pipe during the execution which should be removed and should be handled with care according to the local regulation. While the volume of binder slurry that was not released as slime might cause horizontal displacement and heaving of the ground surface.

7.1.2.3 Quality control during production

To produce stabilized soil columns with guaranteed quality and dimension, it is essential to control and monitor the W/C ratio and density of binder slurry, geometric layout, and operational parameters such as amount of binder, rotation speed of injection pipe (or monitor), shaft speed, etc. The verticality of the mixing tool is usually controlled within about 1/250. In practice, the rotation and withdrawal speeds of the injection pipe are controlled to standard values, while the amount of binder slurry is adjusted by controlling the pumping pressure at the pumping units, as summarized in Table 5.9.

7.1.2.4 Quality assurance

After the construction work, in-situ stabilized soil columns should be investigated in order to verify the design quality, such as continuity, uniformity, strength, permeability or dimension. In Japan, full depth coring and an unconfined compression test on the core samples are most frequently conducted when the improvement purpose is reinforcement or stability. In the case where the improvement purpose is construction of an impermeable zone, continuity of the stabilized soil columns in the successive operation is thought important. The size and strength of the stabilized soil column are dependent on the characteristics of the original ground, the type of binder and operational parameters. The strength and the size of the stabilized soil column listed in Tables 5.9 and 5.10 (Japan Jet Grouting Association, 2011) are guaranteed minimum

Table 5.10 Design values of CCP method (Japan Jet Grouting Association, 2011).

Binder type	soil	W/C ratio	q_u (MN/m²)	cohesion, c (MN/m²)	adhesion (MN/m²)	bending strength (MN/m²)	elastic modulus (MN/m²)
CCP5	sand	90%	1.0				
	clay		0.8				
	organic		0.2				
CCP-6	sand	100%	3.0				
	clay		1.0	$1/6\,q_u$	$c/3$	$2c/3$	$100\,q_u$
CCP-7A	sand	150%	2.0				
	clay		0.5				
CCP-7B	sand	110%	1.0				
	clay		0.5				
CCP-8	organic	100%	0.3				

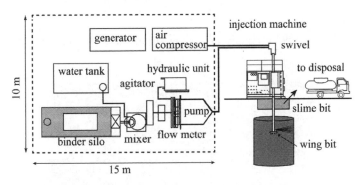

Figure 5.52 Layout of JSG system (Japan Jet Grouting Association, 2011).

values of the technique. In the actual situation, a stabilized soil column having a larger diameter and higher strength than those are constructed in many cases.

7.2   Double fluid technique (JSG method)

7.2.1   Equipment

The double fluid technique is based upon the principles of the single fluid technique, but to enhance its radius of influence it uses a shroud of compressed air concentric about the jet of binder. The two fluid referred to in this technique are binder slurry and air. The binder slurry is injected at a high pressure of 20 MN/m² and is aided by a cone of compressed air of 0.7 MN/m², which shrouds the binder slurry. The air reduces the friction loss, allowing the binder slurry to travel farther from the injection point, thereby producing a larger stabilized soil column diameter than those of the single fluid technique. The air injection produces more slime than the single fluid technique due to the air-lifting effect, which effectively reduce the adverse influence on nearby structures due to horizontal displacement and heaving of ground surface.

The equipment for the technique is the same as that for the single fluid technique, except for the use of two-way coaxial injection pipe and an air compressor, as shown in Figure 5.52. The injection machine consists of a boring machine and an injection

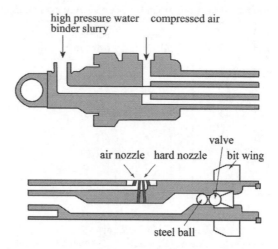

high pressure water   compressed air
binder slurry

valve

air nozzle   hard nozzle   bit wing

steel ball

Figure 5.53 Detail of swievel and tip of JSG method (Japan Jet Grouting Association, 2011).

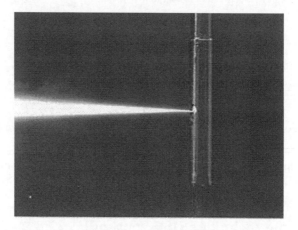

Figure 5.54 Injection of binder from tip of JSG method (Japan Jet Grouting Association, 2011).

pipe. Binder slurry is prepared in the mixer and transferred to the injection machine by the hydraulic pump, and injected into a ground.

Figure 5.53 shows the cross sectional view of the swivel and bottom tip of the injection pipe (Japan Jet Grouting Association, 2011). The injection pipe for the technique is a duplex cylinder with 60.5 mm in outer diameter. An injection nozzle is installed on the side surface of the pipe near its bottom so that the binder slurry is injected in the horizontal direction (Figure 5.54). At the bottom tip, a wing bit of 115 to 150 mm in diameter is installed. Several binders specially designed for the technique are available as shown later in Table 5.13, which are composite of cement as a mother material and chemical additives for achieving various target strengths of the stabilized soil column. The capacities of water tank and pumping unit are typically 5 m³ and 12 m³/h.

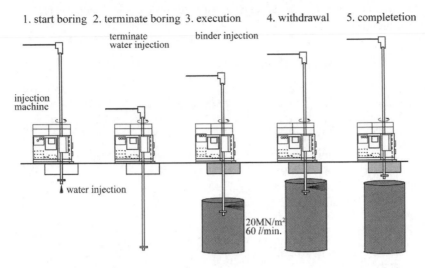

1. start boring 2. terminate boring 3. execution   4. withdrawal   5. completetion

Figure 5.55 Execution process of JSG method (Japan Jet Grouting Association, 2011).

7.2.2 Construction procedure

7.2.2.1 Preparation of site

The field preparation is carried out in accordance with the site specific conditions, which includes suitable access for plant and machinery. The basic layout of the equipment is illustrated in Figure 5.52. The plant usually requires about $150\,m^2$. Before actual operation, execution circumstances should be prepared to assure smooth operation and prevention of environmental impact.

7.2.2.2 Construction work

The construction work of the technique is illustrated in Figure 5.55 (Japan Jet Grouting Association, 2011). After locating the injection machine at the prescribed position, the injection pipe is penetrated into a ground by the help of flushing water of a pressure of about $3\,MN/m^2$ from the flushing port. The injection pipe can penetrate a local stiff layer whose SPT N-value and cohesion are lower than about 50 and $50\,kN/m^2$ respectively. The wing bit installed at the bottom tip of the injection pipe creates a hole of 115 to 150 mm in diameter, larger than that of the injection pipe. When reaching the design depth, the flushing port at the tip of the pipe is closed and the binder slurry at a high pressure of $20\,MN/m^2$ and a cone of compressed air of $0.7\,MN/m^2$ are injected through the nozzle into the soil. Simultaneously, the injection pipe is rotated and withdrawn stepwise by about 25 to 50 mm interval. Table 5.11 summarizes standard operational parameters for sand and clay grounds, which can create a stabilized soil column having the diameter in the table.

The required total volume of binder slurry to create a stabilized soil column can be calculated by Equation (5.2).

$$Q = H \cdot v \cdot q_c \cdot (1 + \beta) \tag{5.2}$$

Table 5.11 Execution conditions for JSG method (Japan Jet Grouting Association, 2011).
(a) For sand layer.

SPT N-value of original soil	$N < 10$	$10 < N < 20$	$20 < N < 30$	$30 < N < 35$	$35 < N < 40$	$40 < N < 50$
Air pressure (MN/m²)	0.7	0.7	0.7	0.7	0.7	0.7
Binder injection pressure (MN/m²)	20.0	20.0	20.0	20.0	20.0	20.0
Withdrawal speed (m/min.)	0.025	0.029	0.03	0.038	0.048	0.059
Rotation speed (rpm)	20	20	20	20	20	20
Flow rate of binder slurry (m³/min.)	0.06	0.06	0.06	0.06	0.06	0.06
Diameter of column (m)	2.0	1.8	1.6	1.4	1.2	1.0

(b) For clay layer.

SPT N-value of original soil	$N < 1$	$N = 1$	$N = 2$	$N = 3$	$N = 4$
Air pressure (MN/m²)	0.7	0.7	0.7	0.7	0.7
Binder injection pressure (MN/m²)	20.0	20.0	20.0	20.0	20.0
Withdrawal speed (m/min.)	0.033	0.037	0.043	0.05	0.063
Rotation speed (rpm)	20	20	20	20	20
Flow rate of binder slurry (m³/min.)	0.06	0.06	0.06	0.06	0.06
Diameter of column (m)	2.0	1.8	1.6	1.4	1.2

where
  H : length of stabilized soil column (m)
  q_c : flow rate of binder slurry (m³/min.)
  Q : total volume of binder slurry (m³)
  v : time required for injection per unit length (min./m)
  β : coefficient (0.06).

Due to the air-lifting effect of the injected air, a large amount of slime is lifted up along the injection pipe during the work which should be removed and handled with care according to the local regulation. The amount of slime during creating a stabilized soil column can be estimated by Equation (5.3).

$$V = V_1 + V_2$$
$$V_1 = (q_c + q_w) \cdot H \cdot v \cdot (1 + \alpha) \tag{5.3}$$
$$V_2 = \sum t \cdot q \cdot \gamma$$

where
  H : length of stabilized soil column (m)
  q : flow rate of drilling pump (m³/min.)
  q_c : flow rate of binder slurry (m³/min.)
  q_w : flow rate of high pressured water injected (m³/min.)
  t : time required for drilling (min.)
  v : time required for injection per unit length (min./m)

Table 5.12 Execution conditions for JSG method and column jet grout method (Japan Jet Grouting Association, 2011).

	JSG method	*column jet grout method*
q	0.04 m³/min.	0.2 m³/min.
q_c	0.06 m³/min.	0.14 or 0.18 m³/min.
q_w	0 m³/min.	0.07 m³/min.
α for sand	0.1	0.1
for clay	0.3	0.15
γ	0.5	0.2

Table 5.13 Design values of JSG method (Japan Jet Grouting Association, 2011).

Binder type	soil	W/C ratio	q_u (MN/m²)	cohesion, c (MN/m²)	adhesion (MN/m²)	bending strength (MN/m²)	elastic modulus (MN/m²)
JG-1	sand	100%	3.0	0.5			300
JG-1	clay	100%	1.0	0.3			100
JG-2	sand	150%	2.0	0.4	$c/3$	$2c/3$	200
JG-3	sand	200%	1.0	0.2			100
JG-4	organic	100%	0.3	0.1			30
JG-5	clay	150%	1.0	0.3			100

V : volume of slime (m³)
V_1 : volume of slime due to column construction (m³)
V_2 : volume of slime due to drilling (m³)
α : coefficient
γ : coefficient

Typical values of the parameters in the Equation are tabulated in Table 5.12 (Japan Jet Grouting Association, 2011).

7.2.2.3 Quality control during production

The quality control is almost the same as the single fluid technique, but the pressure and amount of compressed air are also monitored and controlled.

7.2.2.4 Quality assurance

After the construction work, in-situ stabilized soil columns should be investigated in order to verify the design quality, such as continuity, uniformity, strength, permeability or dimension. The procedure of the quality assurance is the same as the single fluid technique.

The size and strength of the stabilized soil column are dependent on the characteristics of the original ground and type of binder, and operational parameters. The strength and size of the stabilized soil column shown in Tables 5.11 and 5.13 (Japan Jet Grouting Association, 2011) are guaranteed minimum values of the technique. In the actual situation, a stabilized soil column having a larger diameter and higher strength than those are constructed in many cases.

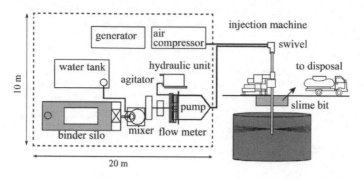

Figure 5.56 Layout of Superjet system (Superjet Association, 2011).

7.3 Double fluid technique (Superjet method)

7.3.1 Equipment

The Superjet Method is one of the double fluid techniques, which can produce a quite large stabilized column of maximum diameter of 5.0 m (Superjet Association, 2011). The binder slurry is injected at a high pressure of 30 MN/m^2 and is aided by a cone of compressed air of 0.7 to 1.05 MN/m^2, which shrouds the binder slurry. The air reduces the friction loss, allowing the binder slurry to travel farther from the injection point, thereby producing a greater diameter of stabilized soil column than that of the single fluid technique. The air injection produces more slime than the single fluid technique due to the air-lifting effect, which effectively reduce the adverse influence on nearby structures due to horizontal displacement and heaving of the ground surface.

The equipment for the technique is shown in Figure 5.56 (Superjet Association, 2011). The injection machine consists of a boring machine and an injection pipe. Binder slurry prepared by the mixer is transferred to the injection machine by the hydraulic pump, and injected into a ground. The injection pipe for the technique is a duplex cylinder with 140 mm in outer diameter. Two injection nozzles are installed on the side surface of the pipe near its bottom so that the binder slurry is injected in the horizontal direction. Several binders specially designed for the technique are available as shown later in Table 5.15, which are a composite of cement as a mother material and chemical additives for achieving various target strengths of the stabilized soil column. The capacities of water tank and pumping unit are typically 60 m^3 and 36 m^3/h.

7.3.2 Construction procedure

7.3.2.1 Preparation of site

The field preparation is carried out in accordance with the site specific conditions, which includes suitable access for plant and machinery. The basic layout of the technique is illustrated in Figure 5.56. The plant usually requires about 200 m^2. Before actual operation, execution circumstances should be prepared to assure smooth operation and prevention of environmental impact.

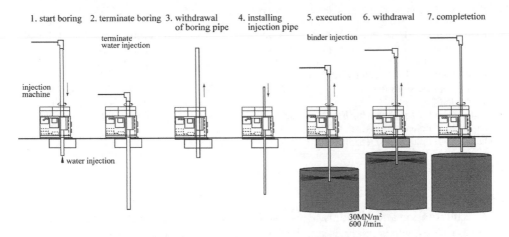

1. start boring   2. terminate boring   3. withdrawal   4. installing   5. execution   6. withdrawal   7. completetion
of boring pipe   injection pipe

Figure 5.57 Execution process of Superjet method (Superjet Association, 2011).

Figure 5.58 Stabilized soil column by Superjet method (Superjet Association, 2011).

7.3.2.2   Construction work

The construction work of the method is illustrated in Figure 5.57 (Superjet Association, 2011). After locating the boring machine at the prescribed position, the boring pipe of about 140 mm in diameter having excavation bits of about 200 to 250 mm in diameter is installed into a ground by the help of flushing water of a pressure of about 5 MN/m². When reaching the design depth, the boring pipe is withdrawn and removed. Then the injection pipe of 140 mm in diameter is installed in the hole to the bottom. At the bottom, the binder slurry is injected about 3 min. during rotating the injection pipe, then the injection pipe is withdrawn stepwise by about 25 to 50 mm interval. During the withdrawal, the binder slurry is injected from the nozzles into the ground at a pressure of about 30 MN/m². The binder slurry injected is around 600 l/min. to create a stabilized soil column of the maximum diameter of 5.0 m, as shown in Figure 5.58 (Superjet Association, 2011).

Table 5.14 Execution conditions for SuperJet method (Superjet Association, 2011).

SPT N-value of original soil for sand	$N < 50$	$50 < N < 100$	$100 < N < 150$	$150 < N$
SPT N-value of original soil for clay	$N < 3$	$3 < N < 5$	$5 < N < 7$	$7 < N < 9$
Air pressure (MN/m^2)	0.7 to 1.05	0.7 to 1.05	0.7 to 1.05	0.7 to 1.05
Binder injection pressure (MN/m^2)	30	30	30	30
Withdrawal speed (m/min.)	0.0625	0.625	0.625	0.625
Rotation speed (rpm)	20	20	20	20
Flow rate of binder slurry (m^3/min.)	0.06	0.06	0.06	0.06
Diameter of column (m) depth <30 m	5.0	4.5	4.0	3.5
Diameter of column (m) depth >30 m	4.5	4.0	3.5	3.0

The rotation and withdrawal speeds of the injection pipe and the amount of binder are controlled accordingly as summarized in Table 5.14 (Superjet Association, 2011). As the amount of binder slurry injected is about 50% of the volume of the stabilized soil column, a large amount of slime is lifted up along the injection pipe during the execution which should be removed and handled with care according to the local regulation.

The required total volume of binder slurry to create a stabilized soil column can be calculated by Equation (5.4).

$$Q = (H \cdot v \cdot q + t_g \cdot q) \cdot \alpha \tag{5.4}$$

where
H : length of stabilized soil column (m)
q : flow rate of binder slurry (m^3/min.)
Q : total volume of binder slurry (m^3)
t_g : injection time of binder slurry at the bottom of stabilization (3 min.)
v : time required for injection per unit length (min./m)
α : coefficient ($=1.06$).

7.3.2.3 Quality control during production

The quality control is almost the same as the single fluid technique, but the pressure and amount of compressed air are also monitored and controlled.

7.3.2.4 Quality assurance

After the construction work, in-situ stabilized soil columns should be investigated in order to verify the design quality, such as continuity, uniformity, strength, permeability or dimension. The procedure of the quality assurance is the same as the single fluid technique.

The size and strength of the stabilized soil column are dependent on and the characteristics of the original ground and the type of binder, and operational parameters. The strength and size of the stabilized soil column shown in Table 5.15 (Superjet Association, 2011) are guaranteed minimum values of the technique. In the actual situation, a stabilized soil column having a larger diameter and higher strength than those are constructed in many cases.

Table 5.15 Design values of Superjet method (Superjet Association, 2011).

Binder type	soil	W/C ratio	q_u (MN/m²)	cohesion, c (MN/m²)	adhesion (MN/m²)	bending strength (MN/m²)	elastic modulus (MN/m²)
SJ-1-H	sand	135%	3.0	0.5			300
SJ-1-H	clay	135%	1.0	0.3			100
SJ-1-L	sand	135%	2.0	0.4			200
SJ-1-L	clay	135%	0.7	0.2			70
SJ-2	sand	150%	3.0	0.5	c/3	2c/3	300
SJ-2	clay	150%	1.0	0.3			100
SJ-3	sand	150%	3.0	0.5			300
SJ3	clay	150%	1.0	0.3			100
SJ-4	organic	100%	0.3	0.1			30

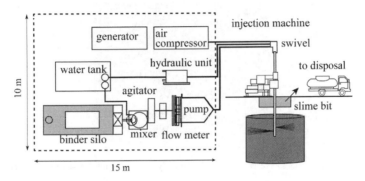

Figure 5.59 Layout of Column jet grount method system (Japan Jet Grouting Association, 2011).

7.4 Triple fluid technique (CJG method)

7.4.1 Equipment

The triple fluids referred to in this technique are binder slurry, air and water. Unlike single fluid and double fluid techniques, water is injected at a high pressure and is aided by a cone of compressed air, which shrouds the water injection. This process produces the air-lifting effect, which evacuates the soil within the intended column diameter. The binder slurry is injected through a separate nozzle below the water and air nozzle to fill the void created by the air-lifting process. The high pressure water of 40 MN/m² and air jet of 0.7 MN/m² are injected to disturb the soil. At the same time, the binder slurry is injected into the soil at a pressure of 2 to 5 MN/m² through a second nozzle positioned just below the air water nozzle. The amount of the binder slurry is typically about 140 to 180 *l*/min. (Japan Jet Grouting Association, 2011).

The equipment for the technique consists of an injection machine, binder silo, water tank, mixer and agitator, hydraulic unit, generator and air compressor as shown in Figure 5.59. The injection machine consists of a boring machine and injection pipe.

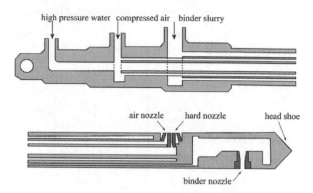

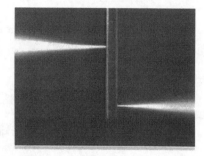

Figure 5.60 Layout of Column jet grout system (Japan Jet Grouting Association, 2011).

Figure 5.61 Injection of binder from tip of Column jet grout method (Japan Jet Grouting Association, 2011).

Binder slurry made in the mixer is transferred to the injection machine by the hydraulic pump, and injected into a ground.

Figure 5.60 shows the cross sectional view of the swivel and tip of the injection pipe (Japan Jet Grouting Association, 2011). The injection pipe for the technique is a triple cylinder with 90 mm in outer diameter. Two injection nozzles are installed on the side surface of the pipe near its bottom, the upper nozzle is for injecting the high pressure water and air jet, and the lower nozzle is for the binder slurry in the horizontal direction respectively (Figure 5.61) (Japan Jet Grouting Association, 2011). Several binders specially designed for the technique are available as shown later in Table 5.17, which are a composite of cement as a mother material and chemical additives for achieving various target strengths of the stabilized soil column. The capacities of water tank and pumping unit are typically 20 m³ and 12 m³/h.

7.4.2 Construction procedure

7.4.2.1 Preparation of site

The field preparation is carried out in accordance with the site specific conditions, which includes suitable access for plant and machinery. The basic layout of the technique is illustrated in Figure 5.59. The plant usually requires about 150 m².

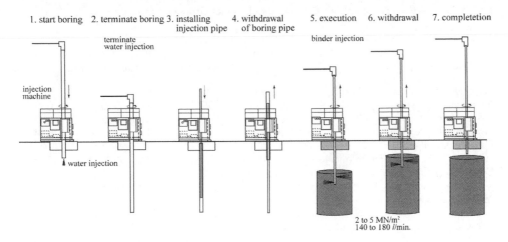

1. start boring   2. terminate boring   3. installing injection pipe   4. withdrawal of boring pipe   5. execution   6. withdrawal   7. completetion

terminate water injection

binder injection

injection machine

water injection

2 to 5 MN/m²
140 to 180 l/min.

Figure 5.62 Execution process of Column jet grout method (Japan Jet Grouting Association, 2011).

Before actual operation, execution circumstances should be prepared to assure smooth operation and prevention of environmental impact.

7.4.2.2   Construction work

The construction work of the method is illustrated in Figure 5.62 (Japan Jet Grouting Association, 2011). After locating the boring machine at the prescribed position, the boring pipe of about 140 mm in diameter is installed into a ground by the help of flushing water of a pressure of about 5 MN/m². When reaching the design depth, the injection pipe of about 90 mm in diameter is installed in the boring pipe, and then the boring pipe is withdrawn and removed. The injection pipe is withdrawn stepwise by about 25 to 50 mm interval. During the withdrawal, the water and air are injected through their respective lines to break up the soil surrounding the injection pipe, while the binder slurry is also injected from the nozzle in the lower level into the ground at a pressure of about 2 to 5 MN/m². The binder slurry injected is around 140 or 180 l/min. to create a stabilized soil column of a diameter of about 1.2 to 2.0 m.

The rotation and withdrawal speeds of the injection pipe and the amount of binder are controlled accordingly as summarized in Table 5.16 (Japan Jet Grouting Association, 2011). The required amount of binder and the volume of slime can be calculated by Equations (5.2) and (5.3) respectively. Quite a large amount of slime is lifted up along the injection pipe during the execution which should be removed and should be handled with care according to the local regulation.

7.4.2.3   Quality control during production

The quality control is almost the same as the single fluid technique, but the pressure and amount of compressed air and pressurized water are also monitored and controlled. The W/C ratio and density of binder slurry is controlled to the design value.

Figure 5.63 shows a typical relationship between the diameter of a stabilized soil column and the withdrawal speed of the injection pipe (Sakata, 1991). The larger

Table 5.16 Execution conditions for CJG method (Japan Jet Grouting Association, 2011).

(a) For sand layer

SPT N-value of original soil	N < 30	30 < N < 50	50 < N < 100	100 < N < 150	150 < N < 175	175 < N < 200
Water injection pressure (MN/m²)	40	40	40	40	40	40
Air pressure (MN/m²)	0.7	0.7	0.7	0.7	0.7	0.7
Binder injection pressure (MN/m²)	2 to 5	2 to 5	2 to 5	2 to 5	2 to 5	2 to 5
Withdrawal speed (m/min.)	0.0625	0.05	0.05	0.04	0.04	0.04
Rotation speed (rpm)	20	20	20	20	20	20
Flow rate of binder slurry (m³/min.)	0.18	0.18	0.16	0.14	0.14	0.14
Diameter of column (m)	2.0	2.0	1.8	1.6	1.4	1.2

(b) For clay layer

SPT N-value of original soil	N < 3	3 < N < 5	5 < N < 7	7 < N < 9
Water injection pressure (MN/m²)	40	40	40	40
Air pressure (MN/m²)	0.7	0.7	0.7	0.7
Binder injection pressure (MN/m²)	2 to 5	2 to 5	2 to 5	2 to 5
Withdrawal speed (m/min.)	0.05	0.05	0.04	0.04
Rotation speed (rpm)	20	20	20	20
Flow rate of binder slurry (m³/min.)	0.18	0.16	0.14	0.14
Diameter of column (m)	2.0	1.8	1.6	1.2

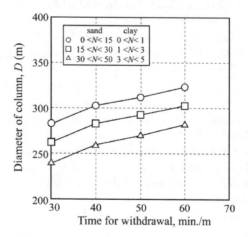

Figure 5.63 Relationship between diameter of stabilized soil column and withdrawal speed of injection pipe (Sakata, 1991).

diameter column can be produced by slowing down the withdrawal speed of the injection pipe.

7.4.2.4 Quality assurance

After the construction work, in-situ stabilized soil columns should be investigated in order to verify the design quality, such as continuity, uniformity, strength, permeability

Table 5.17 Design values of column jet grout method for sandy soil (Japan Jet Grouting Association, 2011).

Binder type	soil	q_u (MN/m²)	cohesion, c (MN/m²)	adhesion (MN/m²)	bending strength (MN/m²)	elastic modulus (MN/m²)
JG-1	sand	3	0.5			300
JG-1	clay	1	0.3			100
JG-2	sand	2	0.4	$c/3$	$2c/3$	200
JG-3	sand	1	0.2			100
JG-4	organic	0.3	0.1			30
JG-5	clay	1	0.3			100

or dimension. The procedure of the quality assurance is the same as the single fluid technique.

The size and strength of the stabilized soil column are dependent on the characteristics of the original ground, the type of binder, and operational parameters. The size and strength of the stabilized column shown in Tables 5.16 and 5.17 (Japan Jet Grouting Association, 2011) are guaranteed minimum values of the technique. In the actual situation, a stabilized soil column having a larger diameter and higher strength than those are constructed in many cases.

7.5 Triple fluid technique (X-jet method)

7.5.1 Equipment

In the X-jet method, the high pressure water of 40 MN/m² and air jets of 0.6 to 1.05 MN/m² are injected at two nozzles on the side surface of the injection pipe to disturb the soil. The two jets are designed to collide each other at a predetermined diameter and to exhaust the jet energy there. At the same time, the binder slurry is injected at a pressure of 4 MN/m² from the other nozzle below the water and air nozzles, which can create the stabilized soil column with uniform diameter of 2.5 m in diameter. The amount of the binder slurry is about 190 to 250 *l*/min. (X-jet Association, 2011).

The equipment for the technique consists of an injection machine, binder silo, water tank, mixer and agitator, hydraulic unit, generator and air compressor as shown in Figure 5.64 (X-jet Association, 2011). The injection machine consists of a boring machine and injection pipe. Binder slurry made in the mixer is transferred to the injection machine by the hydraulic pump, and injected into a ground. The injection pipe for the technique is a triple cylinder with 90 mm in outer diameter. Three injection nozzles are installed on the side surface of the pipe near its bottom, the upper two nozzles are for injecting the high pressure water and air jet (Figure 5.65) (X-jet Association, 2011), and the lower nozzle is for the binder slurry. Several binders specially designed for the technique are available as shown later in Table 5.19, which are a composite of cement as a mother material and chemical additives for achieving various target strengths of the stabilized soil column. The capacities of the water tank and pumping unit are typically 20 m³ and 12 m³/h.

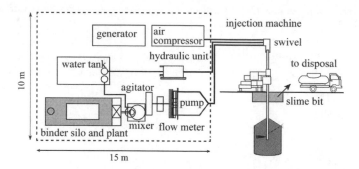

Figure 5.64   Layout of X-jet system (X-jet Association, 2011).

Figure 5.65   Injection of water jets from tip of X-jet method (X-jet Association, 2011).

7.5.2   Construction procedure

7.5.2.1   Preparation of site

The field preparation is carried out in accordance with the site specific conditions, which includes suitable access for plant and machinery. The basic layout of the technique is illustrated in Figure 5.64. The plant usually requires about $150\,\mathrm{m}^2$. Before actual operation, execution circumstances should be prepared to assure smooth operation and prevention of environmental impact.

7.5.2.2   Construction work

The construction work of the technique is illustrated in Figure 5.66 (X-jet Association, 2011). After locating the boring machine at the prescribed position, the boring pipe of about 142 mm in diameter is installed into a ground by the help of flushing water of a pressure of about $5\,\mathrm{MN/m}^2$. When reaching the design depth, the injection pipe of about 90 mm in diameter is installed in the boring pipe, and then the boring pipe is withdrawn and removed. The injection pipe is withdrawn stepwise by about 25 to 50 mm interval. During the withdrawal, the water and air are injected through their respective lines to break up the soil surrounding the injection pipe, while the binder

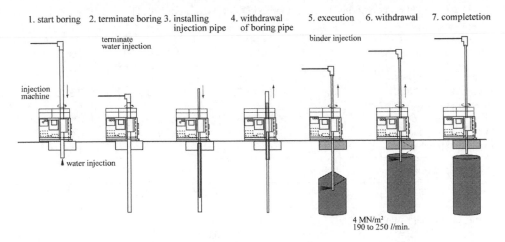

Figure 5.66 Execution process of X-jet method (X-jet Association, 2011).

slurry is also injected from the third nozzle in the lower level into the ground at a pressure of about 4 MN/m². The binder slurry injected is around 190 or 250 *l*/min. to create a stabilized soil column of a uniform diameter of 2.5 m.

The rotation and withdrawal speeds of the injection pipe and the amount of binder are controlled accordingly as summarized in Table 5.18 (X-jet Association, 2011). As the amount of binder slurry injected is about 40 to 90% of the volume of the stabilized soil column, quite a large amount of slime, mixture of soil and binder slurry, is lifted up along the injection pipe during the execution which should be removed and handled with care according to the local regulation.

7.5.2.3 Quality control during production

The quality control is almost the same as the single fluid technique, but the pressure and amount of compressed air and pressurized water are also monitored and controlled. The W/C ratio and density of binder slurry is controlled to the design value.

7.5.2.4 Quality assurance

After the construction work, in-situ stabilized soil columns should be investigated in order to verify the design quality, such as continuity, uniformity, strength, permeability or dimension. The procedure of the quality assurance is the same as the single fluid technique.

The strength of the stabilized soil column is dependent on the characteristics of the original ground, the type of binder, and operational parameters. The strength of stabilized column shown in Tables 5.18 and 5.19 (X-jet Association, 2011) are guaranteed minimum values of the technique. In the actual situation, a stabilized soil column having a higher strength than those are constructed in many cases.

Table 5.18 Execution conditions for X-jet method (X-jet Association, 2011).

(a) For sand layer

SPT N-value of original soil	$N < 50$	$50 < N < 100$	$100 < N < 150$
Water pressure (MN/m^2)	40	40	40
Air pressure (MN/m^2)	0.6 to 1.05	0.6 to 1.05	0.6 to 1.05
Injection pressure (MN/m^2)	4	4	4
Flow rate of binder slurry (m^3/min.)	0.25	0.19	0.19
Withdrawal speed (m/min.)	0.125	0.0625	0.0417
Rotation speed (rpm)	20	20	20

(b) For clay layer

SPT N-value of original soil	$N < 3$	$3 < N < 5$
Water pressure (MN/m^2)	40	40
Air pressure (MN/m^2)	0.6 to 1.05	0.6 to 1.05
Injection pressure (MN/m^2)	4	4
Flow rate of binder slurry (m^3/min.)	0.25	0.19
Withdrawal speed (m/min.)	0.125	0.0625
Rotation speed (rpm)	20	20

Table 5.19 Design values of column jet grout method for sandy soil (X-jet Association, 2011).

Binder type	soil	W/C ratio	q_u (MN/m^2)	cohesion, c (MN/m^2)	adhesion (MN/m^2)	bending strength (MN/m^2)	elastic modulus (MN/m^2)
CROSSSAND	sand	75%	3	0.5			300
CROSSSAND	sand	75%	2	0.4	$c/3$	$2c/3$	200
CROSSNEN	clay	100%	1	0.3			100
CROSSNEN	organic	100%	0.3	0.1			30

8   COMBINED TECHNIQUE

Several techniques combining the mechanical mixing and high pressure injection were developed and available in Japan. There are two types in the technique. One is the combination of mechanical mixing and horizontal jet, in which binder slurry is injected horizontally from the nozzle at the tip of the mixing blade. The other is the combination of mechanical mixing and two inclined jets, in which binder slurry is injected from the two nozzles at the tips of the mixing blades at two different elevation and designed to collide each other at a designated distance (similar to the X-jet method). The combination of the mechanical mixing and high pressure injection can reduce the required power for cutting a large diameter with mechanical mixing alone. When the combined technique is used, it is possible to produce a stabilized soil column in close contacts with underground structures such as piles and sheet pile walls, as shown in Figure 5.67. This provides a large benefit to increase the horizontal resistance of the structure. The two different diameters column can be produced with/without the jet mixing.

Figure 5.67 Stabilized soil column by combined technique in contact with sheet pile wall (by courtesy of Fudo Tetra Corporation).

In this section, the combined technique with mechanical mixing technique and X-jet technique is introduced.

8.1 JACSMAN method

8.1.1 Equipment

8.1.1.1 System and specifications

The combined technique with mechanical mixing and high pressure injection was developed in 1994, which is named JACSMAN (Jet And Churning System MANagement) (Miyoshi and Hirayama, 1993, 1994, 1996; JACSMAN Association, 2011). The system of the method consists of a mixing machine, binder tank and plant, water tank, grout pump and high pressure pump, generator and compressor, as shown in Figure 5.68. The high pressure jets are injected at the nozzles on the tips of the two mixing blades at two different elevations. The two jets were designed to collide each other at a predetermined diameter and to exhaust the jet energy there, which can create a stabilized soil column with uniform diameter (Figure 5.69).

8.1.1.2 Mixing shafts and mixing blades

The JACSMAN machine has two mixing shafts (Figure 5.70) (JACSMAN Association, 2011). The shaft is a square cross section of 250 mm and has triple core barrels in it. They consist of a stack of three mechanical mixing blades. The diameter of the blades is either 1.0 or 1.3 m depending on the type of machine. The outlet of the binder slurry for the mechanical mixing is installed on each shaft between the two mixing blades. The nozzles for high pressure injection are installed on the tips of the lower two mixing blades of each shaft, while the two jets collides each other at 2.3 m from the center. In the technique, the central part of the column of 1.0 or 1.3 m in diameter is produced

Figure 5.68 JACSMAN machine in operation (by courtesy of Fudo Tetra Corporation).

(a) Outlets for high pressure injection.

(b) Outlet for low pressure injection.

Figure 5.69 JACSMAN machine (by courtesy of Fudo Tetra Corporation).

by the mechanical mixing blades while the outer part, 2.3 m in diameter, is produced by the cross-jets (Figure 5.71) (JACSMAN Association, 2011).

8.1.1.3 Plant and pumping unit

A binder plant is prepared for producing and supplying binder slurry to the JACSMAN machine (Figure 5.72). A total of four pump units are installed for supplying the binder slurry to the mixing shafts: two high pressure pumps for the high pressure injection and two low pressure pumps for the mechanical mixing. The high pressure pumps have a capacity of supplying the binder slurry at 0.3 m^3/min. with 30 MN/m^2. The low pressure pumps have a capacity of supplying the binder slurry at 0.21 or 0.4 m^3 per min.

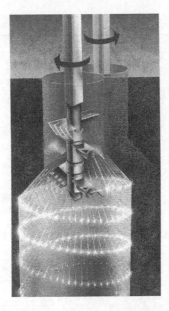

Figure 5.70   Mixing shafts and image of mixing (JACSMAN Association, 2011).

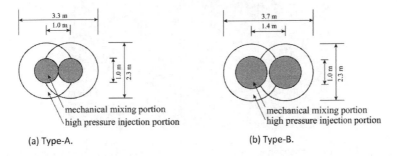

(a) Type-A.   (b) Type-B.

Figure 5.71   Cross section of stabilized soil columns (JACSMAN Association, 2011).

Two types of special binder are provided, named JACSMAN-1 and JACSMAN-2, for the sand, silt and clay soils and for organic soils respectively. The W/C ratio of binder slurry is 100% for both the mechanical mixing and jet mixing.

8.1.1.4   Control unit

A control room is placed in the site, where the admixture condition, quantity of each material, rotation speed of the mixing blades, speed of shafts movement, air pressure, *etc.* are continuously monitored and controlled. These monitoring data can be fed back to the operator for precise construction of the column. Figure 5.73 shows a monitor screen for the method on the machine.

Figure 5.72 Plant and pumping unit (by courtesy of Fudo Tetra Corporation).

Figure 5.73 Monitor screen for JACSMAN (by courtesy of Fudo Tetra Corporation).

8.1.2 Construction procedure

8.1.2.1 Preparation of site

Similarly to the ordinary CDM method, field preparation is carried out in accordance with the site specific conditions, which includes suitable access for plant and machinery, leveling of the working platform. Before actual operation, execution circumstances should be prepared to assure smooth execution and prevention of environmental impact. A sand blanket with about 0.5 to 1.0 m in thickness is usually spread on the ground as aworking platform. Several steel plates with about 1.5 m by 4.0 m are preferably placed on the sand mat so as to assure the bearing capacity of the machine.

8.1.2.2 Field trial test

It is recommended to conduct a field trial test in advance in, or adjacent to the construction site, in order to confirm the smooth execution. In the test, all the equipment monitoring the amount of binder, rotation speed of the mixing blades and penetration

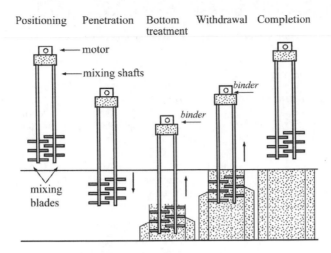

Positioning   Penetration   Bottom   Withdrawal   Completion
treatment

Figure 5.74 Execution process of JACSMAN method (JACSMAN Association, 2011).

and withdrawal speeds of the mixing shafts are calibrated. In the case where the stabilized soil columns should reach and have firm contact with the stiff bearing layer (fixed type improvement), a field trial test should be carried out to measure the change in the electric or hydraulic power required for driving the mixing shafts and the penetration speed of the mixing shafts at the stiff layer so that they can help to detect if the mixing blades have reached the stiff layer in the actual construction.

When there is less experience in similar soil conditions, it is recommended to carry out a field mixing trial and to confirm that the strength and integrity of the trial column/element meet the design requirement.

8.1.2.3   Construction work

The execution process for the JACSMAN is shown in Figure 5.74 (JACSMAN Association, 2011). During the penetration of the mixing shafts, the mixing blades are rotating at 20 rpm to cut and disturb the soil to reduce the strength of the ground so as to make the mixing shafts penetrate by their self-weight. In the early stage of development, the binder slurry was injected to produce the mechanical part in the penetration stage, while the binder slurry was injected to produce the high pressure part in the withdrawal stage. However, the withdrawal injection is adopted currently also for the mechanical mixing. In the withdrawal stage, the direction of blade rotation is reversed. The amount of binder is kept constant while the penetration and withdrawal speeds are controlled so as to assure the design amount of binder should be mixed. The typical execution specifications are summarized in Table 5.20 (JACSMAN Association, 2011).

If the high pressure injection is temporary terminated during the withdrawal stage, a small diameter stabilized column can be produced in the ground.

8.1.2.4   Quality control during production

To produce stabilized soil columns that meet the design requirements, it is essential to control and monitor the quality of binder, geometric layout and operational parameters

Table 5.20 Specifications of JACSMAN method (JACSMAN Association, 2011).

Machine	lifting speed	binder supply		binder content	ground type		application
		jet mix.	mech. mix.		ordinary soil	hard soil	
type-A	0.5 m/min.	600 l/min.	104 l/min.	190 kg/m^3	applicable		underground beam, foundation
type-B	0.5 m/min.	600 l/min.	175 to 80 l/min.	200 to 160 kg/m^3	applicable	applicable	underground beam, foundation, bearing capacity, settlement, liquefaction
type-B	0.5 to 1.0 m/min.	600 l/min.	175 l/min.	200 to 100 kg/m^3	applicable		bearing capacity, settlement, liquefaction

Figure 5.75 Stabilized soil column by JACSMAN method (JACSMAN Association, 2011).

such as quantity of binder, rotation speed of the mixing blade, shaft speed, pressure of binder slurry, *etc.* These monitoring data can be fed back to the operator for precise construction of the column.

8.1.2.5 Quality assurance

After the construction work, the quality of the in-situ stabilized soil columns should be verified in advance of the construction of the superstructure in order to confirm the design quality, such as uniformity, strength, permeability or dimension. Full depth coring, observation of core and testing of selected specimens are conducted for quality assurance in the same way as that for the ordinary CDM.

8.1.2.6 Effect of method

Figure 5.75 shows the stabilized soil column after excavation, which shows the mechanical mixing portion and the jet injection mixing portion (JACSMAN Association, 2011). Figure 5.76 shows the strength distribution of the stabilized soil

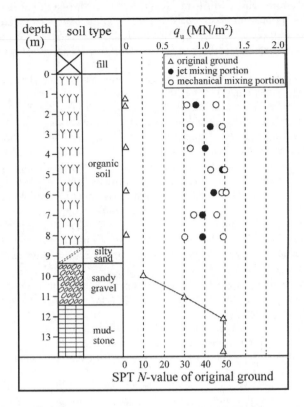

Figure 5.76 Strength distribution along the depth (JACSMAN Association, 2011).

column along the depth (JACSMAN Association, 2011). The equivalent strengths were attained both at the jet injection part and mechanical mixing part, and the strengths at both parts showed a uniform profile along the depth.

REFERENCES

Cement Deep Mixing Method Association (1999) *Cement Deep Mixing Method (CDM)*, *Design and Construction Manual* (in Japanese).

Cement Deep Mixing Method Association (2005) *CDM-Lemni 2/3 Method Technical Manual* (in Japanese).

Cement Deep Mixing Method Association (2006) *CDM-LODIC Method Technical Manual* (in Japanese).

Coastal Development Institute of Technology (2008) *Technical Manual of Deep Mixing Method for Marine Works*. 289p. (in Japanese).

Dry Jet Mixing Association (2010) *Dry Jet Mixing (DJM) Method Technical Manual* (in Japanese).

Endo, S. (1995) New large-diameter deep mixing method combining the advantages of mechanical mixing and jet stirring. *Proc. of the Journal of Japanese Society of Soil Mechanics and Foundation Engineering, Tsuchi to Kiso*. Vol. 448, No. 5, p. 50 (in Japanese).

Horikiki, S., Kamimura, K. & Kurinami, K. (1996) Low displacement deep mixing method (LODIC) and its application. *Kisokou*. Vol. 24. No. 7. pp. 90–94 (in Japanese).

JACSMAN Association (2011) *Technical Data for JACSMAN, Ver. 6*. 21p. (in Japanese).

Japan Jet Grouting Association (2011) *Technical Manual of Jet grouting Method, Ver. 19*. 82p. (in Japanese).

Japanese Society of Soil Mechanics and Foundation Engineering (1985) *Soil Stabilization Techniques*. 389p. (in Japanese).

Kamimura, K., Kami, C., Hara, T., Takahashi, T. & Fukuda, H. (2009a) Application example of deep mixing method with reduced displacement due to mixing (CDM-LODIC). *Proc. of the International Symposium on Deep Mixing and Admixture Stabilization*. pp. 535–540.

Kamimura, K., Kawasaki, H., Hara, T., Takahashi, T. & Fukuda, H. (2009b) Development of triple-axial cement deep mixing method (Lemni 2/3 Method). *Proc. of the International Symposium on Deep Mixing and Admixture Stabilization*. pp. 541–546.

Miyoshi, A. & Hirayama, K. (1993) R&D in soil improvement method by combining water jet and mechanical mixing (Part 1). *Proc. of the 28th Annual Conference of the Japanese Society of Soil Mechanics and Foundation Engineering*. pp. 2519–2520 (in Japanese).

Miyoshi, A. & Hirayama, K. (1994) The job site test of improvement in cohesive soil. *Proc. of the 29th Annual Conference of the Japanese Society of Soil Mechanics and Foundation Engineering*. pp. 2161–2162 (in Japanese).

Miyoshi, A. & Hirayama, K. (1996) Test of solidified columns using a combined system of mechanical churning and jetting. *Proc. of the 2nd International Conference on Ground Improvement Geosystems*. pp. 743–748.

Mizuno, S., Sudou, F., Kawamoto, K. & Endou, S. (1988) Ground displacement due to ground improvement by deep mixing method and countermeasures. *Proc. of the 3rd Annual Symposium of the Japan society of Civil Engineers on experiences in Construction*. pp. 5–19 (in Japanese).

Sakata, M. (1991) A recent execution of RJP method. *Kisoko*. Vol. 6. pp. 80–85 (in Japanese).

Superjet Association (2011) *Technical Manual of Superjet Method*. 44p. (in Japanese).

Terashi, M. (2003) The state of practice in deep mixing method. Grouting and ground treatment, *Proc. of the 3rd International Conference, ASCE Geotechnical Special Publication*. No. 120. Vol. 1. pp. 25–49.

X-jet Association (2011) *Technical Manual of X-jet Method, Ver. 14*. 21p. (in Japanese).

Chapter 6

Design of improved ground by the deep mixing method

I INTRODUCTION

This chapter is intended to introduce the geotechnical design procedure for ground improved by the deep mixing method. The design procedures are formulated by simplified assumptions or idealization of engineering behavior of the improved ground and involve empiricism to some extent, backed up by successful case histories. The simplification/idealization and empiricism are based on the abundant research and experience accumulated by the Japanese mechanical deep mixing system employing vertical rotary shafts and mixing blades such as CDM and DJM. However, the design procedures described in the chapter may be applicable to similar in-situ admixture stabilization including high pressure injection deep mixing if adequate considerations are paid for such characteristics as relation between the strengths of in-situ stabilized soil and laboratory prepared stabilized soil, variability of in-situ stabilized soil, reliability of overlap joint and end bearing.

The ground improved by the deep mixing method is a complicated composite system comprising stiff stabilized soil columns or elements and unstabilized soft soil. The behaviors of the improved ground to various external actions are far different from those found in ordinary relatively uniform ground. Several failure modes including internal and external stabilities may develop depending on the stiffness of improved soil, geometry of deep mixed elements, and external loading conditions. Section 2 is an introductory section intended to provide an overview on the behavior of deep mixed ground to help engineers understand the applicability and limitations of the routine design procedures which are described in Section 4 onward.

In the design process, the geotechnical designer should determine the design parameters of stabilized soil and required level of accuracy of installation taking the capability of the locally available deep mixing technologies into account. Accuracy of installation should cover such items as the location, verticality, depth, reliable contact with bearing layer and overlap of columns (if necessary). Consistency of design and construction is the key to good performance of the improved ground. Section 3 is also an introductory section intended to outline the work flow of a deep mixing project comprising geotechnical design, process design, construction, and QC/QA.

The technical standard for the geotechnical design of improved ground by deep mixing as a foundation of port facilities such as breakwater or revetment by block type and wall type column installation patterns was first established in 1989 by the Ministry of Transport (Ministry of Transport, 1989), which was revised in 1999 and

2007 (Ministry of Transport, 1999; Ministry of Land, Infrastructure, Transport and Tourism, 2007). The Ports and Harbours Association of Japan published the standard and commentaries of the original Japanese version (The Ports and Harbours Association of Japan, 1989, 1999 and 2007) and the Overseas Coastal Area Development Institute of Japan published the English version (The Overseas Coastal Area Development Institute of Japan, 1991, 2002 and 2009). When the deep mixing method expanded its application to various structures as shown in Figures 4.4, several design standards or design guides have been tailored for specific structures by respective organizations which oversee them. The Public Works Research Center published the design method and commentaries of the group column type improved ground for embankment support in 1999 (Public Works Research Center, 1999), which was revised in 2004 (Public Works Research Center, 2004). For applications to building foundation, the Building Center of Japan proposed the Design and Quality Control Guideline of Improved Ground for Building in 1997. The guideline was revised and authorized by the Architectural Institute of Japan in 2006 (Architectural Institute of Japan, 2006). For applications to oil tank foundation, the Fire and Disaster Management Agency gave the official notification on the design procedure for tank foundation in 1995, in which the design procedure of the deep mixing method was specified (Fire and Disaster Management Agency, 1995). The Ministry of Construction proposed a draft design method for liquefaction mitigation (Ministry of Construction, 1999). These design procedures are not identical due to different performance and functional requirements specific to the type of structures.

In this chapter, Section 4 describes the group column type improvement for embankment support, Section 5 describes the traditional design method for block type and wall type improvement for port facilities, Section 6 describes the reliability based design method for block type and wall type improvement for port facilities, and Section 7 deals with the liquefaction mitigation by grid type improvement.

It should be noted that a considerable amount of analyses and even physical modelings have been conducted to supplement these routine design in some instances; for important structures or for the situation much different from the experience.

2 ENGINEERING BEHAVIOR OF DEEP MIXED GROUND

2.1 Various column installation patterns and their applications

When the deep mixing method is used as a solution for problems encountered on a construction project on soft ground, stabilized soil columns/elements are installed by a variety of patterns; block, grid, panel, or group of individual columns as described in Chapter 4. Figure 6.1 illustrates the typical installation patterns.

Table 6.1 shows comparisons of the characteristics of the installation patterns. The block, grid and wall types are manufactured by overlapping deep-mixed stabilized soil columns or elements. The block type is the most stable against both external and internal stability among all the patterns. It may find application in the case of breakwater or a huge earth retaining structure which is subjected to large horizontal forces. The grid type has almost the same function as a block with less stabilized soil volume, which can be applicable when internal stability is less critical compared to a

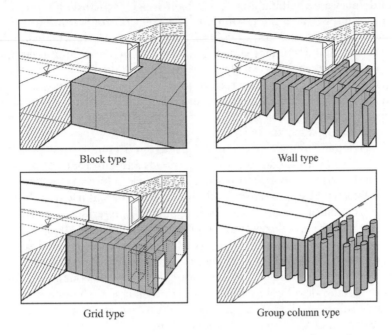

Block type   Wall type

Grid type   Group column type

Figure 6.1 Installation patterns of deep mixing improved ground.

Table 6.1 Characteristics of improvement types.

Type	Stability	Installation	Design consideration
Group column type	Where horizontal loads are small, high stability is obtained.	Overlapping operation is not required.	Requires design on overall stability and on individual column as a pile foundation.
Tangent group column	Where horizontal loads are small, high stability is obtained.	Precise operation is required to achieve firm contact of columns.	Requires design on overall stability and on internal stability of tangent columns.
Wall type	Where all improved walls are linked firmly, high stability is obtained.	Requires precise operation of overlapping of long and short units.	Requires consideration of unimproved soil between walls. Wall spacing and depth of short wall affected by internal stability.
Grid type	Highly stable next to block Type.	Installation sequences are complicated because lattice shape must be formed.	Requires design on three-dimensional internal stress.
Block type	Large solid block resists external loads. Highly stable.	Takes longer time because all columns are overlapped.	Design of size of block is in the same way as the gravity structures.

block. When the instability is dominant to one direction, a panel or wall can effectively improve the stability. The grid type or panel type installation pattern may be selected in order to maintain the stability of embankment slope or foundation support for a retaining structure. When the major concern is the consolidation settlement of soft ground under a low embankment or a light weight structure, a group of individual columns will provide a good solution. The tangent group column is a modified improvement pattern of the group column type improvement, where stabilized soil columns are installed in contact with the adjacent columns without overlapping. As the improvement area ratio is usually of the order of 70 to 80%, larger than that of the group column type improvement, larger effects in bearing capacity and settlement reduction are expected than the group column type improvement. This improvement has frequently applied to embankment slope and small building for increasing stability and bearing capacity respectively.

The behaviour of improved ground depends on a complicated time-dependent interaction between stabilized soil columns/elements and unstabilized soils in the geocomposite system. Modes of deformation leading to failure are governed by such factors as geometry of improvement, relative stiffness of stabilized and unstabilized soils, loading condition typical for specific application, interface properties between structure and stabilized soil/ between stabilized and unstabilized soils. The geometry of improvement include such factors as height/width ratio of stabilized zone, pile installation pattern in the stabilized zone, location of stabilized zone relative to the superstructure (located in the active, transitional, or passive zone), and end bearing condition of stabilized soil columns. As shown in Table 6.1, the interaction between stiff stabilized soil elements and unstabilized soil becomes complicated in the order of block, grid, panel and a group of individual columns.

2.2   Engineering behavior of block (grid and wall) produced by overlap operation

This sub-section deals with the two dimensional behavior of improved ground corresponding to the block type of column installation and explains that the mode of failure changes from external stability to internal stability not only by the strength of deep mixed soil but by the relative strength of improved soil and surrounding soft native soil. This will become a good introduction for engineers to understand why the currently available design procedures involve the examination of different modes of failures. There will be no sub-section for grid and panel, because the overall behavior resembles that of block except for the behavior of unstabilized soil left between panels and grid. The behavior of unstabilized soil between the panels will be discussed as part of the routine design.

2.2.1   Engineering behavior of improved ground leading to external instability

First, let us consider the situation where the stabilized soil columns have *sufficiently large shear strengths and are installed by reliable overlapping procedure*. The words sufficient and reliable are not quantitative but qualitative at this stage of explaining the general idea on the engineering behavior. How sufficient is sufficient or how reliable is reliable will be left for the routine design.

A simple situation of an earth retaining structure (revetment) as shown in Figure 6.2 is considered where the ground surface is flat, soft clay is underlain by a firm bearing stratum and clay thickness is uniform. Before, during and immediately

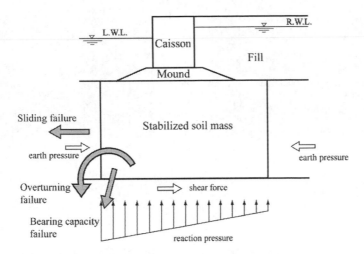

Figure 6.2 Actions and reactions, and modes of external failure of stabilized soil mass underretaining structure.

after the ground improvement work, the forces acting on both sides of the zone of stabilization are the earth pressure at rest, K_0 pressure of the clay ground. At the bottom of the stabilized soil mass, practically uniform reaction forces are acting and balance the weight of the improved soil mass. While fill material is being placed, the fill pressure gradually increases both on top of the stabilized soil mass and on the soft clay behind the stabilized soil mass.

– *Horizontal forces*: The earth pressure of clay under the fill (active side) increases with increasing fill pressure. The horizontal earth pressure in the fill itself also increases. If the fill height is small, these horizontal forces by the fill placement are reacted and balanced by the shear force induced at the bottom of the stabilized soil mass, while the earth pressure acting on passive side of the stabilized soil mass may remain almost constant. The stabilized soil mass starts to displace horizontally outward with increasing fill height (with increasing horizontal force acting on the active side). Then the earth pressure acting on the passive side starts to increase and the shear force at the bottom also increases to maintain the horizontal force equilibrium. When horizontal resistance by passive earth pressure and maximum bottom shear are exceeded, the improved ground may fail by sliding of the stabilized soil mass.
– *Vertical and moment forces*: The increased vertical action by fill placement should also be balanced by the reaction at the bottom of the stabilized soil. Further, the moment equilibrium (for example around the toe of the stabilized soil mass) must be balanced by changing the distribution of the bottom reaction and possibly by the shear forces induced at the side surfaces of the stabilized soil mass. When the vertical force equilibrium and/or the moment equilibrium are violated, two additional modes of failure should be considered. One is the overturning of the stabilized soil mass around the toe of the stabilized soil mass. The other is the

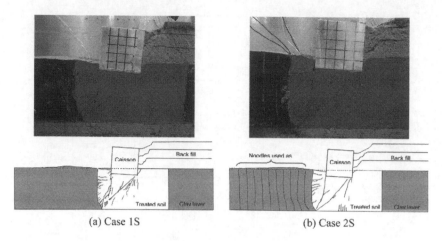

(a) Case 1S   (b) Case 2S

Figure 6.3 Failurepattern of improved soil mass (Ohishi *et al.*, 2005).

bearing capacity failure of the bearing stratum under the stabilized soil mass. The overturning may be a relatively rare mode of failure which may occur when the height/width ratio of stabilized soil mass is excessively large. The bearing capacity of the underlying stratum is the bearing capacity problem of deep foundation under inclined and eccentric loading.

If the stabilized soil columns have *sufficiently large shear strengths and are installed by reliable overlapping procedure* as stated in the beginning of this sub-section, the stabilized soil mass is internally stable and behaves as a rigid body. When the equilibrium is violated, as shown in Figure 6.2 the improved ground fails externally by:

– Sliding failure
– Overturning failure
– Bearing capacity failure.

2.2.2 Engineering behavior of improved ground leading to internal instability

When the strength of a stabilized soil column is not sufficient, there is a risk of excessive deformation or failure of the stabilized soil mass under external actions. Figure 6.3 is two of such example obtained by centrifuge model test conducted by Ohishi *et al.* (2005). In this case, soft clay ground underneath the rigid structure, caisson, was stabilized by the block type improvement. As this is the model test, the stabilized soil block is prepared as 100% intact block without any overlap joints (ideal block). The geometry of the model was designed to assure the external stability. Under the combined action of vertical load due to the concrete caisson and horizontal load due to the placement of backfill, the stabilized soil block is brought to failure by a wedge shaped shear failure of the stabilized soil mass as shown in Figure 6.4 (Ohishi *et al.*, 2005).

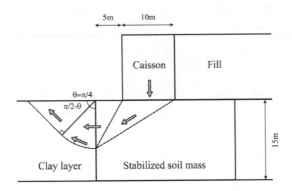

Figure 6.4 Wedge shaped shear failure of stabilized soil block (Ohishi *et al.*, 2005).

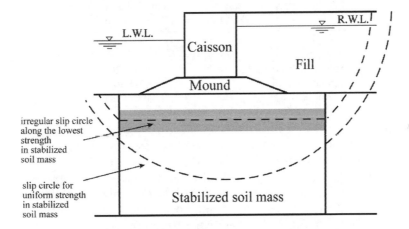

Figure 6.5 Various modes of internal stability of stabilized soil mass.

Figure 6.5 illustrates some of the additional failure modes anticipated for the internal stability of the stabilized soil mass. When a stabilized soil mass is created with relatively uniform strength, general failure as shown by a slip circle may also occur. However, the improvement is not likely to create a uniform material. If the same binder factor is employed throughout the depth in layered ground, a certain soil layer such as a highly organic soil layer may exhibit the lowest strength. Then failure may occur along the lowest strength layer and an irregular slip surface as shown in Figure 6.5.

2.2.3   *Change of failure mode*

As explained above, the engineering behavior of the improved ground depends upon many factors and different modes of failures exist. This is the reason why engineers have to examine the external and internal failures assuming appropriate several modes of failure. Often asked questions are: How strong is strong enough for stabilized soil to avoid the examination of internal stability? Is there any upper limit strength where design can be completed only by slip circle analysis? These are difficult questions to

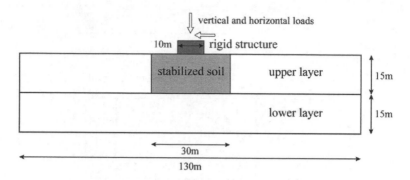

Figure 6.6 Geometry adopted for the numerical simulation (Ohno and Terashi, 2005; Terashi, 2005).

Table 6.2 Strength of stabilized soil, upper layer and lower layer (Ohno and Terashi, 2005; Terashi, 2005).

Soil	Strength
Stabilized soil mass, q_{us}	100–1,000 kN/m²
Upper layer, q_{uu}	30 kN/m²
Lower layer, q_{ub}	100–10,000 kN/m²

answer because the mode of failure depends not only on the strength of stabilized soil but on a variety of factors as explained earlier in this sub-section.

A simple bearing capacity problem of a rigid structure on improved ground is numerically simulated by a large strain elasto-plastic analysis with finite difference method, FLAC to help engineers understand how the mode of failure changes (Ohno and Terashi, 2005; Terashi, 2005). Figure 6.6 shows the model ground for the analyses. The superstructure in the analysis is a rigid structure such as a concrete caisson underlain by a layer of granular material. The original soft ground is a two-layer system. The upper layer is the soft soil to be improved by deep mixing; whose thickness is 15 m. The strength of the upper layer, q_{uu} is a constant value of 30 kN/m². The strength of the lower layer, q_{ub}, whose thickness is 15 m is changed with the calculation case but always stronger than the upper layer. A stabilized soil mass installed in the upper layer by block type are born on the lower layer, whose strength, q_{us} is changed with calculation case. The gravel mound between the concrete caisson and the improved ground is represented by the interface element for which the friction angle is taken as 40°. In the calculations, the strength ratio of the stabilized soil mass, upper layer and lower layer, and the vertical and horizontal loads are changed, as summarized in Table 6.2. Further details of the numerical analyses are found in Ohno and Terashi (2005). The applicability of this numerical simulation was confirmed by centrifuge model testing conducted by Kurisaki *et al.* (2005) and Ohishi *et al.* (2005).

2.2.3.1 Influence of strength ratio q_{ub}/q_{us} on vertical bearing capacity

Figure 6.7 is an example of vertical load and displacement curve in which the strengths of the stabilized soil mass, q_{us} and the lower layer, q_{ub} are 1,000 and 200 kN/m²

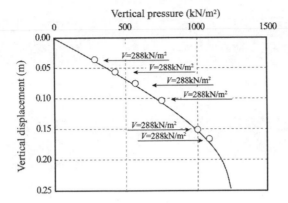

Figure 6.7 Vertical load and displacement curve (Ohno and Terashi, 2005; Terashi, 2005).

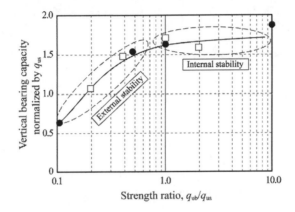

Figure 6.8 Vertical bearing capacity change with q_{ub}/q_{us} (Ohno and Terashi, 2005; Terashi, 2005).

respectively. Figure 6.8 shows the vertical bearing capacity normalized by the strength of stabilized soil, q_{us}. The vertical bearing capacity increases with increasing strength ratio, q_{ub}/q_{us} while q_{ub}/q_{us} is small, which implies that the strength of the bearing layer plays the dominant role in the failure mechanism. When q_{ub}/q_{us} is greater than unity, the bearing capacity becomes almost constant, which implies that the failure takes place in the stabilized soil mass. The failure mechanism of the former may be categorized into external stability and the latter internal stability.

This interpretation on the mode of failure is confirmed by the displacement vector and shear strain distribution at the foundation settlement of 250 mm shown in Figure 6.9. For the case of $q_{ub}/q_{us} = 0.1$, both the displacement vector and strain distribution show that the stabilized soil mass penetrates into the lower layer and show the typical external failure mode. Contrary to this, for the case of $q_{ub}/q_{us} = 10$, all the displacement and strain are concentrated within the stabilized soil mass exhibiting an internal stability. The case for $q_{ub}/q_{us} = 0.5$ seems a transitional case.

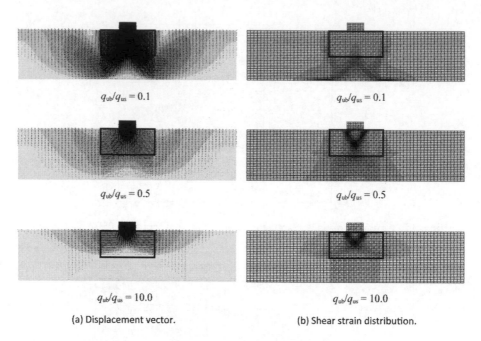

$q_{ub}/q_{us} = 0.1$ $q_{ub}/q_{us} = 0.1$

$q_{ub}/q_{us} = 0.5$ $q_{ub}/q_{us} = 0.5$

$q_{ub}/q_{us} = 10.0$ $q_{ub}/q_{us} = 10.0$

(a) Displacement vector.   (b) Shear strain distribution.

Figure 6.9 Displacement vector and shear strain distribution at 250 mm settlement (Ohno and Terashi, 2005; Terashi, 2005).

Figures 6.8 and 6.9 clearly show that the modal change from the external to internal failure is not governed by the strength of stabilized soil alone but that the relative strength is much more important. It should be noted that the transition of mode at $q_{ub}/q_{ut} = 0.5$ is only valid for this particular geometric condition.

2.2.3.2 Influence of load inclination

When the horizontal load as well as the vertical load apply to the rigid structure, which is the general case for most structures, the failure mode also changes. In this series of calculation, the strengths of the stabilized soil mass, q_{us} and the lower layer, q_{ub} are constant as 1,000 and 200 kN/m². Figures 6.8 and 6.9 show that the external stability is dominant at the ground condition of $q_{ub}/q_{us} = 0.2$ when vertically loaded. The horizontal load component is applied to the rigid structure at five different vertical pressure levels from 288 to 1,008 kN/m² as shown in Figure 6.7. The bearing capacity calculated for the combined vertical and horizontal load components are shown in Figure 6.10. A "cigar" shape of the bearing capacity envelope in the V-H plane is not unique to the deep mixed ground but found for any other uniform ground. In the calculations, the interface element between the rigid foundation and stabilized soil mass was given the friction angle of 40°. The straight line in the figure represents the sliding of the rigid structure on the gravel mound with friction angle of 40°. Interesting is in that the mode of failure is changing while the level of vertical load component is increased.

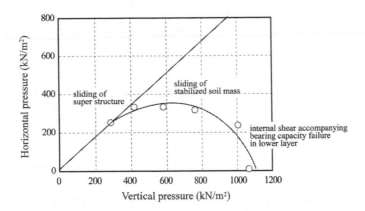

Figure 6.10 Bearing capacity in V-H load component plane (Ohno and Terashi, 2005; Terashi, 2005).

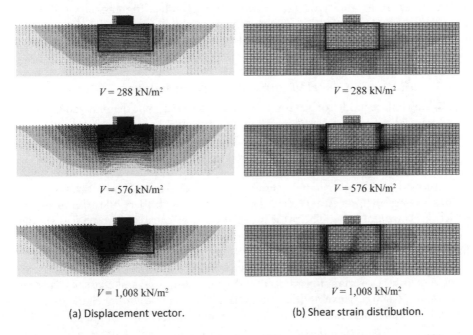

(a) Displacement vector.   (b) Shear strain distribution.

Figure 6.11 Displacement vector and shear strain at 200 mm horizontal displacement (Ohno and Terashi, 2005; Terashi, 2005).

Figure 6.11 shows the displacement and shear strain distributions when the foundation displaces 200 mm horizontally leftward. While the vertical load component is as low as 288 kN/m², no substantial increase in shear strain is found both in the stabilized soil and the surrounding soft soil layer, but the rigid structure horizontally slides over the gravel mound. At intermediate vertical load of 576 kN/m², the rigid structure and deep mixed soil mass move together and passive earth pressure failure is observed in the soft upper layer on the left of the deep mixed soil block, while the stabilized soil

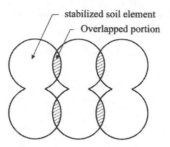

Figure 6.12 Overlapping of adjacent elements to create continuous stabilized soil mass.

exhibits no substantial increase in shear strain. This is the external failure by horizontal sliding mode. Under high vertical load component, a highly sheared zone is observed in the stabilized soil block and it extends down to the lower layer. It is interpreted that the internal instability occurs accompanying the bearing capacity failure in the underlying layer. Figure 6.11 shows that a change of failure modes occurs with the load inclination. Again the modal change is not governed by the strength of stabilized soil alone but the loading conditions also affect the mode of failure.

2.2.3.3 Influence of overlap joint on mode of failure

The various failure modes of a stabilized soil mass (block type deep mixed ground) has been discussed so far for the ideal or simplified situation. The external stability of a deep mixed block was discussed in sub-section 2.2.1 for the situation where each stabilized soil columns has sufficiently large shear strength and a soil block was installed with reliable overlapping procedure. Internal stability of a deep mixed block was discussed in sub-section 2.2.2 by implicitly assuming that the deep mixed soil block is continuous and has no overlap joint. Attention was given to the possible irregular slip surface along the weak horizontal layer. The present subsection, by means of numerical simulation assuming a continuous stabilized soil mass, showed that the mode of failure may change from external to internal depending on various factors.

In real life, however, the deep mixed elements such as block, grid and walls are constructed in-situ by overlapping adjacent individual stabilized soil elements as shown in Figure 6.12. The figure exemplifies the overlap between individual elements produced by a double shafts machine. During the overlapping procedure, the preceding element during initial hardening is partially scraped by the following element. The strength of stabilized soil in the overlapped zone (hatched zone in the figure) is anticipated to be lower than the intact portion of an individual column. Further, it is obvious that the breadth of the overlapped zone is narrower than the individual element.

2.2.3.4 Influence of overlap joint on external stability

The influence of overlap joint on the external stability of deep mixing improved ground was studied by Kitazume et al. (1991) by means of centrifuge modeling. A rigid structure resting on the improved ground was subjected to an increasing horizontal load until the improved ground fails by external stability as shown in Figure 6.13(a). They modeled two extreme cases of column installation patterns. One was the perfect

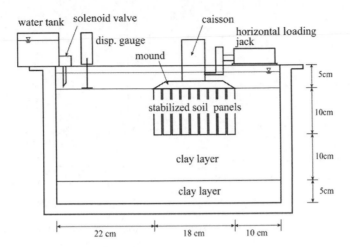

(a) Model setup for floating tangent panel panel.

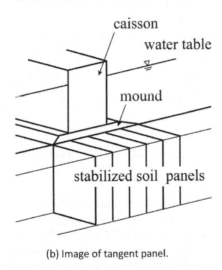

(b) Image of tangent panel.

Figure 6.13 Centrifuge model test on influence of overlap joint (Kitazume *et al.*, 1991).

continuous block of stabilized soil (ideal block). The other extreme was no overlap in which stabilized soil panels were placed in contact each other (referred to tangent panels) as shown in Figure 6.13(b). Both ideal block and tangent panels shared the same width and depth of improvement zone. Also examined was the condition of the bottom end of a stabilized soil. One was a floating type in which the stabilized soil zone does not reach the bearing layer (floating type improvement). In the other case, stabilized soil penetrated through soft soil and had a good contact with the bearing layer (fixed type improvement).

Figures 6.14 and 6.15 show the horizontal load and horizontal displacement relation, and the external failure mode for tangent panels respectively (Kitazume *et al.*,

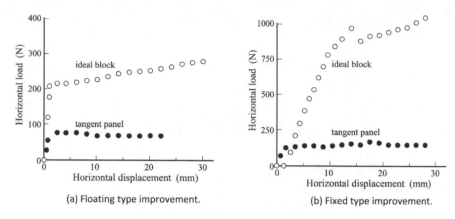

Figure 6.14 Relationship between horizontal load–displacement (Kitazume *et al.*, 1991).

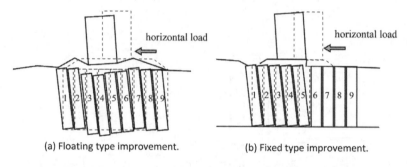

Figure 6.15 External failure of tangent panels under combined vertical and horizontal load (Kitazume *et al.*, 1991).

1991). As shown in Figure 6.14 the ideal block exhibits a much higher horizontal load capacity than the tangent panels both for floating and fixed conditions. For the floating case, the horizontal load capacity of the ideal block is twice as much higher than the tangent panels. The mode of external stability for the floating case in this test conditions is the bearing capacity failure both for the ideal block and the tangent panels. The external failure of the ideal block is determined by the bearing capacity of the floating "block" with full width of improvement. On the contrary, in the case of the tangent panels, only the panels immediately below the rigid structure sustains the load transferred from the rigid structure and the effective width of improvement is reduced to about one half of the ideal block as shown in Figure 6.15. For the fixed type improvement, the horizontal load capacity of the ideal block is four times higher than that of tangent panels. The external failure of the ideal block is determined by the sliding failure of the stabilized soil block as explained earlier in sub-section 2.2.1. On the contrary, in the case of tangent panels, only a part of the panels sustain the load from the rigid structure and the failure takes place by tilting of the panels. The change of failure mode is the major reason for the large difference in the horizontal load capacity in the fixed type improvement. It should be noted that the said quantitative difference of horizontal load capacity is only applicable for this particular geometric condition.

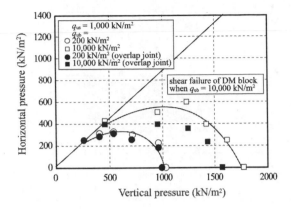

Figure 6.16   Bearing capacity in V-H load component plane (influence of overlap joint) (Ohno and Terashi, 2005; Terashi, 2005).

The description above is based on a comparison of extreme cases of ideal block and tangent panels. However the results demonstrate the importance of a reliable overlap in examining the external stability of a stabilized soil mass with block, grid and panel type of installation. *It should be emphasized that the block, grid and panel type in this design chapter only deals with those created by a careful overlapping procedure at least to the direction perpendicular to the expected failure surface.*

2.2.3.5   Influence of overlap joint on internal stability

The influence of overlap joint on the internal stability of deep mixing improved ground was studied by means of a numerical simulation. The condition of the model ground studied was the same as that explained earlier in sub-section 2.2.3.3. The difference was the introduction of a weak plane model in order to interpret the influence of the overlap joint at least qualitatively. Figure 6.16 shows the horizontal load capacity of the improved ground in the V-H plane. For all the cases the unconfined compressive strength of stabilized soil was kept constant as $1,000 \, \text{kN/m}^2$. The strength of the bearing layer, q_{ub} was changed and those were 200 and $10,000 \, \text{kN/m}^2$. Open circles and open squares in the figure correspond to the ideal block. Solid circles and solid squares correspond to the block with overlap joint. A ubiquitous-joint model was used to study the influence of the overlap joint in which the shear strength on the vertical plane was reduced to 60% and the tensile strength in the horizontal direction was also reduced to 60%.

The results of the calculation shown by open circles for an ideal block are the same as those shown earlier in Figure 6.10. For the cases with a stronger lower layer with $q_{ub} = 10,000 \, \text{kN/m}^2$, the failure modes for an ideal block are shear failure (internal instability) except for the case under a vertical load of $500 \, \text{kN/m}^2$, where the rigid structure slides over the improved ground. It is found that the existence of an overlap joint reduces the horizontal load capacity of the improved ground when the failure is governed by internal stability. Although the shear and tensile strengths on the weak

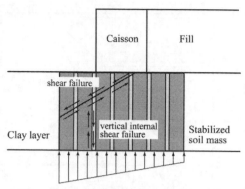

(a) Shear failure and vertical internal shear failure.

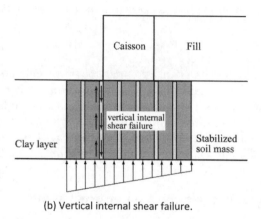

(b) Vertical internal shear failure.

Figure 6.17 Necessity of examining vertical internal shear.

plane were reduced to 60%, the influence on the load capacity was not that much at least for the conditions adopted for this simulation.

A centrifuge model test and numerical simulation conducted to understand the influence of an overlap joint suggest that both the external and internal stabilities of stabilized soil mass depend on the quality of the as-built stabilized soil element especially at the overlap joint. These studies also suggest that the examination of internal stability should involve vertical internal shear of stabilized soil element as shown in Figure 6.17.

2.2.3.6 Summary of failure modes for block type improvement

There exist a variety of failure modes both in the external and internal instability for the simple block type deep mixed ground.

1 External Stability: If the stabilized soil columns have *sufficiently large shear strengths and are installed by reliable overlapping procedure*, the stabilized soil

Figure 6.18 Deformation of clay ground between long walls in extrusion failure (Terashi *et al.*, 1983).

mass behaves as a rigid body. When the equilibrium is violated, as shown in Figure 6.2 the improved ground fails externally either by:

– Sliding failure
– Overturning failure
– Bearing capacity failure

2 Internal Stability: When the strength of the stabilized soil column is not sufficient, there is a risk of excessive deformation or failure of the stabilized soil mass under external actions. While the stabilized soil mass is relatively uniform, the failure pattern may be wedge shear, failure through slip circle or through irregular slip surface. When the overlapping is incomplete, internal failure may develop along the vertical overlap joint faces (Figures 6.5 and 6.17).

3 For the wall type improved ground, there is another important internal failure mode exists. This is the extrusion failure where the unstabilized soil between stiff panels is squeezed out due to the imbalance in the earth pressures acting on the active and passive side of the stabilized zone, as shown in Figure 6.18 (Terashi *et al.*, 1983).

4 The overlap joint face influences both the external and internal stability. When vertical shear along the overlap joint face occurs, the effective width of the stabilized soil block is reduced resulting in reduced external stability (the extreme case of tangent panels was shown in Figures 6.14 and 6.15.

5 The discussion in the sub-section is on a simple situation of horizontal soil layers. When the bearing layer is inclined, external failure may take place along the inclined layer.

6 All the modes of failures described in the present sub-section are considered in the current design procedure (The Ports and Harbours Association of Japan, 2007), in which a simplified calculation procedure is proposed for each mode of failure.

2.3 Engineering behavior of a group of individual columns

Nearly 60% of on-land works in Japan and perhaps roughly 85% of Nordic applications are for settlement reduction and improvement of the stability of embankments by means of the group column type improvement. Routine design practices for the stability of an embankment slope have been carried out following the relatively simple procedure that will be explained later. Although the improvement area ratio (or spacing of individual columns) and required shear strength of the stabilized soil column should be the outcome of a geotechnical design, the improvement area ratio preferred in Japan has been larger than 30% and often exceeds 50%, whereas in Nordic countries the design shear strength has been kept below 150 kN/m^2 regardless of the laboratory and actual field strengths (Swedish Geotechnical Institute, 1997; EuroSoilStab, 2002). The design engineer should understand the reason behind the seemingly conservative approach.

This sub-section intends to explain the actual behavior of the group column improved ground. The geometry of the group column improved ground is three dimensional and naturally its behaviour is much more complicated than simple two dimensional situations for block type improvement discussed in the previous sub-section. However, the following explanation will be given in most cases two-dimensionally. Still this introduction will provide engineers with an insight into real life behavior and enhance their understandings on the applicability and limitation of the routine design practices.

2.3.1 Stability of a group of individual columns

In the early days, a simple failure mode which can be analyzed by a slip circle analysis was imagined both in Japan and Nordic countries (Figure 6.19). This is obviously an analogy from the deep seated failure that may take place in ordinary soft soils. The simple slip circle failure mode is associated with two assumptions. One is that the stabilized soil column and soft soil behave as a composite material which exhibits the weighted average shear strength. The other assumption is that the composite material always fails by shear irrespective of the location along the slip surface.

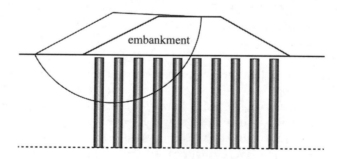

Figure 6.19 Simple circule failure mode.

The slip circle failure mode may not be unrealistic if the strengths of the stabilized soil column and unstabilized soil do not differ too much or if the improvement area ratio is large enough. Terashi and Tanaka (1983) carried out an experimental study to investigate the validity of the assumptions by varying the unconfined compressive strength of stabilized soil from 300 to 1,600 kN/m². The large scale simple shear tests of the composite soil exhibited that the concept of weighted average shear would overestimate the actual shear strength of the composite system. A series of model tests on the bearing capacity of the group column type improved ground exhibited the progressive failure of individual columns and of the overestimation of bearing capacity based on the concept of weighted average shear. In the middle of the 1980s an extensive research program on improved ground by the group of individual columns has been conducted by the initiative of the Public Works Research Institute. Amano et al. (1986) found the possibility of bending failure of the stabilized soil columns by monitoring a test embankment and proposed a design procedure for the embankment support incorporating finite element analysis to examine and avoid the bending failure. Tsukada et al. (1988) compared various column installation patterns and reported the superiority of the buttress type improvement over the group of columns in increasing the passive earth pressure. These early findings have not been incorporated explicitly into the practical design procedure. The current design procedure employed in Japan (Public Works Research Center, 2004) involves two major modes of failure; one is the slip circle analysis to determine the internal stability of stabilized soil columns and the other is the external stability to determine the sliding of a stabilized soil zone as shown in Figure 6.20. In addition to the above examination, the design guide proposes to determine the end bearing capacity of individual columns when the columns are installed to the bearing stratum. The commentary to the design guide, however, emphasizes the importance of learning the previous successful case records in determining the size and location of the improved zone and addresses the following notes; 1) The width to height ratio of the improved zone should be larger than 0.5 at least and preferably larger than 1.0, 2) The range of design strength in terms of the unconfined compressive strength has been between 100 to 600 kN/m², and 3) the most often employed improvement area ratio is larger than 30% and often exceeds 50%.

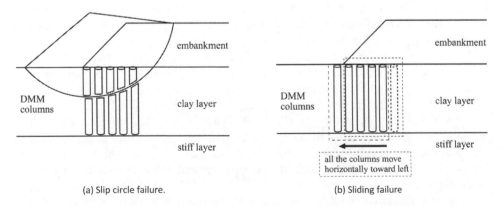

(a) Slip circle failure.　　　　　　　　　　　　(b) Sliding failure

Figure 6.20　Failure modes in the current design procedure (Public Works Research Center, 2004).

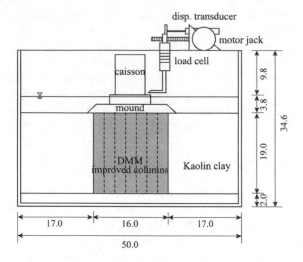

Figure 6.21 Model ground setup (Kitazume *et al.*, 2000).

In Japan, no serious failure was documented nor recorded although numerous applications to embankment support have been designed based on the current design procedure. The lack of information on failure in full scale has prevented an improved understanding of failure modes and the rigorous development of design procedures. In the late 1980s centrifuge model tests to identify the actual modes of failure has started in Japan. Also in Nordic countries, a number of embankments were designed and constructed safely based on the slip circle analysis. However, in the 1990s several failures and/or unexpected large deformations of column stabilized embankments have occurred, which lead Nordic engineers to reconsider the mode of failure (Kivelo, 1998; Broms, 1999). The failure modes described in this sub-section is mostly based on the centrifuge model tests conducted in Japan.

2.3.1.1 Bearing capacity of a group of individual columns

Kitazume *et al.* (1996, 2000) of the Port and Airport Research Institute studied the bearing capacity problem of a rigid concrete structure resting on a group of individual columns. The improvement area ratio of group column type improved ground was 79% where each column is in contact with the adjacent column but without over-lap operations (tangent columns). The strength of stabilized soil columns in terms of unconfined compressive strength varied from 200 to 27,000 kN/m^2. The model setup in Figure 6.21 shows the bearing capacity test under combined vertical and horizontal loadings (Kitazume *et al.*, 2000). The vertical load is given by the self-weight of the structure under the enhanced centrifugal acceleration. A horizontal load was given to the structure by the jack system. For the examination of vertical bearing capacity, a vertical loading system was used instead of the horizontal jack shown in the figure.

Figure 6.22 shows the different modes of failures exhibited by a series of testing (Kitazume *et al.*, 2000). In Case A by vertical loading, an active shear failure zone

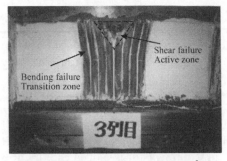

(a) Case A, vertical loading, q_u = 760 kN/m^2.

(b) Case B, vertical and horizontal loading, q_u = 300 kN/m^2.

(c) Case C, vertical and horizontal loading, q_u = 27,000 kN/m^2.

Figure 6.22 Various failure modes for a group of individual columns (Kitazume *et al.*, 2000).

was found for columns immediately below the foundation but the bending failure of the columns was found for the rest of columns in the transitional zone. Cases B and C were conducted under the combined vertical and horizontal loads. The stabilized soil columns with 300 kN/m^2 failed by bending failure in Case B but those columns with an extremely high strength of 27 MN/m^2 did not fail at all and overall failure was governed by the tilting of stabilized soil columns as shown in Case C. The observation suggests that an individual column in the group may fail in a different way according to its location (active zone, transitional zone or passive zone), and that there is an external failure mode such as tilting.

2.3.1.2 Embankment stability on a group of individual columns

The Port and Airport Research Institute also studied the modes of failures for embankment support since the middle of the 1990s and a series of test results has been publicized locally and internationally (Kitazume and Maruyama, 2006, 2007). Kitazume (2008) recently reported all the test results together with design recommendations. In a series of centrifuge tests, soft clay underneath the sloping side of the embankment is improved by a group of individual columns. In all the test cases, soft clay is normally consolidated and underlain by a sand layer. In prototype scale the thickness of clay ground and that of the underlying sand layer is 10 m. The stabilized soil columns having one meter diameters are all bearing on the underlying sand stratum (end bearing column). The embankment fill is placed rapidly until the embankment foundation fails. The slope angle of the embankment constructed during centrifuge flight is around 33.3°. Parameters examined are: strength of stabilized soil column in terms of unconfined compressive strength, q_{us} (varied from 400 kN/m^2 to practically infinite), improvement area ratio a_s (28% and 56%), and number of column rows (varied from 3 to 7).

Figure 6.23 shows some photographs showing different modes of failure (Kitazume, 2008). Photographs shown in the figure are taken after the centrifuge tests in order to observe the final mode of deformation of stabilized soil columns and/or to observe the mode of column failure. Case 6 (q_u of 425 kN/m^2, a_s of 28% and 3 column rows) failed when the fill pressure reached to around 20 kN/m^2. All the stabilized soil columns tilted and bended outwards and exhibited tensile cracks at two different levels. Case 11 (q_u of 1,300 kN/m^2, a_s of 28% and seven rows) failed when the fill pressure reached to around 58 kN/m^2 where stabilized soil columns suffered tilting and bending failure. The columns in Case 3 were modeled by acrylic piles to determine the external failure mode of improved ground. The photograph shows that all the columns tilted outward. The geometry of the test series is typical for embankment support in Japan. No slip circle failure and no sliding failure were observed at least for these tests using end bearing columns. In the tests the time sequence of column failure were detected and found that a group of individual columns did not fail simultaneously but failed progressively. Kitazume proposed three failure modes for embankment stability as shown in Figure 6.24 (Kitazume, 2008) in addition to slip circle failure and sliding failure as shown earlier in Figure 6.20.

Akamoto and Miyake (1989) studied the influence of location of the improved zone relative to the embankment slope. In Case (a) a group of columns are installed only beneath the sloping side of the embankment, in Case (b) the same improved zone is sifted to the location underneath the embankment crest, and finally in Case (c) the same number of stabilized soil columns are installed to a wider zone in comparison to the Cases (a) and (b) as shown in Figure 6.25 (Akamoto and Miyake, 1989). According to the authors, Case (a) failed by tilting and bending failure of columns under the slope. Under the same fill pressure which brought Case (a) to failure, no sign of failure was found in Case (b) except a local small failure of columns at the toe of the embankment slope. Case (c) was most stable among them all. When they increased the column diameter of Case (a) by 1.4 times while keeping the improvement area ratio and the width of improved zone the same as those of Case (a), they reported that they could avoid tilting and bending failure. These results suggest that the group of columns is much more effective in the active zone than in the transient or passive zones, and that

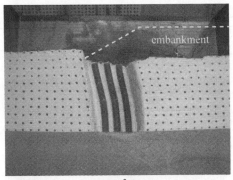

(a) Case 6, q_u = 425 kN/m^2.

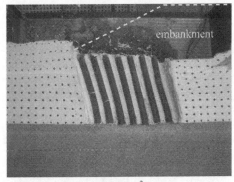

(b) Case 11, q_u = 1,300 kN/m^2.

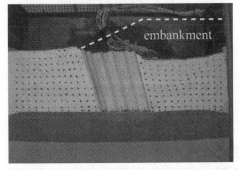

(c) Case 3, q_u = infinitely large (acrylic model piles).

Figure 6.23 Failure modes found in the centrifuge tests (Kitazume, 2008).

the increase in the flexural rigidity of a stabilized soil column is effective in increasing the resistance against tilting and bending mode of failure.

The centrifuge test results referred above were all for the group of stabilized soil columns that reached the bearing stratum. There seems to be limited information on the behavior of the group of floating columns as far as the stability problem is concerned. One example may be found in Figures 6.14(a) and 6.15(a), which were used to explain the behavior of an extremely poorly overlapped block but was actually a behavior of

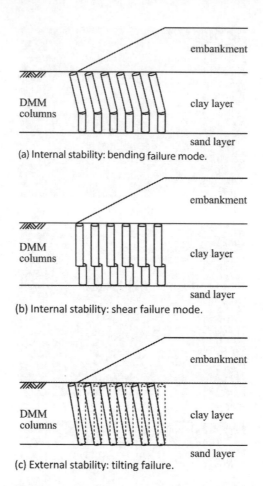

Figure 6.24 Additional failure modes to be examined in group column type improvement (Kitazume, 2008).

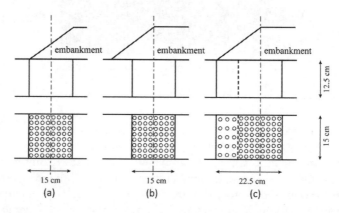

Figure 6.25 Influence of location of improved zone relative to embankment crest (Akamoto and Miyake, 1989).

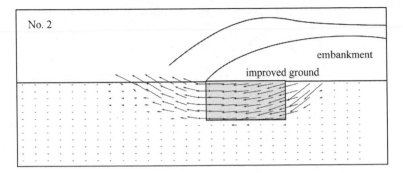

Figure 6.26 Displacement vector during embankment filling (Miyake *et al.*, 1988).

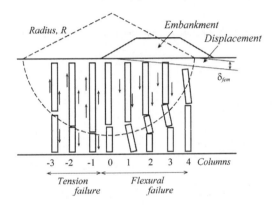

Figure 6.27 Progressive failure (Broms, 1999).

tangent panels. Miyake *et al.* (1988) found an external failure mode resembling the sliding of an improved block by their centrifuge model test as shown in Figure 6.26, although they explained the mode as the slip failure along the bottom end of stabilized soil columns. The unconfined compressive strength of the stabilized soil was around $3\,MN/m^2$, the improvement area ratio was around 30%. All the columns were resting on the overconsolidated clay layer.

Inspired by the centrifuge model tests conducted in Japan and based also on a few cases of full scale failure in Sweden, Nordic engineers also acknowledged the possible overestimation by the assumption of average shear strength. Kivelo (1998) examined the moment capacity of an individual column in the active, transient and passive zone and proposed the methodology of analyzing the stability of embankment slope based on the slip circle analysis. Broms (1999) extended Kivelo's work to explain the progressive failure of the embankment as well. Both Broms and Kivelo assumed a slip circle failure mode that passes through the group of stabilized soil columns and try to incorporate the different column failure patterns in to the limit equilibrium design by slip circle (Figure 6.27).

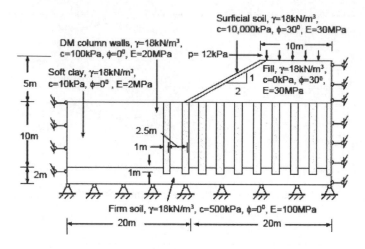

Figure 6.28 Numerical analysis model (Han et al., 2005).

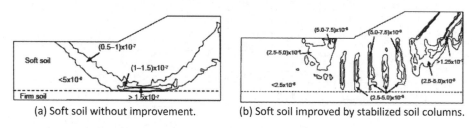

(a) Soft soil without improvement. (b) Soft soil improved by stabilized soil columns.

Figure 6.29 Contours of shear strain rate calculated by 2D FLAC (Han et al., 2005).

2.3.1.3 Numerical simulation of stability of embankment

Han *et al.* (2005) conducted numerical simulations of an embankment supported by a group of stabilized soil columns in order to investigate the modes of deep seated failure. The geometry and material properties of their baseline case are shown in Figure 6.28. A 1 m thick surface layer was used to prevent possible surface failure and examine the deep seated failure. The Mohr-Coulomb failure criteria were used. The stabilized soil was modeled to have a tensile strength equal to 20% of the undrained shear strength. The stabilized soil columns were modeled by continuous stabilized soil walls for ease of calculation by 2D FLAC. Employing the shear strength reduction technique, the factors of safety were obtained for various conditions, and compared with the corresponding factors of safety calculated by the Simplified Bishop's method assuming that the averaged shear strength would develop all along the slip circle.

Figure 6.29 shows the contour of the shear strain rate for soft soil without improvement and that for the deep mixing improved ground (Han *et al.*, 2005). Figure 6.29(a) for an embankment over soft soil without improvement clearly shows the circular slip surface. In Figure 6.29(b) for an embankment supported by stabilized soil walls, higher shear strain rates are found in front and rear of the stabilized soil columns and implies the tilting of stabilized soil columns dominate the mode of failure. The authors

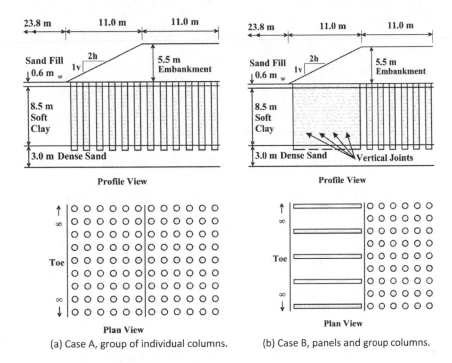

Figure 6.30 Baseline case for examination (Filz and Navin, 2006; Adams et al., 2009).

Table 6.3 Mean material properties (Filz and Navin, 2006; Adams et al., 2009).

Material	γ (kN/m³)	E (kN/m²)	ν	C (kN/m²)	ϕ (deg)
Embankment	19.6	29,900	0.3	0	35
Sand Fill	18.1	12,000	0.33	0	30
Soft Clay	15.1	$200s_u$	0.45	*	0
Dense Sand	22.0	47,900	0.26	0	40
Columns	15.1	207,000	0.45	689	0

*Strength of soft clay varies with depth.

changed the strength of the stabilized soil, wall spacing, and size of wall among others and confirmed that the slip circle analyses generally overestimate the actual factor of safety. The range of improvement area ratio corresponds to 33 to 50%. The undrained shear strength of stabilized soil was changed from 10 kN/m² (equals to original soft clay) to 500 kN/m². The numerical experiments supported the needs of considering various modes of failures which had been pointed out by several researchers by means of model tests.

Filz and Navin (2006) and Adams et al. (2009) have also conducted numerical simulation of the embankment supported by a group of individual columns and compared the factors of safety with those obtained by the ordinary slip circle method. The column installation patterns and the mean material properties used for the study are shown in Figure 6.30 and Table 6.3 respectively. In Case A the stabilized soil columns are 0.9 m

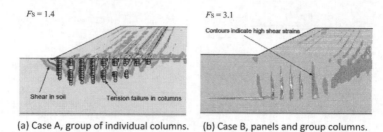

$F_S = 1.4$ $F_S = 3.1$

(a) Case A, group of individual columns. (b) Case B, panels and group columns.

Figure 6.31 Results of numerical analyses (Filz and Navin, 2006; Adams *et al.*, 2009).

in diameter and arranged in a square array with 1.8 m center-to-center spacing both underneath the slope and full height of the embankment, which corresponds to an area replacement of 20%. In Case B the individual columns underneath the side slope of Case A are replaced with panels of stabilized soil which retain an area replacement ratio of 20%. The columns extend from the sand fill, through the clay layer and 0.6 m into the base sand layer. The columns have a cohesion intercept of 689 kN/m² with a total stress friction angle of zero, which corresponds to an unconfined compressive strength of 1.38 MN/m². The shear strength of the soft original clay varies linearly with depth from 10.2 kN/m² at the top to 20.6 kN/m² at the bottom.

The stabilized soil columns and overlapped panels in the improved ground were modeled by two dimensional approximations. A row of stabilized soil columns with 0.9 m diameter was represented by a 0.36 m wide strip with 1.8 m spacing in the 2D analysis. The width of the strip was chosen to match the improvement area ratio of 20%. The original material properties of stabilized soil were given to the strips. The properties of panels and the original soft soil underneath the side slope of Case B were modeled to have the weighted average of stabilized soil and soft soil. The overlapped zone in the panel was given half the shear strength of the intact portion of panels. Numerical simulation of both Cases A and B were conducted and with the strength reduction techniques factors of safety were also obtained.

Figure 6.31 shows the results of numerical analyses (Filz and Navin, 2006, Adams *et al.*, 2009). In the numerical analyses of Case A, the columns bent and broke, and the shear of soil between the columns due to tilting of the columns are observed and confirmed the existence of a variety of failure modes found by centrifuge model tests such by Kitazume *et al.* (1996) and Kitazume (2008). Although the improvement area ratio is the same for both Cases A and B, the factor of safety for the former is 1.4 and that for the latter column installation pattern is 3.1 and much higher than the former. The numerical simulation confirmed the earlier experimental findings of Tsukada *et al.* (1988). Figure 6.31(b) for Case B shows the development of shear strain at the overlap joint faces and confirms the importance of examining vertical shear at the overlap joint in the design. The conventional limit equilibrium analysis assuming circular slip surface together with the assumption of weighted average shear strength produces the same factor of safety for both Case A and Case B. As the slip circle analysis does not take the tilting, bending and other modes of possible failures, the factor of safety calculated is 4.4. As most of the centrifuge modelers pointed out, slip circle analysis overestimate the actual stability of the improved ground supported by a group of individual columns.

In order to justify the two dimensional approximation, Navin and Filz (2006) compared the solution of two dimensional analysis by FLAC and that of three dimensional analysis by FLAC 3D. The examined improved ground was the baseline case shown in Figure 6.30 except for the shear strength of stabilized soil, which was taken as $479 \, kN/m^2$. At least for this particular geometry and soils conditions, the authors reported that the 2D analyses are very close to the 3D analyses.

The most important contribution of Filz and Navin (2006) is the introduction of reliability analysis into the examination of deep mixed ground. The factor of safety for Case A is computed as 1.4 as mentioned above but the probability of failure is found as high as 3.2%. Whereas Case B with the same improvement area ratio but using panels produced a factor of safety of 3.1 and the provability of failure as low as 0.009%. The reliability based design may be extended to provide a rigorous quality assurance scheme for deep mixing.

2.4 Summary of failure modes for a group of stabilized soil columns

There exist a variety of failure modes both in the external and internal instabilities for the group of individual columns and they are far more complicated than block type applications. Modes of instability will be strongly influenced by geometry, location of improved zone relative to the superstructure, end bearing conditions, and as-built quality of deep mixed soil columns. Various failure modes were identified by centrifuge model tests and confirmed also by numerical analyses. Most of the previous studies are focused on the behavior of end-bearing columns. Further study will be necessary for the floating columns.

1   Stabilized soil columns may fail by shear when the strength of stabilized soil does not differ too much from the soft original ground and/or the improvement area ratio is large enough.
2   The limit equilibrium method of slip circle analysis is often used assuming the weighted average shear strength of stabilized soil columns and soft soil. Slip circle analysis generally overestimate the actual stability.
3   Stabilized soil columns with medium to high strengths exhibit shear failure only in the active zone. In the transitional zone, bending failure of the columns dominate. Stabilized soil columns in the passive zone are ineffective in the stability.
4   For the group of stabilized soil columns born on the stiff reliable layer, several failure modes may exist. They are

    – Circular slip
    – Irregular slip surface passing through horizontal failure plane
    – Bending failure
    – Tilting of stabilized soil columns

5   When bending and/or tilting failure modes are anticipated, the capacity against these failure modes depends on the diameter of columns. The overlapped panel is superior to the group of individual columns.
6   When stabilized soil columns are floating, the sliding failure of the improved zone may become an additional external failure mode. Also the end bearing capacity of columns may be one of the external failure modes.

7   The currently available routine design procedures described later does not incorporate all these failure mechanisms. The stability analysis by slip circle method therefore overestimates the actual stability and used with careful consideration on the hidden margin of safety which will be discussed as commentary to the design procedure such as the existence of dry crust, underestimation of original ground and underestimation of column strength in design.

3   WORK FLOW OF DEEP MIXING AND DESIGN

3.1   Work flow of deep mixing and geotechnical design

3.1.1   Work flow of deep mixing

Figure 6.32 shows the work flow common to a project involving deep mixing (Terashi, 2003). This Section is primarily aimed to provide the (geotechnical) design procedure, which correspond to the slight gray frame (double line frame). The design engineer should establish the design parameters for design calculations by assuming the as-built quality of deep mixed ground, which can only be identified after construction. As-built quality highly depends on the site conditions, skill of the contractors, and the capability of locally available deep mixing equipment. In order to guarantee the quality of project, a bench scale test and field trial installation are important. The practicable solution cannot be obtained without understanding the characteristics of the deep mixing project outlined in Figure 6.32. The sequence of work items in the flow may change from a project to another depending on such factors as the size and complexity of the project, the variability of the subsurface conditions, and the anticipated difficulty of deep mixing at the project site. Further details of the work flow and QA/QC related descriptions will be discussed later in Chapter 7.

–   The role of the geotechnical design is to determine, based on the design parameters, the size of improved zone, installation depth and installation pattern so that the improved ground may satisfy the performance criteria of the superstructure. This is an iterative process and the engineer has to change the factors mentioned above until the appropriate solution is reached. The geotechnical designer should establish design parameters and required level of accuracy of installation considering the capability of the current deep mixing technologies. If the engineer intends to use, in his design, end-bearing columns and/or panel or grid element by overlapping individual columns, the designer should consult contractors about the possibility of manufacturing such improved soil element under the expected project site conditions, because the quality of as built stabilized soil element depends highly on the experience of contractors, the capability of their equipment and the availability of skilled operators.
–   The role of the process design is to determine the construction control values to realize the quality of the improved ground specified by the geotechnical design. Specifications may include not only the strength and uniformity of in-situ stabilized soil columns but also the accuracy of installation in order to guarantee the location, depth, stable contact with bearing layer and reliable overlap of columns. Process design is often made possible by the field trial installation using the locally available equipment and materials.

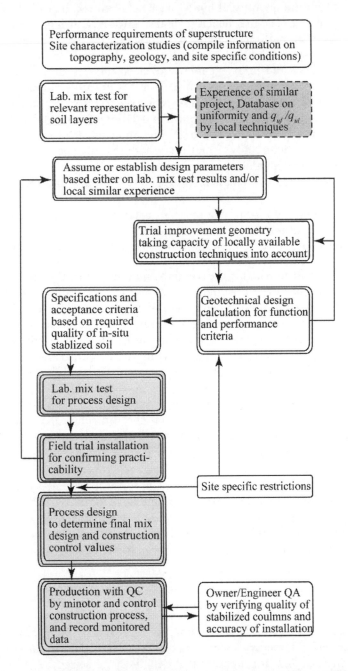

Figure 6.32 Work flow for the project involving deep mixing (Terashi, 2003).

–   The laboratory mix test is often carried out as a bench scale test to determine whether the soft soils at the project site are suitable for deep mixing. The strength of the stabilized soils can be controlled by the amount of binder. However, the cost and the capability of the locally available deep mixing machines may restrict the upper limit for the quantity of binder. It is better to conduct the bench scale test twice in a project. The first bench scale test is for establishing design parameters such as shear strength and should be conducted by owner/engineers. The first bench scale test may be replaced by the engineer's judgement only when there is published data and prior experience of deep mixing for similar soils nearby. However, in many projects, a bench scale test is not undertaken until the design is completed and the specifications for construction are determined because the bench scale test requires soil sampling and laboratory mix tests that normally takes more than four weeks. It is highly recommended for engineers to have an experienced contractor's advice on the strength that may be attainable for the soil at the project site with reasonable cost. The second bench scale test should always be conducted by the contractor before the installation of production columns as a part of QA.

- The properties and uniformity of the in-situ stabilized soil columns are influenced by many factors, among which the capability of the deep mix machine and its operational conditions are important. A field trial test has two aspects.

  - One is to confirm whether the strength and uniformity of columns satisfy those specified in the design document. The other is to determine the criteria for the deep mixing operation. When geotechnical design selected the use of block- or wall-type installation, or when geotechnical design specified the end-bearing columns, machines with poor capability cannot fulfil the requirement. The field trial installation must be carried out with the same machine, same binder, and under the same range of construction control values with those to be used in the production.
  - If the requirements are not met, the geotechnical design should be reconsidered. In this regard, it is recommended that the geotechnical engineer should be involved in the interpretation of the results of the field trial installation. The selection of verification test methods is also important.

As mentioned above, the geotechnical design for deep mixing is iterative. Further, in the worst case, geotechnical and process design have to follow the iteration. *It is the responsibility of the owner/engineer to schedule the sequence of work flow that best suits to the project.* In the case of big projects or a difficult project, it is recommended to carry out the bench scale test and field trial installation in advance to the geotechnical design.

3.1.2 Strategy – selection of column installation pattern

The currently available geotechnical design procedure is different for different column installation pattern. The general characteristics of each installation pattern were briefly summarized in Table 6.1. It is important for engineers to select the most appropriate column installation pattern before conducting analyses. The location of the improved zone relative to the superstructure also influences the performance of deep mixed ground. The engineering behavior described for the block type column installation in Sub-section 2.2 and that for the group of individual columns in Sub-section 2.3 may help the owner/engineer in the selection.

When the stability of an embankment is the major engineering issue, the block or panel type column installation pattern provides a better performance than the group of individual columns. The stability analysis is much simpler and reliable for the block or wall type of column installation. Contrary to this, the construction of a group of individual columns is far simpler and hence construction time and cost are much favorable for a group of individual columns.

In the selection of an appropriate column installation pattern, the owner should consider the experience and capability of the design engineers and the deep mixing specialty contractors available locally.

4   DESIGN PROCEDURE FOR EMBANKMENT SUPPORT, GROUP COLUMN TYPE IMPROVED GROUND

4.1   Introduction

The group column type improvement either by dry or wet method of deep mixing has frequently been applied to embankment support in order to improve stability and to reduce settlement (Figure 6.33). During a quarter century since 1981, the dry method of deep mixing was employed to support at least 2,700 embankments by the group column type improvement. The purpose of improvement was embankment stability for 60% of these case histories and settlement reduction for 40% (Terashi *et al.*, 2009). The design method for the group column type improved ground was proposed by the Public Works Research Center in 1999, and revised in 2004 (Public Works Research Center, 2004).

As mentioned in the previous introductory sections, the block or panel type column installation beneath the sloping side of the embankment provides a better performance than the group of individual columns. Further, the stability analysis is much simpler and reliable for the block or panel type of column installation as will be discussed in Section 5. Nevertheless, the group columns are preferred even for the stability due to the simplicity in construction and cost and time saving.

In this section, the group column type improved ground beneath an embankment is exemplified, where the two dimensional condition is assumed. This section basically introduces the design methodology established by the Public Works Research Center (Public Works Research Center, 2004), but with some additional comments by the authors.

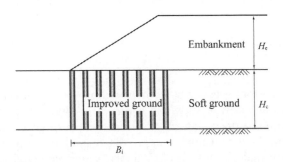

Figure 6.33 Group column type improved ground for embankment support.

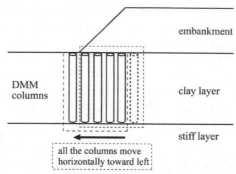

(a) Sliding failure in external stability of improved ground.

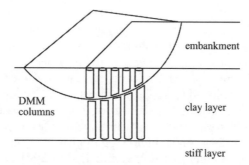

(b) Slip circle failure in internal stability of improved ground.

Figure 6.34 Failure pattern assumed in the current design procedure (Public Works Research Center, 2004).

4.2 Basic concept

In the PWRC design, the group column type improved ground is considered to be a sort of composite ground with an average strength of stabilized soil columns and unstabilized soil between them. In the design, two stabilities are evaluated: external and internal stabilities. The external stability examines the possibility of sliding failure of the improved ground, in which the stabilized soil columns and the unstabilized soil between them moves horizontally as shown in Figure 6.34(a). For the internal stability, the possibility of column failure is evaluated by slip circle analysis (see Figure 6.34(b)).

4.3 Design procedure

4.3.1 Design flow

The design procedure for the group column type improved ground is usually carried out by following the design flow as shown in Figure 6.35 (Public Works Research Center, 2004). After determining the design conditions and dimensions of a superstructure such as an embankment, the dimensions of improved ground are assumed at the first step. The sliding stability analysis and slip circle analysis are conducted for the external and internal stabilities respectively. The horizontal displacement of the improved ground

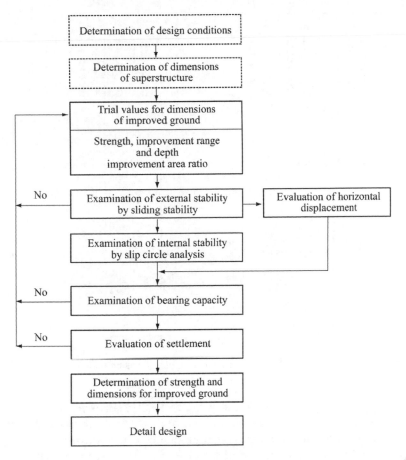

Figure 6.35 Design flow for a group column type improvement (after Public Works Research Center, 2004).

is examined in many cases. The bearing capacity and ground settlement are examined finally, and the details of the improved ground such as strength and dimensions are determined.

4.3.2 Trial values for dimensions of improved ground

The width and depth of improvement, improvement area ratio and strength of stabilized soil column are determined by trial calculations. Trial values for the initial design calculation are established/assumed by considering similar case histories. The width of improvement is usually assumed as the width of embankment side slope for increasing slope stability. For settlement reduction, stabilized soil columns are installed beneath the full height of the embankment. The depth of improvement is classified into two improvement types as schematically shown in Figures 6.36(a) and 6.36(b): fixed type and floating type improvements depending upon whether stabilized

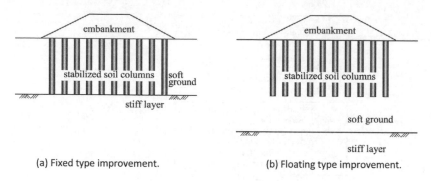

(a) Fixed type improvement.　　　(b) Floating type improvement.

Figure 6.36　Improvement type.

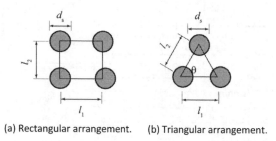

(a) Rectangular arrangement.　　(b) Triangular arrangement.

Figure 6.37　Arrangement of stabilized soil columns.

soil columns reach the stiff layer or not. It can be easily understood that the fixed type improvement is preferable from the viewpoints of increasing stability and reducing settlement. The depth of improvement is usually assumed as the bottom of the soft ground, where the stabilized soil columns reach the stiff layer, the fixed type improvement. In the case where the thickness of the soft ground is quite large, however, the floating type improvement is selected. As the appropriate range for the ratio of the width to the depth of improvement, 0.5 to 1.0 is recommended based on the accumulated experiences.

The improvement area ratio, a_s is represented as the ratio of the sectional area of stabilized soil column to the ground occupied by a single column, as shown in Figure 6.37, and it is calculated by Equations (6.1a) and (6.1b) for rectangular and triangular arrangements respectively. The improvement area ratio, a_s of 0.3 to 0.7 is usually adopted for the foundation of the embankment.

For rectangular arrangement

$$a_s = \frac{\pi d_s^2}{4} \frac{1}{l_1 \cdot l_2} \tag{6.1a}$$

for triangular arrangement

$$a_s = \frac{\pi d_s^2}{4} \frac{1}{l_1 \cdot l_2 \cdot \sin \theta} \tag{6.1b}$$

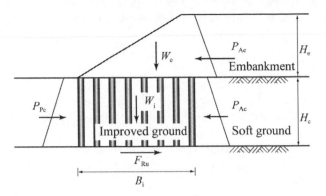

Figure 6.38 External force conditions for sliding failure analysis.

where

a_s : improvement area ratio
d_s : diameter of stabilized soil column (m)
l_1 : spacing between stabilized soil columns (m)
l_2 : spacing between stabilized soil columns (m)
θ : angle of arrangement of stabilized soil columns.

   The designed unconfined compressive strength of a stabilized soil column, q_{uck}, can be assumed at first by Equation (6.2) with a safety factor of 1.0 to 1.2. This equation means that the strength of the stabilized soil column should be higher than the embankment load on the area occupied by the column. As explained later, the strength of the stabilized soil column, however, is recommended to be 200 to 1,000 kN/m² by considering successful case histories.

$$q_{uck} \geq Fs \cdot \frac{\gamma_e \cdot H_e}{a_s} \qquad (6.2)$$

where

a_s : improvement area ratio
Fs : safety factor
H_e : height of embankment (m)
q_{uck} : design unconfined compressive strength of stabilized soil (kN/m²)
γ_e : unit weight of embankment (kN/m³).

4.3.3   Examination of sliding failure

For the external stability, the sliding failure of the improved ground is examined to determine the width and thickness of improved ground. In the design, the stability is evaluated based on the force equilibrium acting on both sides of the improved ground (Figure 6.38), where a two dimensional condition is assumed. The safety factor against sliding failure is calculated by Equation (6.3). In the calculation, the width and thickness of improved ground (mainly the width) are changed to assure the allowable magnitude of Fs_s which is usually 1.3 for the static condition.

$$Fs_s = \frac{P_{pc} + F_{Ri}}{P_{Ac} + P_{Ae}} \tag{6.3}$$

where

B_i : width of improved ground (m)

c_{ub} : undrained shear strength of soft soil beneath improved ground (kN/m^2)

c_{uc} : undrained shear strength of soft soil (kN/m^2)

c_{us} : undrained shear strength of stabilized soil (kN/m^2)

F_{Ri} : total shear force per unit length mobilized on bottom of improved ground (kN/m)

in the case of a sand layer beneath improved ground (fixed type improvement)

$$= \min \begin{cases} (W_e + W_i) \cdot \tan \phi'_b \\ (a_s \cdot c_{us} + (1 - a_s) \cdot c_{uc}) \cdot B_i \end{cases}$$

in the case of a clay layer beneath improved ground (floating type improvement)

$$= \min \begin{cases} c_{ub} \cdot B_i \\ (a_s \cdot c_{us} + (1 - a_s) \cdot c_{uc}) \cdot B_i \end{cases}$$

Fs_s : safety factor against sliding failure of improved ground

P_{Ae} : total static active force per unit length of embankment (kN/m)

$$P_{Ae} = \frac{1}{2} \cdot \gamma_e \cdot H_e^2 \cdot \tan^2 \left(\frac{\pi}{4} - \frac{\phi'_e}{2} \right)$$

P_{Ac} : total static active force per unit length of soft ground (kN/m)

$$P_{Ac} = \frac{1}{2} \cdot \gamma_c \cdot H_c^2 + W_e \cdot H_e - 2 \cdot c_{uc} \cdot H_c$$

P_{Pc} : total static passive force per unit length of soft ground (kN/m)

$$P_{Pc} = \frac{1}{2} \cdot \gamma_c \cdot H_c^2 + 2 \cdot c_{uc} \cdot H_c$$

W_e : weight per unit length of embankment (kN/m)

W_i : weight per unit length of improved ground (kN/m)

ϕ'_b : internal friction angle of soil beneath improved ground

ϕ'_e : internal friction angle of embankment

r_c : unit weight of soft soil (kN/m^3)

4.3.4 Slip circle analysis

The internal stability analysis is evaluated by a slip circle analysis to determine the strength of the stabilized soil column and the improvement area ratio. In the analysis, the composite ground consisting of stabilized soil columns and unstabilized soil is assumed to have an average strength defined by Equation (6.4). As the axial strain of stabilized soil at failure is in many cases smaller than that of the original soil (see Figures 3.8 and 3.9), the shear strength of the original soil doesn't fully mobilize at the failure of

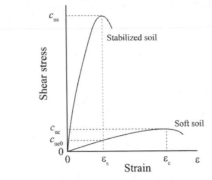

(a) Illustration of stress and strain curves.

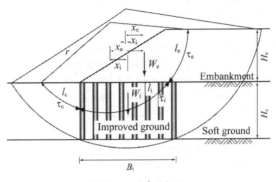

(b) Slip circle analysis.

Figure 6.39 Slip circle analysis.

the stabilized soil. This phenomenon is incorporated in Equation (6.4) by introducing the mobilization factor, k as shown in Figure 6.39(a). However, in cases where the strength of the stabilized soil columns is much higher than that of the original soil, the mobilization factor may provide a negligible influence on the calculation result.

The safety factor against slip circle failure, Fs_{sp} is calculated by the modified Fellenius analysis (see Figure 6.39(b)) with Equation (6.5). The allowable magnitude of safety factor of 1.3 is adopted for the static condition in many cases.

$$\overline{\tau} = a_s \cdot c_{us} + (1 - a_s) \cdot k \cdot c_{uu} \tag{6.4}$$

$$k = \frac{c_{u0}}{c_{uu}}$$

where
    a_s : improvement area ratio
    c_{uu} : undrained shear strength of soft soil (kN/m^2)
    c_{u0} : undrained shear strength of soft soil mobilized at the peak shear strength of stabilized soil (kN/m^2)
    c_{us} : undrained shear strength of stabilized soil (kN/m^2)

k : mobilization factor of soil strength
$\bar{\tau}$: average shear strength of improved ground (kN/m^2).

$$Fs_{sp} = \frac{r \cdot (\tau_c \cdot l_c + \bar{\tau}_i \cdot l_i + \tau_e \cdot l_e)}{W_e \cdot x_e}.$$ (6.5)

where
Fs_{sp} : safety factor against slip circle failure
l_c : length of circular arc in soft ground (m)
l_e : length of circular arc in embankment (m)
l_i : length of circular arc in improved ground (m)
r : radius of slip circle (m)
W_e : weight per unit length of embankment (kN/m)
x_e : horizontal distance of weight of embankment from center of slip circle (m)
τ_c : shear strength of soft ground (kN/m^2)
τ_e : shear strength of embankment (kN/m^2)
$\bar{\tau}_i$: average shear strength of improved ground (kN/m^2).

Equation (6.5) often leads to misunderstanding that the improved ground having a high strength of stabilized soil column and a low improvement area ratio can be an alternative to low strength and a large improvement area ratio to assure the required safety factor. The past experiences, however, have revealed that such an alternative is not suitable because the composite ground concept can't be assured. The improvement area ratio of the improved ground and the strength of the stabilized soil column should be larger than 0.3 and ranging 500 to 1,000 kN/m^2 respectively in order to assure the composite ground concept.

4.3.5 Examination of horizontal displacement

The improved ground consisting of stabilized soil columns and surrounding soil may show horizontal and/or rotational displacement due to the weight of the embankment and the earth pressures acting on the improved ground. When the purpose of improvement includes the reduction of horizontal displacement that may give adverse influence on nearby existing structures, the examination of horizontal displacement is necessary. The PWRC recommends the use of two dimensional finite element analysis. Also recommended is a rough estimation of the horizontal displacement via the magnitude of the minimum safety factor obtained by the slip circle analysis. Figure 6.40 shows an example of the relationship between the horizontal displacement at the toe of the embankment slope and the safety factor against the slip circle failure, which was derived by a series of FEM analyses (Ogawa et al., 1996a, 1996b). According to the figure, the horizontal displacement remains quite small magnitude as long as the safety factor is larger than about 2. The order of 0.2 to 0.3 m in the horizontal displacement takes place when the safety factor becomes lower than about 1.5.

4.3.6 Examination of bearing capacity

The weight of embankment tends to concentrate on the stiff stabilized soil columns. The bearing capacity of the stiff layer at the bottom of the improved ground should be then examined. The PWRC design procedure doesn't specify any particular bearing capacity

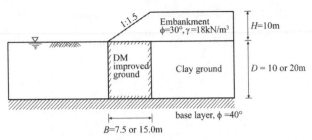

Clay ground, $c_u=5.0 + 1.5z$ (kN/m²) or $c_u=15.0 + 1.5z$ (kN/m²), $\gamma=15$kN/m³
DM improved ground, width =7.5 or 15.0m, $as=0.5$, $q_{uck}=400$kN/m²

(a) Ground condition for FEM analyses.

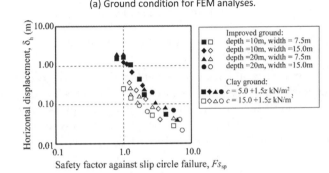

(b) Safety factor and horizontal displacement.

Figure 6.40 Estimation of horizontal displacement by slip circle analysis (Ogawa *et al.*, 1996a and 1996b).

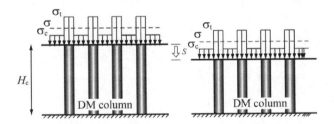

Figure 6.41 Calculation of consolidation settlement.

formula, but left it to the other design standards established by various organizations for specific facilities, such as road, railway, port facility and building.

4.3.7 Examination of settlement

4.3.7.1 Amount of settlement for fixed type improved ground

In the settlement calculation for the fixed type improved ground, it is usually assumed that the stabilized soil columns and the surrounding ground settle uniformly as illustrated in Figure 6.41, where the stress concentration effect is incorporated. This assumption has also been applied to a flexible loading condition such as an

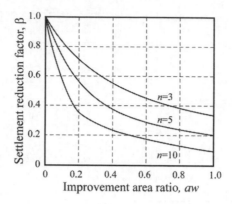

Figure 6.42 Settlement reduction factor along improvement area ratio.

embankment. The final consolidation settlement of improved ground, S, is calculated by multiplying the final consolidation settlement of the original ground without improvement, S_c and a settlement reduction factor, β, as formulated by Equation (6.6).

The final consolidation settlement of the original ground is usually calculated by the Terzaghi's consolidation theory as expressed in Equations (6.7a) to (6.7c). In the case where the original ground consists of multiple layers, the settlement should be calculated as the sum up of the compressive deformations in each layer. The settlement reduction factor, β, is derived by incorporating the stress concentration effect of the stabilized soil columns. The stress concentration ratio, n, can be calculated by a ratio of the coefficient of volume compressibility of the stabilized soil, m_{vs} and that of the unstabilized soil (original soil), m_{vc} as Equation (6.8). The magnitude of m_{vc} is dependent on the strength of the stabilized soil column, but it is usually assumed as 10 to 20 in many cases. The settlement reduction factor, β is shown along the improvement area ratio for various stress concentration ratio in Figure 6.42.

$$S = \beta \cdot S_c$$

$$\beta = \frac{1}{1 + (n-1) \cdot a_s} \tag{6.6}$$

$$S_c = \frac{\Delta e}{1 + e_0} H_c \tag{6.7a}$$

$$S_c = m_{vc} \cdot \sigma \cdot H_c \tag{6.7b}$$

$$S_c = H_c \cdot Cc \cdot \log \frac{\sigma_0 + \sigma}{\sigma_0} \tag{6.7c}$$

$$n = \frac{\sigma_s}{\sigma_c}$$

$$= \frac{m_{vc}}{m_{vs}} \tag{6.8}$$

where
a_s : improvement area ratio
C_c : compression index of soft soil
e_0 : initial void ratio of soil beneath improved ground
H_c : thickness of ground (m)
m_{vc} : coefficient of volume compressibility of unstabilized soil (m^2/kN)
m_{vs} : coefficient of volume compressibility of stabilized soil (m^2/kN)
n : stress concentration ratio (σ_s/σ_c)
S : consolidation settlement of improved ground (m)
S_c : consolidation settlement of soft ground without improvement (m)
β : settlement reduction factor
Δe : increment of void ratio of soft ground
σ : increment of vertical stress (kN/m^2)
σ_0 : initial vertical stress (kN/m^2)
σ_c : vertical stress acting on soft ground between stabilized soil columns (kN/m^2)
σ_s : vertical stress acting on stabilized soil columns (kN/m^2).

4.3.8   Amount of settlement for floating type improved ground

In the case of the floating type improved ground, where a compressible layer is overlain by the improved ground, the ground settlement is calculated as the sum up of the settlement of the improvement portion and that of the unimproved portion. As the PWRC design procedure doesn't specify any design procedure, the design standard specified by the Building Center of Japan is briefly introduced as reference (The Building Center of Japan, 1997), which was derived from the Recommendation for the Design of Building Foundations (Architectural Institute of Japan, 2000).

In the calculation (The Building Center of Japan, 1997), the load equilibrium of three dimensional improved ground is considered (Figure 6.43), in which the stabilized soil columns and the unstabilized soil between them is assumed to behave as a unit. In the design, an imaginary bottom of improved ground, $H_i - H_f$, is calculated at first. Then, the vertical pressure at the imaginary bottom is calculated by assuming a pressure distribution at the imaginary bottom.

For vertical loads equilibrium

$$P = R_u + R_F \tag{6.9}$$

where
B_i : width of improved ground (m)
c_{ub} : undrained shear strength of soil beneath improved ground (kN/m^2)
H_f : height of periphery of improved ground mobilizing cohesion (m)
L_i : length of improved ground (m)
P : vertical load on the top of superstructure (kN)
R_F : cohesive load along periphery of improved column in L_F portion (kN)

$$R_F = \bar{\tau} \cdot H_f \cdot \psi_b$$

R_u : bearing capacity of soil beneath stabilized soil column (kN/m)

$$R_u = 6 \cdot c_{ub} \cdot B_i \cdot L_i$$

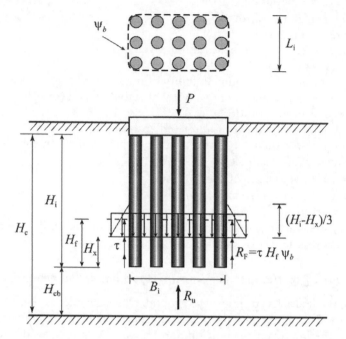

Figure 6.43 Settlement calculation of floating type improvement (The Building Center of Japan, 2000).

$\bar{\tau}$: average cohesion mobilized along H_f (kN/m²)

ψ_b : circumference length of stabilized soil columns (m).

The height of periphery of improved ground, H_f is calculated as Equation (6.10).

$$H_f = \frac{P - R_u}{\tau \cdot \psi_b} \tag{6.10}$$

The imaginary bottom of improved ground, $H_i - H_x$, is calculated as Equation (6.11). In the case of $P < R_u$, H_f should be zero, which indicates the imaginary bottom should be the bottom of the improved ground.

$$H_i - H_x = H_i - \frac{R_F \cdot h_F}{R_u + R_F} \tag{6.11}$$

where

H_i : thickness of improved ground (m)

H_x : distance of imaginary bottom from bottom of improved ground (m)

h_F : point of total R_F force (m).

The vertical stress at the imaginary bottom, p' is calculated by Equation (6.12) with an assumption of the stress distribution. The angle of stress distribution, θ is assumed to be 30°.

$$p' = P \cdot \frac{1}{B_i + 2 \cdot \left(\dfrac{H_i - H_x}{3 \cdot \tan\theta}\right)} \tag{6.12}$$

where

p' : vertical pressure at imaginary bottom of improved ground (kN/m^2)

θ : angle of stress distribution $(°)$.

The amount of settlement in a ground beneath the imaginary bottom, S_{cb}, can be calculated by the Terzaghi's consolidation theory by Equation (6.13).

$$S_{cb} = m_{vc} \cdot (H_{cb} + H_x) \cdot p' \qquad (6.13)$$

where

S_{cb} : settlement in ground beneath imaginary bottom (m)

H_{cb} : thickness of soil beneath improved ground (m)

m_{vc} : coefficient of volume compressibility of soil beneath improved ground (m^2/kN).

The total ground settlement of the floating type improved ground can be calculated by Equation (6.14), which is the sum up of the settlement in the stabilized soil columns' portion and that in a ground beneath the imaginary bottom.

$$S = S_c + S_{cb} \qquad (6.14)$$

4.3.8.1 Rate of settlement

There have been some discussions on the permeability of stabilized soil (Terashi and Tanaka, 1981a, 1981b; Åhnberg, 2003) and whether the stabilized soil column can function as drainage like the vertical drain method or not. The PWRC design standard doesn't specify the design procedure of the rate of consolidation settlement. However, as the accumulated data in Japan have revealed that the permeability of stabilized soil is lower than that of the original soil as shown in Figures 3.43 to 3.45, it is usually assumed in Japan that the stabilized soil column doesn't function as drainage. Therefore the rate of consolidation settlement is usually calculated by a similar manner of the Terzaghi's one dimensional theory with disregarding the stabilized soil columns irrespective of the fixed type and floating type improvements.

4.3.9 Important issues on design procedure

4.3.9.1 Strength of stabilized soil column, improvement area ratio and width of improved ground

In the design, the geometry and strength of stabilized soil columns can be obtained by trial and error manner. In each iteration, the designer should select appropriate design parameters considering the site condition, skill of the contractors, and the capability of locally available deep mixing equipment as discussed in Section 3. The assumption of composite ground adopted in the PWRC design procedure generally overestimates the stability as discussed in sub-section 2.3. To avoid the different failure modes such as tilting and bending failures, it is recommended to determine the improvement geometry based on the successful case histories. The PWRC design recommends an improvement area ratio, a_s, larger than 0.3 or 0.5 for preventing instability under the sloping side of an embankment and/or large horizontal deformation. According to a recent survey, 80% of case histories selected an improvement area ratio larger than 0.5 (Terashi *et al.*, 2009).

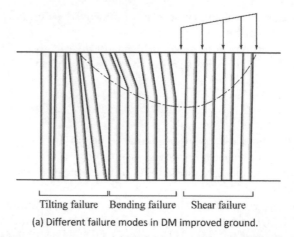

Tilting failure   Bending failure   Shear failure

(a) Different failure modes in DM improved ground.

(b) Centrifuge test results on failure modes in DM improved ground (Kitazume *et al.*, 1999)

Figure 6.44 Different failure modes in DM improved ground.

Recently a group column type improvement with relatively small improvement area ratio has been adopted for the settlement reduction purpose, named 'Alicc method', where the improvement area ratio is about 0.1 to 0.2 (Public Works Research Institute, 2007).

For the improvement of retaining wall foundations and horizontal resistance of bridge abutment foundation piles, the improvement area ratio of 0.6 to 0.8 has often been applied.

4.3.9.2 Limitation of design procedure based on slip circle analysis

The slip circle passes through the columns as far as its strength is relatively low, but passes out of the improved ground when the strength exceeds a certain value. In this case, the slip circle analysis provides the minimum width and depth of improvement but it will not provide the solution for strength and improvement area ratio (Kitazume, 2008).

As section 2.3 has revealed, there exist several failure modes for group columns such as shearing, bending and tilting. Figure 6.44 exemplifies the different failure modes (Kitazume *et al.*, 2000). When the width to depth ratio of the improved zone is small or when the improvement area ratio is smaller than 0.3, more sophisticated

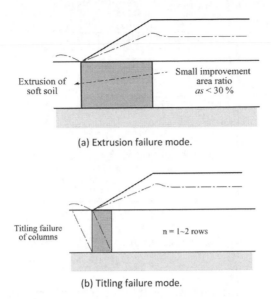

(a) Extrusion failure mode.

(b) Titling failure mode.

Figure 6.45 Failure pattern in case of small improvement width.

analysis should be undertaken (Kitazume, 2008). The extrusion failure of unstabilized soil (Figure 6.45(a)) and tilting of stabilized soil columns (Figure 6.45(b)) and bending failure of stabilized soil columns should be considered.

5   DESIGN PROCEDURE FOR BLOCK TYPE AND WALL TYPE IMPROVED GROUNDS

5.1   Introduction

The deep mixing method was originally developed in the 1970s for in-water works in order to improve foundation of port facility such as quay wall, sea revetment and breakwater. In such applications the improved ground is subjected to not only large vertical loads due to self-weight of the superstructure and surcharge but also a large horizontal wave force in breakwater, earth pressures of backfilled ground in quay wall and sea revetment and seismic inertia forces. Therefore the block, wall or grid type improvements have been applied to port facility. The design standard was specified by Ministry of Transport in 1989 (Ministry of Transport, 1989) and revised by incorporating the accumulated research results and field experiences on the soil properties and the interaction of improved and unimproved ground (Ministry of Transport, 1999). In 2007, the design standard was fully revised based on the reliability design concept. In the revised design procedure, the average and variation of the soil parameters and the external forces are incorporated by the partial safety factors to evaluate the stability of improved ground. The design method of the previous version (Ministry of Transport, 1999) is described in this section and the reliability based design (Ministry of Land, Infrastructure, Transport and Tourism, 2007) will be described in Section 6.

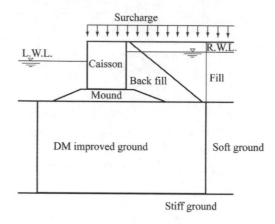

Figure 6.46 Caisson type quay wall on block type improved ground.

The standard and commentaries were published by the Ports and Harbours Association of Japan for the Japanese version (The Ports and Harbours Association of Japan, 1989, 1999 and 2007) and by the Overseas Coastal Area Development Institute of Japan for the English version (The Overseas Coastal Area Development Institute of Japan, 1991, 2002 and 2009).

In this section, the design procedure is described for a caisson type quay wall on a block type and wall type improved ground as shown in Figure 6.46, where the two dimensional condition is assumed. The quay wall is consisted of a caisson, gravel mound, backfill, fill and DM improved ground.

5.2 Basic concept

As port facilities are subjected to large horizontal loads, large magnitude of tensile and bending stresses may develop if the group column type improvement were selected and progressive failure of individual columns is anticipated due to the low bending and tensile strength of stabilized soil columns. Therefore, massive improved ground such as block, wall or grid type improvements have been applied to port and harbor facilities to improve the foundation ground. When the stabilized soil columns are overlapped to make a continuous stabilized soil mass, the boundary surfaces between adjacent columns (a sort of construction joint) may become weak points in the improved ground. Therefore, sufficiently high safety factors were applied to the strength of in-situ stabilized soil; this in turn results in quite a high strength of stabilized soil of the order of $1 \, MN/m^2$. Due to the extraordinary difference between the engineering characteristics of the stabilized soil and unstabilized surrounding soft soil, the stabilized soil block and wall are not considered to be a part of the ground, but rather to be a rigid structural member buried in a ground to transfer external forces to a deeper reliable stratum.

In the seismic design, the seismic coefficient analysis is applied in Japan where the dynamic cyclic loads are converted to quasi-static load by multiplying the unit weight of the structure by the seismic coefficient. The design seismic coefficient, k_h, is obtained by Equation (6.15).

$$k_h = k_{h0} \cdot C_g \cdot C_s \tag{6.15}$$

where
k_h : design seismic coefficient
k_{h0} : regional seismic coefficient
C_g : subsoil condition factor
C_s : importance factor.

The magnitude of the regional seismic coefficient, k_{h0}, is determined as 0.08 to 0.15 corresponding to the possibility of occurrence of earthquakes. The subsoil condition factor, C_g, is determined either as 0.8, 1.0 or 1.2 according to the properties and thickness of subsoil strata. As it is found that the improved ground has better seismic characteristics than the original (unstabilized) ground, the subsoil condition factor of 0.8 for design of the superstructure can be adopted in the case where the improved ground has sufficient extent. The importance factor, C_s, is determined either as 0.8, 1.0, 1.2 or 1.5 for characteristics and importance of structure. The design seismic coefficient values calculated by Equation (6.15) are rounded off to the second decimal place.

5.3   Design procedure

5.3.1   Design flow

As discussed in Section 2.2, the engineering behavior of the improved ground depends upon many factors and different modes of failures exist both in external and internal stability. The routine design is iterative and each mode of failure is examined independently until the most appropriate geometry of improvement and strength of stabilized soil are determined. The design flow for the block type and wall type improved grounds is shown in Figure 6.47 (The Ports and Harbours Association of Japan, 1999; The Overseas Coastal Area Development Institute of Japan, 2002). The design concept is, for the sake of simplicity, derived by analogy with the design procedure for a gravity type structure such as a concrete retaining structure. For the wall type improvement composed of long and short walls as shown in Figure 6.1, the basic design concept can be assumed to be similar to the block type improvement.

The first step of the design procedure is the stability analysis of the superstructure to assure the superstructure and improved ground can behave as a unit against external loadings. The second step is an "external stability analysis" of improved ground in which the sliding failure, the overturning failure and the bearing capacity of improved ground are evaluated. The third step is an "internal stability analysis" of improved ground, in which the induced stresses due to the external forces are calculated and confirmed to be lower than the allowable values. In the wall type improved ground, additionally the extrusion failure is also examined, where unstabilized soil between the long walls might be squeezed out (Figure 6.18). The fourth step is the examination of displacement of the improved ground.

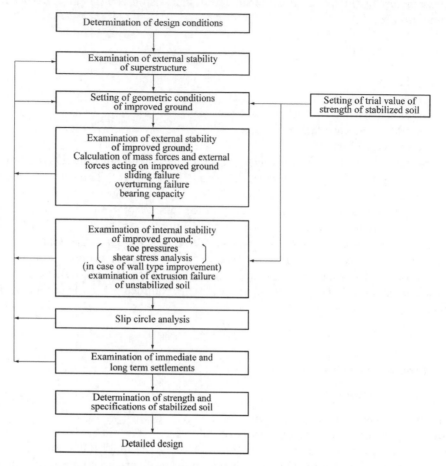

Figure 6.47 Flow of design procedure for block and wall types improvement (The Ports and Harbours Association of Japan, 1999; The Overseas Coastal Area Development Institute of Japan, 2002).

5.3.2 Examination of the external stability of a superstructure

For the external stability analysis of the superstructure at the first step of the design procedure, the improved ground whose size and strength are not determined yet is assumed to be stiff enough to have sufficient bearing capacity to support the super-structure. The sliding and overturning failures of the superstructure are examined at this step in order to determine its size and weight. In the calculation of the sliding fail-ure, it is assumed that the superstructure (caisson) moves horizontally on the mound due to the active earth pressure of the backfill and its seismic inertia force (Figure 6.48). In the overturning failure, it is assumed that the superstructure rotates about its front bottom edge. The safety factor against sliding and overturning failures are calculated by Equations (6.16) and (6.17) respectively. The minimum safety factors against sliding and overturning are specified as 1.3 in many cases respectively.

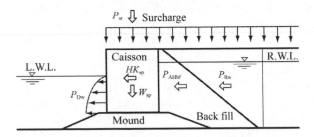

Figure 6.48 Stability calculation of superstructure (The Ports and Harbours Association of Japan, 1999; The Overseas Coastal Area Development Institute of Japan, 2002).

For static condition of sliding failure

$$Fs_s = \frac{(W_{sp} + P_{su}) \cdot \tan \phi'_m}{P_{AHbf} + P_{Rw}} \tag{6.16a}$$

for seismic condition of sliding failure

$$Fs_s = \frac{(W_{sp} + P_{su}) \cdot \tan \phi'_m}{P_{DAHbf} + P_{Rw} + P_{Dw} + HK_{sp}} \tag{6.16b}$$

for static condition of overturning failure

$$Fs_o = \frac{W_{sp} \cdot x_{sp} + P_{su} \cdot x_{su}}{P_{AHbf} \cdot y_{AHbf} + P_{Rw} \cdot y_{Rw}} \tag{6.17a}$$

for seismic condition of overturning failure

$$Fs_o = \frac{W_{sp} \cdot x_{sp} + P_{su} \cdot x_{su}}{P_{DAHbf} \cdot y_{DAHbf} + P_{Rw} \cdot y_{Rw} + P_{Dw} \cdot y_{Dw} + HK_{sp} \cdot y_{sp}} \tag{6.17b}$$

where
Fs_s : safety factor against sliding failure of superstructure
Fs_o : safety factor against overturning failure of superstructure
HK_{sp} : total seismic inertia force per unit length of superstructure (kN/m)
P_{AHbf} : total static active force per unit length of backfill (kN/m)
P_{DAHbf} : total dynamic active force per unit length of backfill (kN/m)
P_{Dw} : total dynamic water force per unit length (kN/m)
P_{su} : total surcharge force per unit length (kN/m)
P_{Rw} : total residual water force per unit length (kN/m)
W_{sp} : weight per unit length of superstructure (kN/m)

x_{sp} : horizontal distance of weight of superstructure from its edge (m)

x_{su} : horizontal distance of total surcharge force from front edge of superstructure (m)

y_{AHbf} : vertical distance of horizontal component of static active force of backfill from bottom of superstructure (m)

y_{DAHbf} : vertical distance of horizontal component of total dynamic active force of backfill from bottom of superstructure (m)

y_{Dw} : vertical distance of total dynamic water force from bottom of superstructure (m)

y_{Rw} : vertical distance of total residual water force from bottom of superstructure (m)

y_{sp} : vertical distance of weight of superstructure from its bottom (m)

ϕ'_m : internal friction angle of the gravel mound.

The P_{Dw} is the total dynamic water force acting on the caisson in the case of an earthquake, whose magnitude can be calculated by Equation (6.18) according to the Westergaard equation.

$$
\begin{aligned}
P_{Dw} &= \int \frac{7}{8} \cdot \gamma_w \cdot k_h \cdot \sqrt{H_w \cdot h} \, dh \\
&= \frac{7}{12} \cdot \gamma_w \cdot k_h \cdot H_w^2
\end{aligned}
\tag{6.18}
$$

where

h : depth from water surface (m)

H_w : water depth (m)

k_h : seismic coefficient

γ_w : unit weight of water (kN/m^3).

5.3.3 Trial values for the strength of stabilized soil and geometric conditions of improved ground

The field strength of stabilized soil, improvement type, and width and thickness, are assumed. The initial trial value for the width of improved ground is usually assumed as the sum of the widths of the gravel mound and backfill as the minimum. The thickness of the improved ground is usually assumed as the thickness of the soft ground because the fixed type improved ground is desirable from the view point of stability and displacement. When laboratory mix test results are available, an appropriate field strength is assumed considering the economy and the construction aspects. If laboratory mix test data is not available, 2,000 to 3,000 kN/m^2 in terms of unconfined compressive strength is ordinarily adopted as the field strength in the case of in-water works.

5.3.4 Examination of the external stability of improved ground

In the "external stability analyses," three failure modes are examined for the assumed improved ground: sliding, overturning and bearing capacity failures. The design loads adopted in the external stability analysis are schematically shown in Figure 6.49 (The Ports and Harbours Association of Japan, 1999; The Overseas Coastal Area

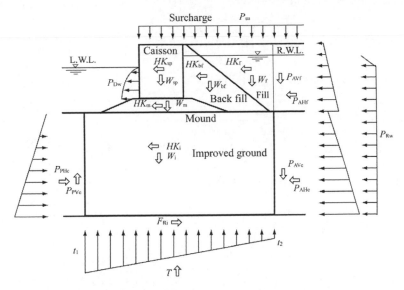

Figure 6.49 Schematic diagram of design loads (The Ports and Harbours Association of Japan, 1999; The Overseas Coastal Area Development Institute of Japan, 2002).

Development Institute of Japan, 2002). They include the active and passive earth pressures, surcharge and external forces acting on the boundary of the improved ground, the mass forces generated by gravity, and the seismic inertia forces.

In the stability analysis of the wall type improved ground, it is sometimes necessary to assume the magnitudes of external forces acting on unstabilized soil and stabilized soil independently. In general, it can be assumed that the active and passive earth pressures act uniformly on the long wall, short wall and unstabilized soil between the long walls. For vertical loads, it is assumed that the self-weight of the superstructure, and the surcharge and external forces acting on the superstructure and the weight of stabilized soil are concentrated on the long wall.

5.3.4.1 Sliding and overturning failures

In the calculation of sliding failure, it is assumed that the improved ground and the superstructure move horizontally at the bottom boundary of improved soil due to the unbalance of the earth pressures and/or the seismic inertia forces. In the overturning failure, it is assumed that the improved ground and the superstructure rotate about the front bottom edge of the improved ground. The sliding and overturning stabilities are calculated by the equilibrium of the horizontal and the moment forces, and the safety factors against these failures are calculated by Equations (6.19) and (6.20) respectively. The minimum safety factors are usually 1.3 and 1.0 for the static and dynamic conditions respectively.

For static condition of sliding failure

$$Fs_s = \frac{P_{PHc} + F_{Ri}}{P_{AHc} + P_{Rw}} \tag{6.19a}$$

for seismic condition of sliding failure

$$Fs_s = \frac{P_{DPHc} + F_{Ri}}{P_{DAHc} + P_{Rw} + P_{Dw} + HK_{sp} + HK_m + HK_{bf} + HK_f + HK_i} \qquad (6.19b)$$

for static condition of overturning failure

$$Fs_o = \frac{P_{PHc} \cdot y_{PHc} + P_{AVc} \cdot x_{AVc} + P_{su} \cdot x_{su} + \Sigma W \cdot x}{P_{AHc} \cdot y_{AHc} + P_{Rw} \cdot y_{Rw}} \qquad (6.20a)$$

$$\Sigma W \cdot x = W_{sp} \cdot x_{sp} + W_m \cdot x_m + W_{bf} \cdot x_{bf} + W_f \cdot x_f + W_i \cdot x_i$$

for seismic condition of overturning failure

$$Fs_o = \frac{P_{DPHc} \cdot y_{DPHc} + P_{DAVc} \cdot x_{DAVc} + P_{su} \cdot x_{su} + \Sigma W \cdot x}{P_{DAHc} \cdot y_{DAHc} + P_{Rw} \cdot y_{Rw} + P_{Dw} \cdot y_{Dw} + \Sigma HK \cdot y} \qquad (6.20b)$$

$$\Sigma W \cdot x = W_{sp} \cdot x_{sp} + W_m \cdot x_m + W_{bf} \cdot x_{bf} + W_f \cdot x_f + W_i \cdot x_i$$

$$\Sigma HK \cdot y = HK_{sp} \cdot y_{sp} + HK_m \cdot y_m + HK_{bf} \cdot y_{bf} + HK_f \cdot y_f + HK_i \cdot y_i$$

where
- B_i : width of improved ground (m)
- c_{uc} : undrained shear strength of soft soil (kN/m^2)
- F_{Ri} : shear force per unit length mobilized on bottom of improved ground (kN/m)

  for block type improvement resting on sandy layer (fixed type)

  $$F_{Ri} = F_{Rs}$$

  for wall type improvement resting on sandy layer (fixed type)

  $$F_{Ri} = F_{Rs} + F_{Ru}$$

  for block and wall type improvements resting on clay (floating type)

  $$F_{Ri} = c_{uc} \cdot B_i$$

- F_{Rs} : total shear force per unit length mobilized by sand layer at the bottom of improved ground (kN/m)

  $$F_{Rs} = (W_{sp} + W_m + W_{bf} + W_f + W_s + P_{su} + P_{AVc} - P_{PVc}) \cdot \tan \phi_b'$$

- F_{Ru} : total shear force per unit length mobilized by unstabilized soil between long walls at the bottom of improved ground (kN/m)
  in the case of a sand layer beneath improved ground,

  $$= \min \begin{cases} W_u \cdot \tan \phi_b' \cdot \dfrac{L_s}{L_s + L_\ell} \\[2ex] c_{uc} \cdot B_i \cdot \dfrac{L_s}{L_s + L_\ell} \end{cases}$$

Fs_0 : safety factor against overturning failure of improved ground
Fs_s : safety factor against sliding failure of improved ground
HK_{bf} : total seismic inertia force per unit length of backfill (kN/m)
HK_f : total seismic inertia force per unit length of fill (kN/m)
HK_i : total seismic inertia force per unit length of improved ground (kN/m)
HK_m : total seismic inertia force per unit length of mound (kN/m)
HK_{sp} : total seismic inertia force per unit length of superstructure (kN/m)
L_l : thickness of long wall of improved ground (m) as shown later in
             Figure 6.56
L_s : thickness of short wall of improved ground (m) as shown later in
             Figure 6.56
P_{AHc} : horizontal component of total static active force per unit length of soft
             ground (kN/m)
P_{AVc} : vertical component of total static active force per unit length of soft
             ground (kN/m)
P_{DAHc} : horizontal component of total dynamic active force per unit length of
             soft ground (kN/m)
P_{DAVc} : vertical component of total dynamic active force per unit length of soft
             ground (kN/m)
P_{DPHc} : horizontal component of total dynamic passive force per unit length of
             soft ground (kN/m)
P_{Dw} : total dynamic water force per unit length (kN/m)
P_{PIIc} : horizontal component of total static passive force per unit length of
             soft ground (kN/m)
P_{PVc} : vertical component of total static passive force per unit length of
             soft ground (kN/m)
P_{Rw} : total residual water force per unit length (kN/m)
P_{su} : total surcharge force per unit length (kN/m)
W_{bf} : weight per unit length of backfill (kN/m)
W_f : weight per unit length of fill (kN/m)
W_i : weight per unit length of improved ground (kN/m)
W_m : weight per unit length of mound (kN/m)
W_s : weight per unit length of stabilized soil (kN/m)
W_{sp} : weight per unit length of superstructure (kN/m)
W_u : weight per unit length of unstabilized soil (in case of wall type
             improvement) (kN/m)
x_{AVc} : horizontal distance of vertical component of total static active force
             from front edge of improved ground (m)
x_{bf} : horizontal distance of weight of backfill from front edge of
             improved ground (m)
x_{DAVc} : horizontal distance of vertical component of total dynamic active force
             from front edge of improved ground (m)
x_f : horizontal distance of weight of fill from front edge of improved
             ground (m)
x_i : horizontal distance of weight of improved ground from its front
             edge (m)
x_m : horizontal distance of weight of mound from front edge of improved
             ground (m)

x_{sp} : horizontal distance of weight of superstructure from front edge of improved ground (m)

x_{su} : horizontal distance of total surcharge force from front edge of improved ground (m)

y_{AHc} : vertical distance of horizontal component of total static active force from bottom of improved ground (m)

y_{bf} : vertical distance of total seismic inertia force of backfill from bottom of improved ground (m)

y_{DAHc} : vertical distance of horizontal component of total dynamic active force from bottom of improved ground (m)

y_{DPHc} : vertical distance of horizontal component of total dynamic passive force from bottom of improved ground (m)

y_{Dw} : vertical distance of total dynamic water force from bottom of improved ground (m)

y_f : vertical distance of total seismic inertia force of fill from bottom of improved ground (m)

y_i : vertical distance of total seismic inertia force of improved ground from bottom of improved ground (m)

y_m : vertical distance of total seismic inertia force of mound from bottom of improved ground (m)

y_{PHc} : vertical distance of horizontal component of total static passive force from bottom of improved ground (m)

y_{Rw} : vertical distance of total residual water force from bottom of improved ground (m)

y_{sp} : vertical distance of total seismic inertia force of superstructure from bottom of improved ground (m)

ϕ'_b : internal friction angle of soil beneath improved ground.

5.3.4.2 Bearing capacity

The bearing capacity of improved ground is evaluated by the classical Terzaghi's bearing capacity theory which can incorporate the effects of loading condition and embedment condition. In the design, the subgrade reactions at the front edge and the rear edge of the bottom of the improved ground are examined so as to satisfy the allowable bearing capacity through Equations (6.21) to (6.23).

$$t_1 \leq q_f$$
$$t_2 \leq q_f \tag{6.21}$$

For static condition

$$e = \frac{B_i}{2} - \frac{(P_{PHc} \cdot y_{PHc} + P_{AVc} \cdot x_{AVe} + P_{su} \cdot x_{su} + \Sigma W \cdot x) - (P_{AHc} \cdot y_{AHc} + P_{Rw} \cdot y_{Rw})}{W_{sp} + W_m + W_{bf} + W_f + W_i + P_{AVc} - P_{PVc}}$$

$$\Sigma W \cdot x = W_{sp} \cdot x_{sp} + W_m \cdot x_m + W_{bf} \cdot x_{bf} + W_f \cdot x_f + W_i \cdot x_i \tag{6.22a}$$

for seismic condition

$$(P_{DPHc} \cdot y_{DPHc} + P_{DAVc} \cdot x_{DAVe} + P_{su} \cdot x_{su} + \Sigma W \cdot x)$$

$$e = \frac{B_i}{2} - \frac{- (P_{DAHc} \cdot y_{DAHc} + P_{Rw} \cdot y_{Rw} + P_{Dw} \cdot y_{Dw} + \Sigma HK \cdot y)}{W_{sp} + W_m + W_{bf} + W_f + W_i + P_{DAVc} - P_{DPVc}} \qquad (6.22b)$$

$$\Sigma W \cdot x = W_{sp} \cdot x_{sp} + W_m \cdot x_m + W_{bf} \cdot x_{bf} + W_f \cdot x_f + W_i \cdot x_i$$

$$\Sigma HK \cdot x = HK_{sp} \cdot y_{sp} + HK_m \cdot y_m + HK_{bf} \cdot x_{bf} + HK_f \cdot y_f + HK_i \cdot y_i$$

In the case of $e <= B_i/6$

$$t_1 = \frac{W_{sp} + W_m + W_{bf} + W_f + W_i + P_{su} + P_{AVc} - P_{PVc}}{B_i} \cdot \left(1 + \frac{6 \cdot e}{B_i}\right) \cdot \frac{L_s + L_l}{L_l} \Bigg\}$$

$$t_2 = \frac{W_{sp} + W_m + W_{bf} + W_f + W_i + P_{su} + P_{AVc} - P_{PVc}}{B_i} \cdot \left(1 - \frac{6 \cdot e}{B_i}\right) \cdot \frac{L_s + L_l}{L_l}$$

$$(6.22c)$$

In the case of $e >= B_i/6$

$$t_1 = \frac{2 \cdot (W_{sp} + W_m + W_{bf} + W_f + W_i + P_{su} + P_{AVc} - P_{PVc})}{3 \cdot B_i} \cdot \frac{L_s + L_l}{l_l} \qquad (6.22d)$$

where

e : eccentricity (m)
L_l : thickness of long wall of improved ground (m) as shown later in Figure 6.52
L_s : thickness of short wall of improved ground (m) as shown later in Figure 6.52
t_1 : subgrade reaction at front edge of improved ground (kN/m^2)
t_2 : subgrade reaction at rear edge of improved ground (kN/m^2).

$$q_f = \frac{1}{Fs}\left(\frac{1}{2}\gamma \cdot B_i \cdot N_\gamma + c_{ub} \cdot N_c + q \cdot (N_q - 1)\right) + q \qquad (6.23)$$

where

c_{ub} : undrained shear strength of soil beneath improved ground (kN/m^2)
Fs : safety factor
N_c : bearing capacity factor of soil beneath improved ground
N_q : bearing capacity factor of soil beneath improved ground
N_γ : bearing capacity factor of soil beneath improved ground
q : effective overburden pressure at bottom of improved ground (kN/m^2)
q_f : bearing capacity of soil beneath improved ground (kN/m^2)
γ : unit weight of soil beneath improved ground (kN/m^3).

In the wall type improved ground, the bearing capacity at the bottom of the long wall is the bearing capacity problem of a deep rectangular foundation interfered by the adjacent foundations. The B_i is the width of improved ground, L_l is the thickness of the long wall, and S_i is the center to center spacing of long walls as shown in Figure 6.50

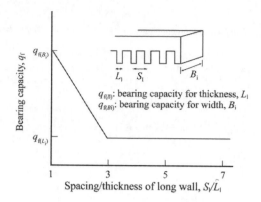

Figure 6.50 Simple design of bearing capacity of wall type improvement (Terashi and Kitazume, 1987).

(Terashi and Kitazume, 1987). When S_i/L_1 is unity, the bearing capacity of a strip foundation of width B_i applies. When S_i/L_1 is large, the bearing capacity of a strip foundation of width L_1 applies. The increase of the bearing capacity of stabilized long walls caused by the interference of adjacent long walls has been demonstrated in a series of centrifuge model tests and the simple design shown in Figure 6.50 and Equation (6.24) have been proposed by Terashi and Kitazume (1987).

$$q_f = q_{f(L1)} + \frac{1}{2} \cdot (q_{f(Bi)} - q_{f(L1)}) \cdot (3 - S_1/L_1) \qquad (6.24)$$

$$q_{f(L1)} = \frac{1}{Fs} \cdot \left(\frac{1}{2} \cdot \gamma \cdot L_1 \cdot N_\gamma + c_{ub} \cdot N_c \right) + q \cdot N_q$$

$$q_{f(Bi)} = \frac{1}{Fs} \cdot \left(\frac{1}{2} \cdot \gamma \cdot B_i \cdot N_\gamma + c_{ub} \cdot N_c \right) + q \cdot N_q$$

where
  L_1 : thickness of long wall of improved ground (m)
  $q_{f(L1)}$: bearing capacity of strip foundation with thickness of long wall, L_1 (kN/m²)
  $q_{f(Bi)}$: bearing capacity of strip foundation with width of improved ground, B_i (kN/m²)
  S_1 : center to center spacing of long walls of improved ground (m).

5.3.5 Examination of the internal stability of improved ground

In the "internal stability analysis," the induced stresses in the improved ground are calculated based on the elastic theory. The shape and size of the improved ground are determined so that the induced stresses are lower than the allowable strengths of the stabilized soil. In the calculation, the stabilized soil is generally assumed to have a uniform property for the sake of simplicity even it contains possibly weaker zones due to construction process such as overlap joints. The effect of the strength at the

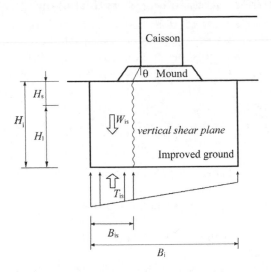

Figure 6.51 Internal stability of improved ground.

overlapping portion is taken into account when determining the allowable strengths of stabilized soil, which will be described later.

Instead of running numerical analysis such as the finite element method, simple calculation methods have been generally applied as routine design. The loading conditions applied to the internal stability analysis are generally assumed to be same as those for the external stability analysis, as already shown in Figure 6.49. At this stage of calculation, however, the external stability is already satisfied with a certain safety margin and hence horizontal resisting forces exceeds the driving forces. Earth pressure at the passive side may be chosen appropriately between the passive and at rest pressures. According to the accumulated experiences in design, the internal stability evaluation at the two critical parts as shown in Figure 6.51 is considered sufficient as long as the shape of stabilized soil is within the experiences: (a) subgrade reactions at the front edge and rear edge of the improved ground, and (b) average shear stress along a vertical shear plane at the front edge of superstructure.

5.3.5.1 Subgrade reaction at the front edge of improved ground

The subgrade reactions at the front edge and rear edge of the improved ground should satisfy the criteria as shown in Equation (6.25). The subgrade reactions are calculated by Equation (6.22).

$$t_1 - p_{PHc} \le \sigma_{ca}$$
$$t_2 - p_{AHc} \le \sigma_{ca}$$

(6.25)

where
σ_{ca}: allowable compressive strength of stabilized soil (kN/m^2).

5.3.5.2 Average shear stress along a vertical plane

The average shear stress induced along the vertical shear plane at the front face of the superstructure should satisfy the criteria as shown in Equations (6.26). In the case where a mound underlies the superstructure, the stress distribution at an angle of around 30° can be taken into account to find the vertical shear plane (Figure 6.51).

$$\tau \leq \tau_{ca} \tag{6.26a}$$

$$\tau = \frac{L_l + L_s}{L_l \cdot H_i + L_s \cdot H_s} \left(\frac{L_l}{L_l + L_s} \int_0^{B_{is}} t_{is} \, db - W_{is} \right) \tag{6.26b}$$

where

B_{is} : distance of vertical shear plane from toe of improved ground (m)
H_i : height of improved ground (m)
H_s : height of short wall of improved ground (m)
L_l : thickness of long wall of improved ground (m) as shown later in Figure 6.52
L_s : thickness of short wall of improved ground (m) ($L_s = 0$ for block type improved ground)
t_{is} : subgrade reaction at bottom of improved ground (kN/m²)
W_{is} : weight per unit length of improved ground at part of B_{is} (kN/m)
τ : average shear stress along vertical shear plane (kN/m²)
τ_{ca} : allowable shear strength of stabilized soil (kN/m²).

In the case of the wall type improved ground, the external load acting on the short wall should be transferred to the long wall. The shear stress along the lap joint between long walls and short wall is also examined (see Figure 6.52). The induced shear stress along the vertical lap joint should satisfy the criteria by Equation (6.27). Induced stress, τ in the equation should be examined appropriately by considering the load distribution as shown in Figure 6.52, both for the static and seismic conditions.

$$\tau \leq \tau_{ca} \tag{6.27a}$$

$$\tau = \frac{L_s}{2 \cdot H_s \cdot B_s} \left(\int_0^{B_s} t_{ls} \, db + W_{ss} \right) \tag{6.27b}$$

where
B_s : width of short wall of improved ground (m)
H_s : height of short wall of improved ground (m)

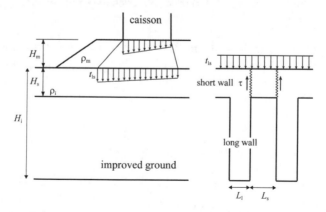

Figure 6.52 Vertical shear of short wall in wall type improved ground.

t_{ls} : induced stress at top of short wall of improved ground (kN/m^2).
W_{ss} : weight per unit length of short wall of improved ground (kN/m)

5.3.5.3 Allowable strengths of stabilized soil

The allowable strengths of stabilized soil are defined by Equations (6.28) to (6.30).

$$\sigma_{ca} = \frac{1}{Fs}\alpha \cdot \beta \cdot \gamma \cdot \overline{q_{uf}}$$

$$= \frac{1}{Fs}\alpha \cdot \beta \cdot \gamma \cdot \lambda \cdot \overline{q_{ul}} \tag{6.28}$$

$$\tau_{ca} = \frac{1}{2}\sigma_{ca} \tag{6.29}$$

$$\sigma_{ta} = 0.15 \cdot \sigma_{ca} \le 200 \text{ kN/m}^2 \tag{6.30}$$

where
Fs : safety factor
$\overline{q_{uf}}$: average unconfined compressive strength of in-situ stabilized soil (kN/m^2)
$\overline{q_{ul}}$: average unconfined compressive strength of stabilized soil manufactured in laboratory (kN/m^2)
α : coefficient of effective width of stabilized soil column
β : reliability coefficient of overlapping
γ : correction factor for strength variability
λ : ratio of $\overline{q_{uf}}/\overline{q_{ul}}$ (usually 0.5 to 1 according to past experience)
σ_{ca} : allowable compressive strength of stabilized soil (kN/m^2)
σ_{ta} : allowable tensile strength of stabilized soil (kN/m^2)
τ_{ca} : allowable shear strength of stabilized soil (kN/m^2).

- *safety factor, Fs*: As all the allowable strengths are based on the unconfined compressive strength, q_u, in which no effect of creep and cyclic loading are incorporated. In the practical design procedure, the safety factors of 3.0 and 2.0 for the static and dynamic conditions respectively are usually adopted to incorporate these effects, and also to incorporate the importance of the structure, the type of loads, the design method, and the reliability of the materials.
- *coefficient of effective width of stabilized soil column, α*: The block type and wall type of improved grounds manufactured by overlapping execution are in general composed of stabilized soil columns and unstabilized soil between the columns, as shown in Figure 6.53 for the case of a two mixing shafts machine. The coefficient of the effective width of the stabilized soil column, α is calculated by Equation (6.12) to compensate for the unstabilized part. As the tolerance of overlapping is adopted around 200 mm in many cases, the α value is usually 0.8 to 0.9 (Cement Deep Mixing Method Association, 1999).

$$\alpha = \min\left(\frac{lx}{Dx}, \frac{ly}{Dy}\right) \tag{6.31}$$

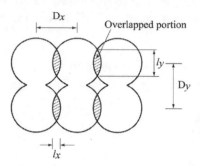

Figure 6.53 Effective width formed by improvement machine.

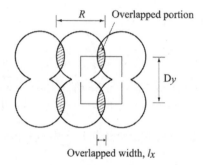

Figure 6.54 Connecting surfaces of stabilized soil columns.

- *reliability coefficient of overlapping, β*: In the overlapping execution, a stabilized soil column during hardening is partially scraped by the following column, as shown in Figure 6.54 for the case of a two mixing shafts machine. The strength in the overlapped portion is anticipated to be lower than that of other parts of the column. The reliability coefficient of overlapping is the ratio of the strength of overlapped and non-overlapped portions. Its magnitude is influenced by various factors, such as the time interval until overlapping, the execution capacity of the DM machine, and the type of binder. The β value of 0.8 to 0.9 had been adopted in early days. However with increasing successful applications, the influence of overlap explained in b) and c) are currently considered together by employing $\alpha \cdot \beta = 0.8$ to 0.9.
- *correction factor for strength variability, γ*: It is generally known that the unconfined compressive strength of in-situ stabilized soil exhibits variability to some extent (Section 6 in Chapter 3). The correction factor for strength variability is a coefficient used to account for the variability. The γ value of 0.5 to 0.6 is usually adopted.
- *ratio of field strength and laboratory strength, $\overline{q_{uf}/q_{ul}}$, λ*: The accumulated data clearly shows that the average unconfined compressive strength of in-situ stabilized

Figure 6.55 Deformation of clay ground between long walls in extrusion failure (Terashi *et al.*, 1983).

soil, $\overline{q_{uf}}$ is lower than laboratory stabilized soil, $\overline{q_{ul}}$ in on-land constructions but almost equivalent in in-water works as already shown in Section 6 in Chapter 3. The value of λ can be taken as 1.0 in the design of in-water applications.

Since the coefficients, α, β, γ and λ are not independent but are actually closely related, it is difficult to determine their magnitudes individually. According to the successful previous projects, the adopted ratio of the allowable compressive strength to the unconfined compressive strength in laboratory stabilized soil has been between 1/6 and 1/10.

The in-situ strength of a stabilized soil column for in-water works is preferably determined to be 2,000 to 3,000 kN/m², rough estimations of σ_{ca}, τ_{ca} and σ_{ta} are 260 to 670, 160 to 225 and 50 to 75 kN/m² respectively.

5.3.5.4 Extrusion failure

For the wall type improvement (Figure 6.1), the extrusion failure must also be examined. The extrusion failure is a failure mode considered for the unstabilized soil remaining between the long walls which is subjected to the unbalanced active and passive earth pressures, as shown in Figure 6.55 (Terashi *et al.*, 1983). In the design procedure, the soft soil between the long walls is assumed to move as a rigid body in the shape of a rectangular prism, where the width and length of the prism are taken as the width of improved ground and the length of short wall respectively (Figure 6.56). The minimum safety factor is calculated by Equation (6.32) by changing the height of the prism, H_{PR} and it should be higher than the allowable value. The minimum safety factor is usually specified as 1.2 and 1.0 for the static and seismic conditions respectively.

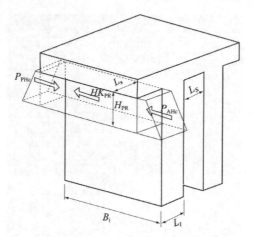

Figure 6.56 Conceptual diagram of extrusion of unimproved soil.

For the static condition

$$Fs_e = \frac{2(L_s + H_{pr}) \cdot B_i \cdot c_{uc} + P_{PHc} \cdot L_s}{(P_{AHc} + P_{Rw}) \cdot L_s} \qquad (6.32a)$$

for the seismic condition

$$Fs_e = \frac{2(L_s + H_{pr}) \cdot B_i \cdot c_{uc} + P_{DPHc} \cdot L_s}{(P_{DAHc} + P_{Rw} + HK_{pr}) \cdot L_s} \qquad (6.32b)$$

where
B_i : width of improved ground (m)
c_{uc} : undrained shear strength of soft soil at assumed prism (kN/m^2)
Fs_e : safety factor against extrusion failure
H_{pr} : height of assumed prism (m)
HK_{pr} : total seismic inertia force per unit length of soil prism (kN)
L_s : thickness of short wall of improved ground (m)
P_{AHc} : horizontal component of total static active force per unit length acting
        on the prism (kN/m)
P_{DAHc}: horizontal component of total dynamic active force per unit length acting
        on the prism (kN/m)
P_{DPHc}: horizontal component of total dynamic passive force per unit length acting
        on the prism (kN/m)
P_{PHc} : horizontal component of total static passive force per unit length acting
        on the prism (kN/m)
P_{Rw} : total residual water force per unit length acting on the prism (kN/m).

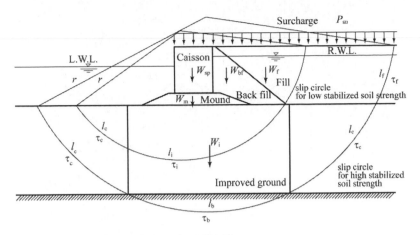

Figure 6.57 Slip circle analysis.

5.3.6 Slip circle analysis

Slip circle analysis is carried out to evaluate the overall stability of the improved ground, the superstructure and the surrounding soil by Equation (6.33). As the strength of stabilized soil is a very high value, a slip circle analysis passing through the improved ground is not necessary in many cases, as shown in Figure 6.57. In the case where sufficient bearing capacity is assured, slip circle analysis is not necessary in many cases. The slip circle analysis in seismic condition is not specified in the design standard. Safety factors obtained on slip circles that pass outside stabilized soil mass are useful for evaluting the validity of the improvement geometry with respect to the external stability.

$$Fs_{sp} = \frac{r \cdot (\tau_c \cdot l_c + \tau_f \cdot l_f + \tau_i \cdot l_i)}{\begin{array}{c} W_{sp} \cdot x_{sp} + W_m \cdot x_m + W_{bf} \cdot x_{bf} + W_f \cdot x_f + W_c \cdot x_c + W_i \cdot x_i \\ + P_{su} \cdot x_{su} + P_{Rw} \cdot y_{Rw} \end{array}} \quad (6.33)$$

where

Fs_{sp} : safety factor against slip circle failure
l_c : length of circular arc in soft ground (m)
l_f : length of circular arc in fill (m)
l_i : length of circular arc in improved ground (m)
r : radius of slip circle (m)
x_{bf} : horizontal distance of weight of backfill from center of slip circle (m)
x_c : horizontal distance of weight of soft ground from center of slip circle (m)
x_f : horizontal distance of weight of fill from center of slip circle (m)
x_i : horizontal distance of weight of improved ground from center of slip circle (m)
x_m : horizontal distance of weight of mound from center of slip circle (m)
x_{sp} : horizontal distance of weight of superstructure from center of slip circle (m)
x_{su} : horizontal distance of total surcharge force from center of slip circle (m)

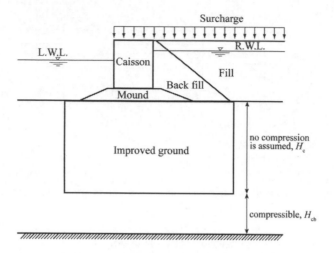

Figure 6.58 Ground settlement of floating type improved ground.

y_{Rw} : vertical distance of total residual water force from center of slip circle (m)
τ_c : shear strength of soft ground (kN/m^2)
τ_i : average shear strength of improved ground (kN/m^2)
τ_f : shear strength of fill (kN/m^2).

5.3.7 Examination of immediate and long term settlements

After the optimum cross section of the improved ground is determined by the above procedures, the immediate and the long term settlements of the improved ground should be examined. Usually, the deformation of the stabilized soil itself can be negligible because of its high rigidity and large consolidation yield pressure. Therefore, the settlement of the improved ground is calculated as the deformation of the soft ground beneath the improved ground. In the case of the fixed type improvement where the stabilized soil columns reach the stiff layer (Figure 6.36(a)), the settlement can be assumed to be negligible. In the case of the floating type improvement (Figure 6.36(b)), the consolidation settlement beneath the improved ground is calculated by the Terzaghi's one dimensional consolidation theory, as Equations (6.34) and (6.35) (see Figure 6.58).

$$p = \frac{W_{sp} + W_m + W_{bf} + W_f + W_i + P_{su}}{B_i} \tag{6.34}$$

$$S = \frac{\Delta e}{1 + e_0} H_{cb} \tag{6.35a}$$

$$S = m_{vc} \cdot (p - p_0) \cdot H_{cb} \tag{6.35b}$$

$$S = H_{cb} \cdot C_c \cdot \log \frac{p}{p_0} \tag{6.35c}$$

where

B_i : width of improved ground (m)

C_c : compression index of soft soil

H_{cb} : thickness of soil beneath improved ground (m)

m_{vc} : coefficient of volume compressibility of soil beneath improved ground (m^2/kN)

p : subgrade reaction at bottom of improved ground (kN/m^2)

p_0 : initial vertical stress at bottom of improved ground (kN/m^2), before improvement

S : settlement (m)

e_0 : initial void ratio of soil beneath improved ground

Δe : increment of void ratio of soil beneath improved ground.

5.3.8   Determination of strength and specifications of stabilized soil

Design engineer is responsible for writing specifications on strength and geometric layout of stabilized soil columns including end bearing condition and minimum required overlapping width. The design of deep mixing involves examination of several failure modes both in external and internal stability. Only the designer knows which mode is the most critical one. These information should better be reflected in the acceptance criteria.

5.4   Sample calculation

An example of calculations for the most common deep mixing application is shown in Figure 6.59 (Terashi *et al.*, 1985). In this example, the superstructure is a revetment composed of a gravel mound and a concrete caisson supporting the earth pressure induced by backfill. The superstructure is to be constructed on a soft clay layer underlain by a reliable bearing stratum of dense sand as shown in the upper left corner of the figure.

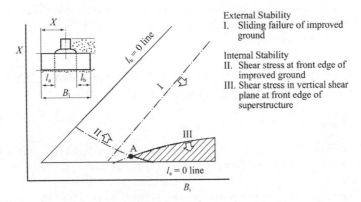

Figure 6.59 Determination of optimum design (Terashi *et al.*, 1985).

The initial approximation of the width of the improved ground in this trial calculation is shown by two vertical dotted lines at both ends of the gravel mound in the figure. To assure the required safety factor, the width of the improved ground, B_i is increased by increasing distances l_a and/or l_b. The three curves in the figure denote: the minimum extent of the improved ground that satisfies the requirement for the sliding failure of improved ground (Curve I), the induced shear stress at the front edge of the improved ground (II), and the shear stress in the vertical shear plane at front edge of the superstructure (III). The arrow added to each curve shows the direction toward a higher safety factor against each mode of failure. The hatched zone in the figure satisfies all the requirements and the dimension at point "A" is the optimum one. In this particular example, overturning, bearing capacity, and extrusion failures are not the governing factors. As is shown, usually a couple of failure modes become critical factors in determining the shape and extent of the improved ground.

5.5 Important issues on design procedure

The design loads acting on the improved ground for the bearing capacity analysis were originally considered to be the ultimate active and passive earth pressures, the same as in the "external stability analysis." However, in a case where the improved ground is sufficiently stable with the safety factor against sliding and overturning failures, it is easily understood that the earth pressure on the passive side of the improved ground and the shear strength on the bottom of the improved ground are not fully mobilized to the ultimate value. Many research efforts have revealed this phenomenon experimentally and analytically (e.g. Terashi *et al.*, 1989, Kitazume, 1994). Terashi *et al.* (1989) proposed the design loading conditions, based on their centrifuge tests, in which the earth pressures acting on the improved ground should be close to the pressures at rest in the internal stability analysis as long as the safety factor against external stability is relatively large. The design loading conditions should therefore be carefully determined by considering the margin of the safety factor against the external stabilities. According to the investigation, the design load on the passive side for the bearing capacity analysis should be determined by considering the force equilibrium of loads acting on the active side and modified shear force on the bottom.

Since the magnitude and distribution of the earth pressures up to failure are still not well determined, detailed analysis such as FEM analysis should be conducted to achieve a more reliable and precise design.

6 DESIGN PROCEDURE FOR BLOCK TYPE AND WALL TYPE IMPROVED GROUNDS, RELIABILITY DESIGN

6.1 Introduction

In 2007, the design standard of deep mixing improved ground for port facilities was fully revised in which the reliability design concept was adopted. In the revised design method, the average and variation of soil parameters and external forces are incorporated by partial factors in the performance verifications. Here the design standard specified by the Ministry of Land, Infrastructure, Transport and Tourism (Ministry of

Land, Infrastructure, Transport and Tourism, 2007; The Ports and Harbours Association of Japan, 2007; The Overseas Coastal Area Development Institute of Japan, 2009) is briefly introduced, where the caisson type quay wall on the block type and wall type improved ground is shown as an example (see Figure 6.46). The background of the standard and details on the partial safety factors are presented by Kitazume and Nagao (2007).

6.2 Basic concept

The basic concept, the design procedure and the assumed failure patterns of the revised design method are the same as the previous one introduced in Section 5. In the design method for port facilities, the stabilized soil of block or wall is not considered to be part of a ground, but rather to be a rigid structural member buried in a ground to transfer external forces to a reliable stratum. The average and variation of soil parameters and external forces are incorporated by partial factors in the performance verifications.

The Hyogoken-Nambu earthquake caused serious damages to many kinds of infrastructures and required to revise the seismic designs. The Japan Society of Civil Engineers proposed a new design concept for civil engineering infrastructures, in which the seismic design of infrastructures should be evaluated under the Level 1 and Level 2 earthquake ground motions. The design assumes the Level 1 earthquake has a similar magnitude to those targeted in the previous design, which is estimated to take place once or twice in the life span of the infrastructure. The Level 2 earthquake, on the other hand, is categorized into a huge earthquake like the Hyogoken-Nambu earthquake. Its magnitude should be estimated by identifying the fault line and mechanism of anticipated earthquakes. Any infrastructures should be assured the seismic stability in the Level 1 earthquake ground motion. For the level 2 earthquake ground motion, any infrastructures should be assured the sustainability incorporating their importance.

The performance verification of variable states in respect of the Level 1 earthquake ground motion can be conducted, equivalent to gravity type quay walls, by either a simplified method (seismic coefficient method), or by a detailed method (nonlinear seismic response analysis considering dynamic interaction of the ground and structures). Examination of accidental states in respect of the Level 2 earthquake ground motion may also be necessary depending on the performance requirements of facilities.

6.3 Design procedure

6.3.1 Design flow

The design procedure for the block type and wall type improved grounds of port facilities is shown in Figure 6.60 (The Ports and Harbours Association of Japan, 2007; The Overseas Coastal Area Development Institute of Japan, 2009). The design concept is, for the sake of simplicity, derived by analogy with the design procedure for a gravity type structure such as a concrete retaining structure. In the wall type improvement composed of long and short walls as shown in Figure 6.1, the basic design concept can be assumed to be similar to the block type improvement.

The first step is evaluation of actions including setting of seismic coefficient for verification. The second step of the procedure is examination of the external stability

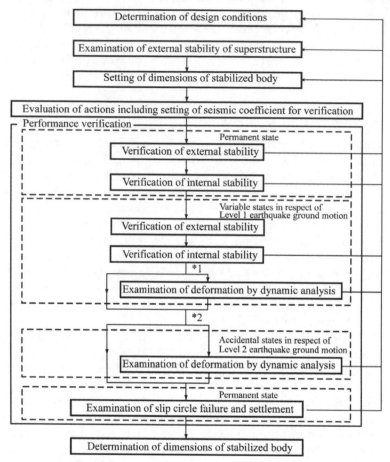

Figure 6.60 Flow of the current design procedure (The Ports and Harbours Association of Japan, 2007; The Overseas Coastal Area Development Institute of Japan, 2009).

of the superstructure to assure the superstructure and improved ground can behave as a unit. The third step is verification in the permanent state, which includes verification of "external stability" and "internal stability" of improved ground. In the verification of the external stability, sliding failure, overturning failure and bearing capacity of the improved ground are evaluated. In the verification of the internal stability, the induced stresses due to the external forces are calculated and confirmed to be lower than the allowable values. The wall type improved ground is also examined for extrusion failure, where unstabilized soil between the long walls might be squeezed out.

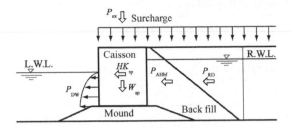

Figure 6.61 Force conditions of superstructure for verufucation (The Ports and Harbours Association of Japan, 2007; The Overseas Coastal Area Development Institute of Japan, 2009).

The fourth step is verification in the Level 1 earthquake ground motion, which includes verification of "external stability" and "internal stability" of improved ground. In some cases, the same verification is required for the accidental state in respect of the Level 2 earthquake ground motion. Then, slip circle failure and settlement of improved ground are examined.

In this section, the design procedure is described for a caisson type quay wall on a block type and wall type improved ground as shown in Figure 6.46, where the two dimensional condition is assumed. The quay wall is consisted of a caisson, gravel mound, backfill, fill and block type or wall type improved ground. The permanent state and variable state (Level 1 earthquake ground motion) for each verification are explained together.

6.3.2 Examination of external stability of a superstructure

For the external stability analysis of a superstructure at the first step of the design procedure, the improved ground whose size and strength are not determined yet is assumed to stiff enough to have sufficient bearing capacity to support the superstructure (see Figure 6.61). The sliding and overturning failures of the superstructure are verified at this step in order to determine its size and weight.

6.3.2.1 Sliding failure

The performance verification for the sliding failure is calculated by Equation (6.36), where γ is the partial factor, and the subscripts k and d denote the characteristic value and design value respectively, and $f_d = \gamma_f \cdot f_k$. The partial factors are summarized in Table 6.4. In the table, "earthquake resistant port facilities" is particularly strong and reinforced facilities to transport evacuees and emergency supplies for large scale earthquake.

For permanent state

$$f_{m_d} \cdot (W_{sp_d} + P_{su_d}) \geq \gamma_a \cdot \gamma_i \cdot (P_{AHbf_d} + P_{Rw_d}) \tag{6.36a}$$

for variable states in respect of the Level 1 earthquake ground motion

$$f_{m_d} \cdot (W_{sp_d} + P_{su_d}) \geq \gamma_a \cdot \gamma_i \cdot (P_{DAHbf_d} + P_{Rw_d} + P_{Dw_d} + HK_{sp_d}) \tag{6.36b}$$

Table 6.4 Partial factors for sliding failure.

(a) For the permanent state for earthquake resistant port facilities.
target reliability index, β_T 3.1
target system failure probability, P_{fT} 1.0×10^{-3}
reliability index, β in calculation of γ 3.31

	γ	α	μ/X_k	V
friction coefficient, γ_f	0.55	0.946	1.06	0.15
horizontal component of total active force, γ_{PAH}	1.15	−0.288	1.00	0.12
total residual water force, γ_{PRw}	1.00	−0.024	1.00	0.05
unit weight of reinforced concrete, γ_{WRC}	0.95	0.026	0.98	0.02
unit weight of no-reinforced concrete, γ_{WNC}	1.00	0.009	1.02	0.02
unit weight of sand filled in caisson, γ_{WSAND}	1.00	0.143	1.02	0.04
structural analysis factor, γ_a	1.00	−	−	−

(b) For the permanent state for ordinary port facilities.
target reliability index, β_T 2.7
target system failure probability, P_{fT} 4.0×10^{-3}
reliability index, β in calculation of γ 2.89

	γ	α	μ/X_k	V
friction coefficient, γ_f	0.60	0.935	1.06	0.15
horizontal component of total active force, γ_{PAH}	1.15	−0.316	1.00	0.12
total residual water force, γ_{PRw}	1.00	−0.027	1.00	0.05
unit weight of reinforced concrete, γ_{WRC}	0.95	0.028	0.98	0.02
unit weight of no-reinforced concrete, γ_{WNC}	1.00	0.010	1.02	0.02
unit weight of sand filled in caisson, γ_{WSAND}	1.00	0.157	1.02	0.04
structural analysis factor, γ_a	1.00	−	−	−

(c) For variable states in respect of the Level I earthquake ground motion.
target reliability index, β_T
target system failure probability, P_{fT}
reliability index, β in calculation of γ

	γ	α	μ/X_k	V
friction coefficient, γ_f	1.00	−	−	−
horizontal component of total active earth, γ_{PAH}	1.00	−	−	−
total residual water force, γ_{PRw}	1.00	−	−	−
unit weight of reinforced concrete, γ_{WRC}	1.00	−	−	−
unit weight of no-reinforced concrete, γ_{WNC}	1.00	−	−	−
unit weight of sand filled in caisson, γ_{WSAND}	1.00	−	−	−
structural analysis factor, γ_a	1.00	−	−	−

where

f_m : coefficient of friction of mound
HK_{sp} : total seismic inertia force per unit length of superstructure (kN/m)
P_{AHbf} : total static active force per unit length of backfill (kN/m)
P_{DAHbf} : total dynamic active force per unit length of backfill (kN/m)
P_{Dw} : total dynamic water force per unit length (kN/m)
P_{su} : total surcharge force per unit length (kN/m)
P_{Rw} : total residual water force per unit length (kN/m)

Table 6.5 Partial factors for overturning failure.

(a) For the permanent state for earthquake resistant port facilities.

target reliability index, β_T	3.1			
target system failure probability, P_{fT}	1.0×10^{-3}			
reliability index, β in calculation of γ	3.31			

	γ	α	μ/X_k	V
horizontal component of total active force, γ_{PAH}	1.35	−0.832	1.00	0.12
total residual water force, γ_{PRw}	1.05	−0.092	1.00	0.05
unit weight of reinforced concrete, γ_{WRC}	0.95	0.097	0.98	0.02
unit weight of no-reinforced concrete, γ_{WNC}	1.00	0.035	1.02	0.02
unit weight of sand filled in caisson, γ_{WSAND}	0.95	0.538	1.02	0.04
structural analysis factor, γ_a	1.00	−	−	−

(b) For the permanent state for ordinary port facilities.

target reliability index, β_T	2.7			
target system failure probability, P_{fT}	4.0×10^{-3}			
reliability index, β in calculation of γ	2.89			

	γ	α	μ/X_k	V
horizontal component of total active force, γ_{PAH}	1.30	−0.842	1.00	0.12
total residual water force, γ_{PRw}	1.05	−0.092	1.00	0.05
unit weight of reinforced concrete, γ_{WRC}	0.95	0.094	0.98	0.02
unit weight of no-reinforced concrete, γ_{WNC}	1.00	0.034	1.02	0.02
unit weight of sand filled in caisson, γ_{WSAND}	0.95	0.521	1.02	0.04
structural analysis factor, γ_a	1.00	−	−	−

(c) For variable states in respect of the Level I earthquake ground motion.

target reliability index, β_T	3.1			
target system failure probability, P_{fT}	1.0×10^{-3}			
reliability index, β in calculation of γ	3.31			

	γ	α	μ/X_k	V
horizontal component of total active force, γ_{PAH}	1.35	−0.832	1.00	0.12
total residual water force, γ_{PRw}	1.05	−0.092	1.00	0.05
unit weight of reinforced concrete, γ_{WRC}	0.95	0.097	0.98	0.02
unit weight of no-reinforced concrete, γ_{WNC}	1.00	0.035	1.02	0.02
unit weight of sand filled in caisson, γ_{WSAND}	0.95	0.538	1.02	0.04
structural analysis factor, γ_a	1.00	−	−	−

W_{sp} : weight per unit length of superstructure (kN/m)
γ_a : structural analysis factor (generally assumed to be 1.0)
γ_i : structural factor (generally assumed to be 1.0).

6.3.2.2 Overturning failure

The performance verification for the overturning failure is calculated by Equation (6.37), and the partial factors are summarized in Table 6.5.

For permanent state

$$W_{sp_d} \cdot x_{sp} + P_{su_d} \cdot x_{su} \geq \gamma_a \cdot \gamma_i \cdot (P_{AHbf_d} \cdot y_{AHbf} + P_{Rw_d} \cdot y_{Rw}) \tag{6.37a}$$

for variable states in respect of the Level l earthquake ground motion

$$W_{sp_d} \cdot x_{sp} + P_{su_d} \cdot x_{su} \geq \gamma_a \cdot \gamma_i \cdot (P_{DAHbf_d} \cdot y_{DAHbf} + P_{Rw_d} \cdot y_{Rw} + P_{Dw_d} \cdot y_{Dw} + HK_{sp_d} \cdot y_{sp})$$

(6.37b)

where

x_{sp} : horizontal distance of weight of superstructure from its edge (m)

x_{su} : horizontal distance of total surcharge force from front edge of superstructure (m)

y_{AHbf} : vertical distance of horizontal component of total static active force of backfill from bottom of superstructure (m)

y_{DAHbf} : vertical distance of horizontal component of total dynamic active force of backfill from bottom of superstructure (m)

y_{Dw} : vertical distance of total dynamic water force from bottom of superstructure (m)

y_{Rw} : vertical distance of total residual water force from bottom of superstructure (m)

y_{sp} : vertical distance of weight of superstructure from its bottom (m)

γ_a : structural analysis factor (generally assumed to be 1.0).

γ_i : structural factor (generally assumed to be 1.0).

6.3.3 Setting of geometric conditions of improved ground

The field strength of stabilized soil, improvement type, and width and thickness, are assumed. The initial trial value for the width of improved ground is usually assumed as the sum of the widths of the gravel mound and backfill as the minimum. The thickness of the improved ground is usually assumed as the thickness of soft ground because the fixed type improved ground is desirable from the view point of stability and displacement. When laboratory mix test results are available, an appropriate field strength is assumed considering the economy and the construction aspects. If laboratory mix test data is not available, 2,000 to 3,000 kN/m² in terms of unconfined compressive strength is ordinarily adopted as the field strength in the case of in-water work.

6.3.4 Evaluation of seismic coefficient for verification

6.3.4.1 For level I performance verification

The seismic coefficient of the Level 1 performance verification for a superstructure on DM improved ground (e.g. caisson, mound, backfill and fill) can be obtained by Equation (6.38), which incorporates the allowable displacement of the superstructure. The allowable displacement is specified in the standard depending on the type of structure (The Ports and Harbours Association of Japan, 2007; The Overseas Coastal Area Development Institute of Japan, 2009), but should be specified depending upon its type and importance. In the case of a gravity type quay wall, the D_a value of 100 mm is specified. The magnitude of the modified maximum seismic acceleration, α_c is obtained by seismic response analyses incorporating the maximum acceleration at bed rock, the ground conditions, and the time duration of an earthquake (Kitazume and Nagao, 2007).

$$k_{h1k} = 1.78 \cdot \left(\frac{D_a}{D_r}\right)^{-0.55} \cdot \frac{\alpha_c \cdot 0.64}{g} + 0.04 \tag{6.38}$$

where

D_a : allowable displacement (mm)
D_r : reference displacement ($= 100$ mm)
g : gravity ($= 9.8$ m/s^2)
k_{h1k} : seismic coefficient for superstructure
α_c : modified maximum seismic acceleration (m/s^2).

The seismic coefficient for the external forces acting on the improved ground, k_{h2k}, the seismic coefficient for dynamic earth pressures acting on the superstructure, k'_{h2k}, and the seismic coefficient for dynamic earth pressures acting on the improved ground, k_{h3k}, can be calculated by Equation (6.39) (Kitazume and Nagao, 2007).

$$k_{h2k} = k_{h1k} \cdot 0.65 \tag{6.39a}$$

$$k'_{h2k} = k_{h1k} \tag{6.39b}$$

$$k_{h3k} = 1.78 \cdot \left(\frac{D_a}{D_r}\right)^{-0.55} \cdot \frac{\alpha_c}{g} + 0.04 \tag{6.39c}$$

where

k_{h2k} : seismic coefficient for external forces acting on DM improved ground
k'_{h2k} : seismic coefficient for dynamic force acting on superstructure
k_{h3k} : seismic coefficient for dynamic force acting on DM improved ground.

6.3.4.2 For level 2 performance verification

The Level 2 performance verification should be carried out by dynamic analyses which can incorporate the effect of liquefaction on the displacement of the superstructure and the improved ground.

6.3.5 *Examination of the external stability of improved ground*

In the "external stability analysis," three failure modes are examined for the assumed improved ground: sliding, overturning and bearing capacity failures. The design loads adopted in the external stability analysis are schematically shown in Figure 6.62. They include the active and passive earth pressures, surcharge and external forces acting on the boundary of improved ground, the mass forces generated by gravity, and the seismic inertia forces.

In the stability analysis of the wall type improved ground, it is sometimes necessary to assume the magnitudes of external forces acting on unstabilized soil and stabilized soil independently. In general, it can be assumed that the active and passive earth pressures act uniformly on the long wall, short wall and unstabilized soil between the long walls. For vertical loads, it is assumed that the self-weight of the superstructure,

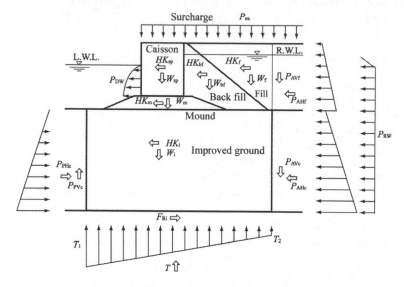

Figure 6.62 Schematic diagram of design loads (The Ports and Harbours Association of Japan, 2007; The Overseas Coastal Area Development Institute of Japan, 2009).

and the surcharge and external forces acting on the superstructure and the weight of stabilized soil are concentrated on the long wall.

6.3.5.1 Sliding failure

In the calculation of sliding failure, it is assumed that the improved ground and the superstructure move horizontally at the bottom boundary of improved ground due to the unbalance of the earth pressures and/or the seismic inertia forces. The performance verification for the sliding failure is calculated by Equation (6.40). For the wall type improved ground, the two sliding patterns are assumed: frictional shear strength is mobilized at the bottom of the improved ground, and frictional shear strength at the bottom of the long wall and cohesive strength mobilized at the bottom of unstabilized soil between the long walls. The partial factors determined on this basis are as shown in Table 6.6.

For permanent state

$$P_{\mathrm{PHc_d}} + F_{\mathrm{Ri_d}} \geq \gamma_a \cdot \gamma_i \cdot (P_{\mathrm{AHc_d}} + P_{\mathrm{RW_d}}) \tag{6.40a}$$

for variable states in respect of the Level 1 earthquake ground motion

$$P_{\mathrm{DPHc_d}} + F_{\mathrm{Ri_d}} \geq \gamma_a \cdot \gamma_i \cdot (P_{\mathrm{DAHc_d}} + P_{\mathrm{RW_d}} + P_{\mathrm{Dw_d}} + HK_{\mathrm{sp_d}} + HK_{\mathrm{m_d}}$$
$$+ HK_{\mathrm{bf_d}} + HK_{\mathrm{f_d}} + HK_{\mathrm{i_d}}) \tag{6.40b}$$

Table 6.6 Partial factors for sliding failure.

(a) For the permanent state.
i) In the case of sand layer beneath improved ground
target reliability index, β_T 2.9
target system failure probability, P_{fT} 2.1×10^{-3}
reliability index, β in calculation of γ 3.0

	γ	α	μ/X_k	V
weight, $\gamma_{Wb}, \gamma_{Wc}, \gamma_{Wf}, \gamma_{Wm}, \gamma_{Ws}, \gamma_{Wu}$	1.00	0.131	1.00	0.03
horizontal component of total active force, γ_{PAH}	1.15	−0.519	1.00	0.10
vertical component of total active force, γ_{PAV}	1.00	0.000	1.00	−
horizontal component of passive earth pressure, γ_{PPH}	0.90	0.277	1.00	0.10
vertical component of passive earth pressure, γ_{PPV}	1.00	0.000	1.00	−
undrained shear strength at active side, γ_{Cua}	1.00	0.000	1.00	−
undrained shear strength at passive side, γ_{Cup}	1.00	0.000	1.00	−
coefficient of friction, γ_μ	0.70	1.000	1.00	0.10
structural analysis factor, γ_a	1.00	−	−	−

ii) In the case of a clay layer beneath improved ground.
target reliability index, β_T 2.9
target system failure probability, P_{fT} 2.1×10^{-3}
reliability index, β in calculation of γ 3.0

	γ	α	μ/X_k	V
weight, $\gamma_{Wb}, \gamma_{Wc}, \gamma_{Wf}, \gamma_{Wm}, \gamma_{Ws}, \gamma_{Wu}$	1.00	0.000	1.00	−
horizontal component of total active force, γ_{PAH}	1.15	−0.461	1.00	0.10
vertical component of total active force, γ_{PAV}	1.00	0.000	1.00	−
horizontal component of passive earth pressure, γ_{PPH}	0.85	0.454	1.00	0.10
vertical component of passive earth pressure, γ_{PPV}	1.00	0.000	1.00	−
undrained shear strength at active side, γ_{Cua}	1.00	0.000	1.00	−
undrained shear strength at passive side, γ_{Cup}	1.00	0.000	1.00	−
coefficient of friction, γ_μ	0.75	0.831	1.00	0.10
structural analysis factor, γ_a	0.80	0.202	1.00	0.33
weight, $\gamma_{Wb}, \gamma_{Wc}, \gamma_{Wf}, \gamma_{Wm}, \gamma_{Ws}, \gamma_{Wu}$	1.00	−	−	−

(b) For variable states in respect of the Level I earthquake ground motion.
i) In the case of a sand layer beneath improved ground.
target reliability index, β_T 2.9
target system failure probability, P_{fT} 2.1×10^{-3}
reliability index, β in calculation of γ 3.0

	γ	α	μ/X_k	V
weight, $\gamma_{Wb}, \gamma_{Wc}, \gamma_{Wf}, \gamma_{Wm}, \gamma_{Ws}, \gamma_{Wu}$	1.00	−	−	−
horizontal component of total active force, γ_{PAH}	1.00	−	−	−
vertical component of total active force, γ_{PAV}	1.00	−	−	−
horizontal component of passive earth pressure, γ_{PPH}	1.00	−	−	−
vertical component of passive earth pressure, γ_{PPV}	1.00	−	−	−
undrained shear strength at active side, γ_{Cua}	1.00	−	−	−
undrained shear strength at passive side, γ_{Cup}	1.00	−	−	−
coefficient of friction, γ_μ	1.00	−	−	−
structural analysis factor, γ_a	1.00	−	−	−

(Continued)

Table 6.6 Continued.

ii) In the case of a clay layer beneath improved ground.

target reliability index, β_T	2.9			
target system failure probability, P_{fT}	2.1×10^{-3}			
reliability index, β in calculation of γ	3.0			

	γ	α	μ/X_k	V
weight, $\gamma_{Wb}, \gamma_{Wc}, \gamma_{Wf}, \gamma_{Wm}, \gamma_{Ws}, \gamma_{Wu}$	1.00	–	–	–
horizontal component of total active force, γ_{PAH}	1.00	–	–	–
vertical component of total active force, γ_{PAV}	1.00	–	–	–
horizontal component of passive earth pressure, γ_{PPH}	1.00	–	–	–
vertical component of passive earth pressure, γ_{PPV}	1.00	–	–	–
undrained shear strength at active side, γ_{Cua}	1.00	–	–	–
undrained shear strength at passive side, γ_{Cup}	1.00	–	–	–
coefficient of friction, γ_μ	1.00	–	–	–
structural analysis factor, γ_a	1.00	–	–	–
weight, $\gamma_{Wb}, \gamma_{Wc}, \gamma_{Wf}, \gamma_{Wm}, \gamma_{Ws}, \gamma_{Wu}$	1.00	–	–	–

where

F_{Ri} : total shear force per unit length mobilized on bottom of improved ground (kN/m)

for block type improvement resting on sandy layer (fixed type)

$$F_{Ri} = F_{Rs}$$

for wall type improvement resting on sandy layer (fixed type)

$$F_{Ri} = F_{Rs} + F_{Ru}$$

for block and wall type improvements resting on clay (floating type)

$$F_{Ri} = c_{uc} \cdot B_i$$

F_{Rs} : total shear force per unit length mobilized by sand layer at the bottom of improved ground (kN/m)

$$R_s = \gamma_\mu \cdot \mu_k \cdot (W_{sp} + W_m + W_{bf} + W_f + W_s + \gamma_{Psu} \cdot P_{su} + \gamma_{P_{PVc}} \cdot P_{PV_c} + \gamma_{P_{AVc}} \cdot P_{AV_c})$$

F_{Ru} : total shear force per unit length mobilized by unstabilized soil between long walls at the bottom of improved ground (kN/m)
in the case of a sand layer beneath improved ground,

$$= \min \cdot \begin{cases} \gamma_\mu \cdot \gamma_{W_u} \cdot W_u \cdot \mu_k \cdot \dfrac{L_s}{L_s + L_\ell} \\ \gamma_{c_{uc}} \cdot c_{uc} \cdot B_i \cdot \dfrac{L_s}{L_s + L_\ell} \end{cases}$$

HK_{bf} : total seismic inertia force per unit length of backfill (kN/m)

HK_f : total seismic inertia force per unit length of fill (kN/m)

HK_i : total seismic inertia force per unit length of improved ground (kN/m)

HK_m : total seismic inertia force per unit length of mound (kN/m)
HK_{sp} : total seismic inertia force per unit length of superstructure (kN/m)
P_{AHc} : horizontal component of total static active force per unit length (kN/m)
P_{AVc} : vertical component of total static active force per unit length (kN/m)
P_{DAHc} : horizontal component of total dynamic active force per unit length (kN/m)
P_{DAVc} : vertical component of total dynamic active force per unit length (kN/m)
P_{DPHc} : horizontal component of total dynamic passive force per unit length (kN/m)
P_{DPVc} : vertical component of total dynamic passive force per unit length (kN/m)
P_{Dw} : total dynamic water force per unit length (kN/m)
P_{PHc} : horizontal component of total static passive force per unit length (kN/m)
P_{PVc} : vertical component of total static passive force per unit length (kN/m)
P_{Rw} : total residual water force per unit length (kN/m)
W_{sp} : weight per unit length of superstructure (kN/m)
W_f : weight per unit length of fill (kN/m)
W_i : weight per unit length of improved ground (kN/m)
W_m : weight per unit length of mound (kN/m)
W_s : weight per unit length of stabilized soil (kN/m)
W_u : weight per unit length of unstabilized soil (in case of wall type improvement) (kN/m)
γ : partial factor
γ_a : structural analysis factor (generally assumed to be 1.0)
γ_i : structural factor (generally assumed to be 1.0)
μ_k : coefficient of friction of soil beneath improved ground

6.3.5.2 Overturning failure

In the overturning failure, it is assumed that the improved ground and the superstructure rotate about the front bottom edge of the improved ground. The performance verification for the overturning failure is calculated by Equation (6.41), where the symbol γ is the partial factor for its subscript, and the subscripts k and d denote the characteristic value and design value, respectively. The partial factors for the overturning failure are summarized in Table 6.7.

For permanent state

$$
\begin{aligned}
P_{PHc_d} \cdot y_{PHc} + P_{AVc_d} \cdot x_{AVc} + P_{su_d} \cdot x_{su} + W_{sp_d} \cdot x_{sp} + W_{m_d} \cdot x_m \\
+ W_{bf_d} \cdot x_{bf} + W_{f_d} \cdot x_f + W_{i_d} \cdot x_i \geq \gamma_i \cdot \gamma_a \cdot (P_{AHc_d} \cdot y_{AHc} + P_{RW_d} \cdot y_{Rw})
\end{aligned}
\tag{6.41a}
$$

for variable states in respect of the Level 1 earthquake ground motion

$$
\begin{aligned}
P_{DPHc_d} \cdot y_{DPHc} + P_{DAVc_d} \cdot x_{DAVc} + P_{su_d} \cdot x_{su} + W_{sp_d} \cdot x_{sp} + W_{m_d} \cdot x_m \\
+ W_{bf_d} \cdot x_{bf} + W_{f_d} \cdot x_f + W_{i_d} \cdot x_i \geq \gamma_i \cdot \gamma_a \cdot (P_{DAHc_d} \cdot y_{DAHc} \\
+ P_{RW_d} \cdot y_{Rw} + P_{Dw_d} \cdot y_{Dw} + \Sigma HK \cdot y)
\end{aligned}
\tag{6.41b}
$$

$$
\Sigma HK \cdot y = HK_{sp_d} \cdot y_{sp} + HK_{m_d} \cdot y_m + HK_{bf_d} \cdot y_{bf} + HK_{f_d} \cdot y_f + HK_{i_d} \cdot y_i
$$

where
x_{AVc} : horizontal distance of vertical component of total static active force from bottom of improved ground (m)

Table 6.7 Partial factors for overturning failure.

(a) For the permanent state. target reliability index, β_T 2.9
target system failure probability, P_{fT} 2.1×10^{-3}
reliability index, β in calculation of γ 3.0

	γ	α	μ/X_k	V
horizontal component of total active force, γ_{PAH}	1.25	−0.882	1.00	0.10
vertical component of total active force, γ_{PAV}	1.00	0.029	1.00	0.10
horizontal component of passive earth pressure, γ_{PPH}	0.85	0.382	1.00	0.10
undrained shear strength at active side, γ_{Cua}	1.00	0.102	1.00	0.10
weight of mound, γ_{Wm}	1.00	0.030	1.00	0.03
weight of backfill, γ_{Wb}	1.00	0.055	1.00	0.03
weight of stabilized soil, γ_{Ws}	1.00	0.102	1.00	0.03
weight of unstabilized soil, γ_{Wu}	1.00	0.074	1.00	0.03
structural analysis factor, γ_a	1.00	−	−	−

(b) For variable states in respect of the Level I earthquake ground motion.
target reliability index, β_T 2.9
target system failure probability, P_{fT} 2.1×10^{-3}
reliability index, β in calculation of γ 3.0

	γ	α	μ/X_k	V
horizontal component of total active force, γ_{PAH}	1.00	−	−	−
vertical component of total active force, γ_{PAV}	1.00	−	−	−
horizontal component of passive earth pressure, γ_{PPH}	1.00	−	−	−
undrained shear strength at active side, γ_{Cua}	1.00	−	−	−
weight of mound, γ_{Wm}	1.00	−	−	−
weight of backfill, γ_{Wb}	1.00	−	−	−
weight of stabilized soil, γ_{Ws}	1.00	−	−	−
weight of unstabilized soil, γ_{Wu}	1.00	−	−	−
structural analysis factor, γ_a	1.00	−	−	−

x_{bf} : horizontal distance of weight of backfill from front edge of improved ground (m)

x_{DAV_c} : horizontal distance of vertical component of total dynamic active force from front edge of improved ground (m)

x_f : horizontal distance of weight of fill from front edge of improved ground (m)

x_i : horizontal distance of weight of improved ground from its front edge (m)

x_m : horizontal distance of weight of mound from front edge of improved ground (m)

x_{sp} : horizontal distance of weight of superstructure from front edge of improved ground (m)

x_{su} : horizontal distance of total surcharge force from front edge of improved ground (m)

y_{AHc} : vertical distance of horizontal component of total static active force from bottom of improved ground (m)

y_{AVc} : vertical distance of vertical component of total static active force from bottom of improved ground (m)

y_{bf} : vertical distance of total seismic inertia force of backfill from bottom of improved ground (m)

y_{DAHc} : vertical distance of horizontal component of total dynamic active force from bottom of improved ground (m)

y_{Dw} : vertical distance of total dynamic water force from bottom of improved ground (m)

y_f : vertical distance of total seismic inertia force of fill from bottom of improved ground (m)

y_i : vertical distance of total seismic inertia force of improved ground from bottom of improved ground (m)

y_m : vertical distance of total seismic inertia force of mound from bottom of improved ground (m)

y_{DPHc} : vertical distance of horizontal component of total dynamic passive force from bottom of improved ground (m)

y_{Rw} : vertical distance of total residual water force from bottom of improved ground (m)

y_{sp} : vertical distance of total seismic inertia force of superstructure from bottom of improved ground (m)

y_{PHc} : vertical distance of horizontal component of total static passive force from bottom of improved ground (m)

γ_i : structural factor (generally assumed to be 1.0)

γ_a : structural analysis factor.

6.3.5.3   Bearing capacity

As the deep mixing improved ground is assumed as a buried structure in this design procedure, its bearing capacity is evaluated by the classical bearing capacity theory which can incorporate the effects of loading condition and embedded condition. In the design, the subgrade reactions at the front edge and the rear edge of the bottom of the improved ground are calculated by Equations (6.22). The performance verification for the bearing capacity is calculated as Equation (6.42), while the bearing capacity of the improved ground is calculated by Equation (6.43).

$$t_1 \leq q_{ar_d} \tag{6.42a}$$

$$t_2 \leq q_{ar_d} \tag{6.42b}$$

$$q_{ar_d} = \gamma_R \left(\gamma_d \cdot \frac{B_i}{2} \cdot N_{\gamma_d} + c_{ub} \cdot N_{c_d} + q \cdot \left(N_{q_d} - 1 \right) \right) + q \tag{6.43}$$

where

B_i : width of improved ground (m)

c_{ub} : undrained shear strength of soil beneath improved ground (kN/m^2)

q : effective overburden pressure at bottom of improved ground (kN/m^2)

q_{ar} : bearing capacity (kN/m^2)

γ_d : unit weight of soil beneath improved ground (kN/m^3)

N_c : bearing capacity factor of soil beneath improved ground

N_q : bearing capacity factor of soil beneath improved ground

N_γ : bearing capacity factor of soil beneath improved ground.

The bearing capacity of a row of stabilized soil walls in the wall type improved ground is a problem of mutual interference of the bearing capacities of deep rectangular foundations. The increase of the bearing capacity of stabilized soil walls caused by the interference of adjacent walls has been demonstrated in a series of centrifuge model tests and the simple design shown in Figure 6.50 has been proposed by Terashi and Kitazume (1987). For the bearing capacity of wall type improved ground, the performance verification is calculated as Equation (6.44) incorporating the effect of mutual interference between the long walls, where γ is the partial factor, and the subscripts k and d denote the characteristic value and design value, respectively.

$$q_{f_d} = q_{f(L_1)_d} + \frac{1}{2}(q_{f(B_i)_d} - q_{f(L_1)_d}) \cdot \left(3 - \frac{S_1}{L_1}\right) \tag{6.44}$$

$$q_{f(L_1)_d} = \gamma_R \left(\gamma_d \frac{L_1}{2} N_{\gamma_d} + c_{u_b} \cdot N_{c_d} + q(N_{q_d} - 1)\right) + q$$

$$q_{f(B_i)_d} = \gamma_R \left(\gamma_d \frac{B_i}{2} N_{\gamma_d} + c_{u_b} \cdot N_{c_d} + q(N_{q_d} - 1)\right) + q$$

where
L_l : thickness of long wall of improved ground (m)
$q_{f(Ll)}$: bearing capacity of strip foundation with thickness of long wall, L_l (kN/m^2)
$q_{f(Bi)}$: bearing capacity of strip foundation with width of improved ground, B_i (kN/m^2)
S_l : center to center spacing of long walls of improved ground (m).

6.3.6 Examination of internal stability of improved ground

In the "internal stability analysis," the induced stresses in the improved ground are calculated based on the elastic theory. The shape and size of the improved ground are determined so that the induced stresses are lower than the allowable strengths of the stabilized soil. In the calculation, the stabilized soil is generally assumed to have a uniform property for the sake of simplicity even it contains possibly weaker zones due to construction process such as overlap joints. The effect of the strength at the overlapping portion is taken into account when determining the allowable strengths of stabilized soil, which will be described later.

Instead of running numerical analysis such as the finite element method, simple calculation methods have been generally applied as routine design. The loading conditions applied to the internal stability analysis are generally assumed to be the same as those for the external stability analysis, as already shown in Figure 6.62. At this stage of calculation, however, the external stability is already satisfied with a certain safety margin and hence horizontal resisting forces exceeds the driving forces. Earth pressure at the passive side may be chosen appropriately between the passive and at rest pressures. According to the accumulated experiences in design, the internal stability evaluation at the two critical parts as shown in Figure 6.51 is considered sufficient as long as the shape of stabilized soil is within the experiences: (a) subgrade reactions at the front edge and rear edge of improved ground, and (b) average shear stress along a vertical shear plane at the front edge of the superstructure.

Table 6.8 Partial factors for subgrade reactions.

(a) For the permanent state.
target reliability index, β_T 2.9
target system failure probability, P_{fT} 2.1×10^{-3}
reliability index, β in calculation of γ 3.0

	γ	α	μ/X_k	V
design strength of stabilized soil, γ_{qus}	0.55	–	–	–
subgrade reaction, γ_{t1}, γ_{t2}	1.05	−0.116	1.00	0.03
weight of unstabilized soil, γ_{Wc}	1.00	0.001	1.00	0.03
structural analysis factor, γ_a	1.00	–	–	–

(b) For variable states in respect of the Level I earthquake ground motion.
target reliability index, β_T 2.9
target system failure probability, P_{fT} 2.1×10^{-3}
reliability index, β in calculation of γ 3.0

	γ	α	μ/X_k	V
design strength of stabilized soil, γ_{qus}	0.67	–	–	–
subgrade reaction, γ_{t1}, γ_{t2}	1.05	–	–	–
weight of unstabilized soil, γ_{Wc}	1.00	–	–	–
structural analysis factor, γ_a	1.00	–	–	–

6.3.6.1   Subgrade reactions at front edge of improved ground

The subgrade reactions at the front edge and rear edge of the improved ground, t_1 and t_2, should be smaller than the design value as shown in Equation (6.45). The partial factors for the subgrade reactions are summarized in Table 6.8.

For permanent state

$$f_{c_d} \geq \gamma_a \cdot \gamma_i (t_{1_d} - P_{AHc_d})$$
$$f_{c_d} \geq \gamma_a \cdot \gamma_i (t_{2_d} - P_{AHc_d})$$
(6.45a)

for variable states in respect of the Level 1 earthquake ground motion

$$f_{c_d} \geq \gamma_a \cdot \gamma_i (t_{1_d} - P_{DPHc_d})$$
$$f_{c_d} \geq \gamma_a \cdot \gamma_i (t_{2_d} - P_{DAHc_d})$$
(6.45b)

where
  f_c : design compressive strength (kN/m^2)
  t_1 : subgrade reaction at front edge (kN/m^2)
  t_2 : subgrade reaction at rear edge (kN/m^2)
  γ_i : structural factor (generally assumed to be 1.0)
  γ_a : structural analysis factor (generally assumed to be 1.0).

6.3.6.2   Average shear stress along a vertical shear plane

The average shear stress induced along a vertical shear plane at the front face of the superstructure (Figure 6.51) should satisfy the criteria as shown in Equations (6.46). In

Table 6.9 Partial factors for average shear stress along a vertical shear plane.

(a) For the permanent state.

target reliability index, β_T	2.9			
target system failure probability, P_{fT}	2.1×10^{-3}			
reliability index, β in calculation of γ	3.0			

	γ	α	μ/X_k	V
design strength of stabilized soil, γ_{qus}	0.55	–	–	–
subgrade reaction, γ_{t1}, γ_{t2}	1.05	−0.115	1.00	0.03
weight of unstabilized soil, γ_{Wc}	1.00	0.005	1.00	0.03
structural analysis factor, γ_a	1.00	–	–	–

(b) For variable states in respect of the Level I earthquake ground motion.

target reliability index, β_T	2.9			
target system failure probability, P_{fT}	2.1×10^{-3}			
reliability index, β in calculation of γ	3.0			

	γ	α	μ/X_k	V
design strength of stabilized soil, γ_{qus}	0.67	–	–	–
subgrade reaction, γ_{t1}, γ_{t2}	1.00	–	–	–
weight of unstabilized soil, γ_{Wc}	1.00	–	–	–
structural analysis factor, γ_a	1.00	–	–	–

the case where a mound underlies the superstructure, the stress distribution at an angle of around 30° can be taken into account to find the vertical shear plane. In the case where gravel mound underlies the superstructure, the stress distribution at an angle of around 30° can be taken into account to find the shear failure plane. The partial factors for the vertical shear failure of the long wall part are summarized in Table 6.9.

$$f \le f_{sh_d} \tag{6.46a}$$

$$f = \gamma_a \cdot \gamma_i \cdot \frac{L_l + L_s}{L_l \cdot H_i + L_s \cdot H_s} \left(\frac{L_l}{L_l + L_s} \int_0^{B_{is}} t_{is} \, db - W_{is_d} \right) \tag{6.46b}$$

where

B_{is} : distance of vertical shear plane from toe of improved ground (m)
H_i : height of improved ground (m)
H_s : height of short wall of improved ground (m)
L_l : thickness of long wall of improved ground (m) as shown in Figure 6.52
L_s : thickness of short wall of improved ground (m) as shown in Figure 6.52
t_{is} : subgrade reaction at bottom of improved ground (kN/m^2)
W_{is} : weight per unit length of improved ground at part of B_{is} (kN/m)
f : average shear stress along vertical shear plane (kN/m^2)
f_{sh} : design shear strength of stabilized soil (kN/m^2).

In the case of the wall type improved ground, the external load acting on the short wall should be transferred to the long wall. The shear stress along the lap joint between long wall and short wall is also examined (see Figure 6.52). The performance verification for the short wall is calculated by Equation (6.47). The partial factors for

Table 6.10 Partial factors for vertical shear failure of short wall.

(a) For the permanent state.

target reliability index, β_T	2.9	
target system failure probability, P_{fT}	2.1×10^{-3}	
reliability index, β in calculation of γ	3.0	

	γ	α	μ/X_k	V
design strength of stabilized soil, γ_{quc}	0.55	–	–	–
subgrade reaction, γ_{tl}	1.05	−0.091	1.00	0.03
weight of stabilized soil, γ_{Wt}	1.00	−0.006	1.00	0.03
weight of mound, γ_{Wm}	1.00	−0.006	1.00	0.03
structural analysis factor, γ_a	1.00	–	–	–

(b) For variable states in respect of the Level I earthquake ground motion.

target reliability index, β_T	2.9	
target system failure probability, P_{fT}	2.1×10^{-3}	
reliability index, β in calculation of γ	3.0	

	γ	α	μ/X_k	V
design strength of stabilized soil, γ_{quc}	0.67	–	–	–
subgrade reaction, γ_{tl}	1.00	–	–	–
weight of stabilized soil, γ_{Wt}	1.00	–	–	–
weight of mound, γ_{Wm}	1.00	–	–	–
structural analysis factor, γ_a	1.00	–	–	–

the vertical shear failure of the lap joint are summarized in Table 6.10. Induced stress f in the equation should be examined appropriately by considering the load distribution as shown in Figure 6.52, both for the static and seismic conditions.

$$f_{sh_d} \geq \gamma_a \cdot \gamma_i \cdot \frac{L_s}{2 \cdot H_s \cdot B_s} \left(\int_0^{B_s} t_{ls} \, db + W_{ss} \right) \tag{6.47}$$

where

H_s : height of short wall of improved ground (m)
L_s : thickness of short wall of improved ground (m)
t_{ls} : induced stress at top of short wall of improved ground (kN/m²)
W_{ss} : weight per unit length of short wall of improved ground (kN/m)
γ_i : structural factor (generally assumed to be 1.0)
γ_a : structural analysis factor (generally assumed to be 1.0).

6.3.6.3 Allowable strengths of stabilized soil

The design strengths of stabilized soil are defined by Equations (6.48) to (6.50). The coefficients in the equations can be referred in Section 5.3.5.3.

$$f_c = \alpha \cdot \beta \cdot q_{uc_d} \tag{6.48}$$

$$= \alpha \cdot \beta \cdot \gamma_{quc} \cdot q_{uc_k} \tag{6.49}$$

$$f_{sh} = \frac{1}{2}f_c$$

$$f_t = 0.15 \cdot f_c \leq 200 \text{ kN/m}^2$$

(6.50)

where

f_c : design compressive strength of stabilized soil (kN/m^2)

f_{sh} : design shear strength of stabilized soil (kN/m^2)

f_t : design tensile strength of stabilized soil (kN/m^2)

q_{uck} : design unconfined compressive strength of stabilized soil (kN/m^2)

α : coefficient of effective width of stabilized soil column

β : reliability coefficient of overlapping.

6.3.6.4   Extrusion failure

For the wall type improvement (Figure 6.1), the extrusion failure must also be examined. The extrusion failure is a failure mode considered for the unstabilized soil remaining between the long walls which is subjected to the unbalanced active and passive earth pressures, as shown in Figure 6.55 (Terashi *et al.*, 1983). In the design procedure, the soft soil between the long walls is assumed to move as a rigid body in the shape of a rectangular prism, where the width and length of the prism are taken as the width of improved ground and the length of short wall respectively (see Figure 6.56). In the examination, the performance verification calculated by Equation (6.51) by changing the height of the prism, H_{PR} should be higher than the allowable value. The partial factor for the extrusion failure are summarized in Table 6.11.

For permanent state

$$2(L_s + H_{pr}) \cdot B_i \cdot c_{uc_d} + P_{PHc_d} \cdot L_s \geq \gamma_a \cdot \gamma_i \cdot (P_{AHc_d} + P_{Rw_d}) \cdot L_s \tag{6.51a}$$

for variable states in respect of the Level 1 earthquake ground motion

$$2(L_s + H_{pr}) \cdot B_i \cdot c_{uc_d} + P_{DPHc_d} \cdot L_s \geq \gamma_a \cdot \gamma_i \cdot (P_{DAHc_d} + P_{Rw_d} + HK_{pr}) \cdot L_s$$

(6.51b)

where

B_i : width of improved ground (m)

c_{uc} : undrained shear strength of soft soil at assumed prism (kN/m^2)

H_{pr} : height of assumed prism (m)

HK_{pr} : total seismic inertia force per unit length of soil prism (kN)

L_s : thickness of short wall of improved ground (m)

P_{AHc} : horizontal component of total static active force per unit length acting on the prism (kN/m)

P_{DAHc} : horizontal component of total dynamic active force per unit length acting on the prism (kN/m)

P_{DPHc} : horizontal component of total dynamic passive force per unit length acting on the prism (kN/m)

P_{PHc} : horizontal component of total static passive force per unit length acting on the prism (kN/m)

Table 6.11 Partial factors for extrusion failure.

(a) For the permanent state.

target reliability index, β_T 2.9
target system failure probability, P_{fT} .... 2.1×10^{-3}
reliability index, β in calculation of γ 3.0

	γ	α	μ/X_k	V
strength of unstabilized soil, γ_{cu}	0.75	0.955	1.00	0.10
horizontal component of total active force acting on unstabilized soil between long wall, γ_{PAH}	1.05	−0190	1.00	0.10
horizontal component of passive earth pressure acting on unstabilized soil between long wall, γ_{PPH}	0.95	0.182	1.00	0.10
unit weight of unstabilized soil, γ_{wu}	1.00	0.000	1.00	0.10
structural analysis factor, γ_a	1.00	–	–	–

(b) For variable states in respect of the Level I earthquake ground motion.

target reliability index, β_T 2.9
target system failure probability, P_{fT} .... 2.1×10^{-3}
reliability index, β in calculation of γ 3.0

	γ	α	μ/X_k	V
strength of unstabilized soil, γ_{cu}	1.00	–	–	–
horizontal component of total active force acting on unstabilized soil between long wall, γ_{PAH}	1.00	–	–	–
horizontal component of passive earth pressure acting on unstabilized soil between long wall, γ_{PPH}	1.00	–	–	–
unit weight of unstabilized soil, γ_{wu}	1.00	–	–	–
structural analysis factor, γ_a	1.00	–	–	–

P_{Rw} : total residual water force per unit length acting on the prism (kN/m)
γ_i : structural factor (generally assumed to be 1.0)
γ_a : structural analysis factor (generally assumed to be 1.0).

6.3.7 Slip circle analysis

Slip circle analysis is carried out to evaluate the overall stability of the improved ground, the superstructure and the surrounding soil. As the strength of stabilized soil is a very high value, a slip circle analysis passing through the improved ground is not necessary in many cases, as shown in Figure 6.57. In the case where sufficient bearing capacity is assured, slip circle analysis is not necessary in many cases. The slip circle analysis in seismic condition is not specified in the design standard. The performace verification on slip circles that pass outside stabilized soil mass are useful for evaluating the validity of the improvement geometry with respect to the external stability.

$$
\begin{aligned}
r \cdot (\tau_{c_d} \cdot l_c + \tau_{f_d} \cdot l_f + \tau_{i_d} \cdot l_i) \\
\geq \gamma_a \cdot \gamma_i \cdot (W_{sp} \cdot x_{sp} + W_m \cdot x_m + W_{bf} \cdot x_{bf} + W_f \cdot x_f + W_c \cdot x_c \\
+ W_i \cdot x_i + P_{su} \cdot x_{su} + P_{Rw} \cdot y_{Rw})
\end{aligned}
\tag{6.52}
$$

where

l_c : length of circular arc in soft ground (m)

l_f : length of circular arc in fill (m)

l_i : length of circular arc in improved ground (m)

r : radius of slip circle (m)

x_{bf} : horizontal distance of weight of backfill from center of slip circle (m)

x_c : horizontal distance of weight of soft ground from center of slip circle (m)

x_f : horizontal distance of weight of fill from center of slip circle (m)

x_i : horizontal distance of weight of improved ground from center of slip circle (m)

x_m : horizontal distance of weight of mound from center of slip circle (m)

x_{sp} : horizontal distance of weight of superstructure from center of slip circle (m)

x_{su} : horizontal distance of total surcharge force from center of slip circle (m)

y_{Rw} : vertical distance of total residual water force from center of slip circle (m)

τ_c : shear strength of soft ground (kN/m^2)

τ_i : average shear strength of improved ground (kN/m^2)

τ_f : shear strength of fill (kN/m^2).

6.3.8 Examination of immediate and long term settlements

After the optimum cross section of the improved ground is determined by the above procedures, the immediate and the long term settlements of the improved ground should be examined. Usually, the deformation of the stabilized soil itself can be negligible because of its high rigidity and large consolidation yield pressure. Therefore, the displacement of the improved ground is calculated as the deformation of the soft layers surrounding or beneath the stabilized soil. In the case of the fixed type improvement where the stabilized soil reaches the stiff layer, the settlement can be assumed to be negligible. In the case of the floating type improvement (Figure 6.58), the consolidation settlement beneath the improved ground is calculated by the Terzaghi's one dimensional consolidation theory, as shown in Equations (6.34) and (6.35).

6.3.9 Determination of strength and specifications of stabilized soil

The design engineer is responsible for writing the specifications on strength and geometric layout of stabilized soil columns including end bearing condition and minimum required overlapping width. The design of deep mixing involves examination of several failure modes both in external and internal stability. Only the designer knows which mode is the most critical. These information should better be reflected in the acceptance criteria.

7 DESIGN PROCEDURE OF GRID TYPE IMPROVED GROUND FOR LIQUEFACTION PREVENTION

7.1 Introduction

The block and grid types of deep mixing improvement have been applied to liquefaction prevention. In the block type improvement, a whole area of liquefiable soil is stabilized in order to increase liquefaction resistance. Required cohesion of stabilized soil for liquefaction prevention is relatively small of the order of 100 kN/m^2 (Zen *et al.*, 1987).

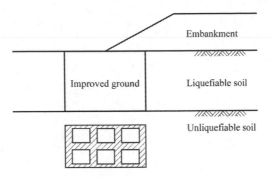

Figure 6.63 Grid type improved ground for liquefaction prevention.

In the grid type improvement, the grid wall is expected to restrict the shear deformation of soil between the walls during an earthquake, which in turn prevents liquefaction. The Ministry of Construction carried out a series of researches together with construction firms to investigate the effect of the grid improvement on liquefaction prevention, and published a technical report in 1999 in which a draft design procedure is proposed (Ministry of Construction, 1999).

In this section, the grid type improved ground beneath an embankment is exemplified, where the two dimensional condition is assumed. Of course, it is preferable to improve the whole area of potentially liquefiable layer in order to minimize adverse influence due to liquefaction. However, it is not seldom to improve only part of the layer for some reasons such as the economic limitation and magnitude of anticipated damage. Here the grid type improved ground beneath the embankment slope is discussed, as shown in Figure 6.63.

7.2 Basic concept

As the grid type improved ground is a rigid structure with high strength, the improved ground is assumed as a rigid structure in the design procedure, similar to the block type and wall type improved grounds. After determining the width of grid (spacing of grid walls) for liquefaction prevention, the width and thickness of improved ground are determined by examination of the external and internal stability analyses, which is a quite similar concept to the design of block type and wall type improved grounds.

7.3 Design procedure

7.3.1 Design flow

As the grid wall of improved ground functions to restrict the shear deformation of the soil between the grid walls, the width of grid (spacing of grid walls) is an essential parameter for liquefaction prevention. In the design, the spacing is determined at first based on the design earthquake and the thickness of the liquefiable soil. Then the external and internal stabilities are examined to obtain the width, height and strength of improved ground, and the thickness of the grid wall. The design procedure for the

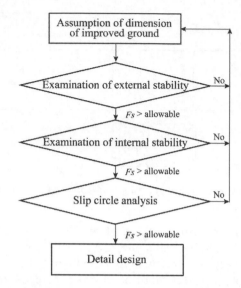

Figure 6.64 Design flow of grid type improvement for liquefaction prevention (Ministry of Construction, 1999).

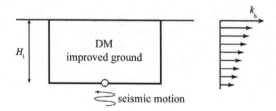

Figure 6.65 Design seismic coefficient.

grid type improvement for liquefaction prevention is shown in Figure 6.64 (Ministry of Construction, 1999).

7.3.2 Design seismic coefficient

The design seismic coefficient at the ground surface, k_{h0}, is used to evaluate the possibility of liquefaction, the earth pressures and pore water pressure acting on the improved ground, and the seismic inertia forces. This seismic coefficient is also used to evaluate the seismic inertia force of the embankment on the improved ground. As the seismic coefficient for an underground structure is usually smaller than that at the ground surface, the design seismic coefficient at the bottom of the improved ground, k_h is used to evaluate the seismic inertia force of improved ground as shown in Figure 6.65, which can be calculated by Equation (6.53).

$$k_h = \gamma_d \cdot k_{h0}$$
$$\gamma_d = 1 - 0.015 \cdot z \tag{6.53}$$

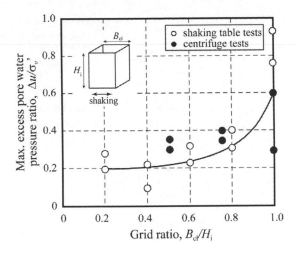

Figure 6.66 Relationship between maximum excess pore water pressure ratio and grid ratio.

where
  k_h : seismic coefficient of improved ground at its bottom
  k_{h0} : seismic coefficient at the surface of ground
  z : depth (m)
  γ_d : reduction factor.

7.3.3 Determination of width of grid

At first, the width of grid (spacing of grid walls) is determined. The width of grid, B_{cl} is determined by Figure 6.66 which was derived from the shaking table tests and centrifuge model tests. In the figure, the maximum excess pore pressure ratio, $\Delta u/\Delta\sigma'_v$ at the mid depth of the potentially liquefiable layer is plotted. According to the figure, the grid ratio, B_{cl}/H_i should be smaller than about 0.8 to assure the effect of liquefaction prevention. As the fixed type improved ground is desirable from the view point of stability and displacement, the thickness of the potentially liquefiable layer is assumed to be the full thickness of the soft layer, H_c.

7.3.4 Assumption of specifications of improved ground

The width and height of improved ground are assumed. As the fixed type improved ground is desirable, the thickness of improved ground is assumed to be H_c. The width of improved ground is determined by considering the external and internal stabilities as described in the following sections.

7.3.5 Examination of the external stability of improved ground

7.3.5.1 Sliding and overturning failures

In the "external stability analysis" of the improved ground, three failure modes are examined: sliding, overturning and bearing capacity failures. The design loads considered in the external stability analysis are schematically shown in Figure 6.67. They

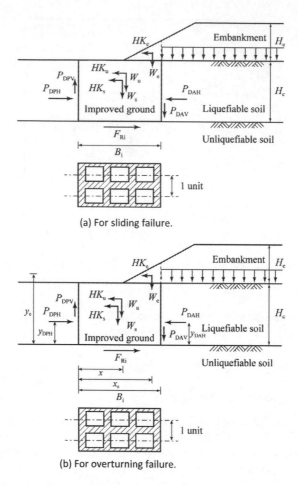

(a) For sliding failure.

(b) For overturning failure.

Figure 6.67 Schematic diagram of design loads.

include the active and passive earth pressures, the other external forces acting on the improved ground, the mass forces generated by gravity, and the seismic inertia forces. In the analyses, the stabilized soil walls and unstabilized soil between them are assumed to behave as a unit.

In the examination of sliding failure, it is assumed that the improved ground moves horizontally by the active earth pressure due to the embankment and the seismic inertia force of the improved ground. In the overturning failure, it is assumed that the improved ground rotates about its front edge. The sliding and overturning stabilities are calculated by the equilibrium of horizontal and moment forces, and the safety factors against these failures in seismic condition are calculated by Equations (6.54) and (6.55) respectively. The minimum safety factors are usually specified as 1.0 and 1.1 respectively.

For sliding failure

$$Fs_s = \frac{P_{DPH} + F_{Ri}}{P_{DAH} + HK_s + HK_u + HK_e} \tag{6.54}$$

for overturning failure

$$Fs_o = \frac{P_{DPH} \cdot y_{DPH} + P_{DAV} \cdot x_{DAV} + W_s \cdot x_s + W_u \cdot x_u + W_e \cdot x_e}{P_{DAH} \cdot y_{DAH} + HK_s \cdot y_s + HK_u \cdot y_u + HK_e \cdot y_e} \qquad (6.55)$$

where

a_s : improvement area ratio

B_i : width of improved ground (m)

c_{uc} : undrained shear strength of soft soil (kN/m^2)

c_{us} : undrained shear strength of stabilized soil (kN/m^2)

F_{Ri} : total shear force per unit length mobilized on bottom of improved ground (kN/m)

  in the case of sand layer beneath improved ground (fixed type improvement)

$$= \min \begin{cases} (W_e + W_s + W_u) \cdot \tan \phi_b' \\ c_{us} \cdot B_i \cdot a_s \end{cases}$$

  in the case of clay layer beneath improved ground (floating type improvement)

$$= c_{uc} \cdot B_i$$

Fs_o : safety factor against overturning failure of improved ground

Fs_s : safety factor against sliding failure of improved ground

HK_e : total seismic inertia force per unit length of embankment on improved ground (kN/m)

HK_s : total seismic inertia force per unit length of stabilized soil (kN/m)

HK_u : total seismic inertia force per unit length of unstabilized soil between grid walls (kN/m)

P_{DAH} : horizontal component of total dynamic active and pore water forces per unit length of soft ground and embankment (kN/m)

P_{DPH} : horizontal component of total dynamic passive and pore water forces per unit length of soft ground (kN/m)

P_{DAV} : vertical component of total dynamic active and pore water forces per unit length of soft ground and embankment (kN/m)

W_e : weight per unit length of embankment on improved ground (kN/m)

W_s : weight per unit length of stabilized soil (kN/m)

W_u : weight per unit length of unstabilized soil between grid walls (kN/m)

x_{DAV} : horizontal distance of vertical component of total dynamic active force from front edge of improved ground (m)

x_e : horizontal distance of weight of embankment on improved ground from front edge of improved ground (m)

x_s : horizontal distance of weight of stabilized soil from front edge of improved ground (m)

x_u : horizontal distance of weight of unstabilized soil between grid walls from front edge of improved ground (m)

y_{DAH} : vertical distance of horizontal component of total dynamic active force from bottom of improved ground (m)

y_{DPH} : vertical distance of horizontal component of total dynamic passive force from bottom of improved ground (m)

y_e : vertical distance of weight of embankment on improved ground from bottom of improved ground (m)

y_s : vertical distance of weight of stabilized soil from bottom of improved ground (m)

y_u : vertical distance of weight of unstabilized soil between grid walls from bottom of improved ground (m)

ϕ'_b : internal friction angle of soil beneath improved ground.

The magnitude of the horizontal component of total dynamic active forces, P_{DAH} can be calculated by Equation (6.56) for various liquefaction resistance values, F_L.

a)   in the case the ground is liquefiable and $F_L < 1.0$, the soil behind improved ground is assumed to fully liquefy

$$P_{DAH} = \frac{1}{2} \cdot \gamma'_c \cdot H_c^2 + w \cdot H_c + P_{Rw} + P_{Dw} \qquad (6.56a)$$

b)   in the case the ground is liquefiable and $F_L > 1.0$, the soil behind improved ground is assumed to partially liquefy

$$P_{DAH} = \max \begin{cases} K_A \cdot \left(\frac{1}{2}\gamma'_c \cdot H_c^2 + w \cdot H_c\right) + P_{Rw} + r_u \cdot (1 - K_A) \cdot \\ \left(\frac{1}{2}\gamma'_c \cdot H_c^2 + w \cdot H_c\right) + P_{Dw} \\ K'_{EA} \cdot \left(\frac{1}{2}\gamma'_c \cdot H_c^2 + w \cdot H_c\right) + P_{Rw} \end{cases} \qquad (6.56b)$$

c)   in the case the ground is assumed not to liquefy in the soil behind improved ground

$$P_{DAH} = K_{EA} \cdot \left(\frac{1}{2}\gamma'_c \cdot H_c^2 + w \cdot H_c\right) + P_{Rw} \qquad (6.56c)$$

Similarly, the magnitude of the horizontal component of total dynamic passive force, P_{DPH} can be calculated by Equation (6.57) for various liquefaction resistance values, F_L.

a)   in the case the ground is liquefiable and $F_L < 1.0$, the soil in front of improved ground is assumed to fully liquefy

$$P_{DPH} = \frac{1}{2}\gamma'_c \cdot H_c^2 + P_{Rw} - P_{Dw} \qquad (6.57a)$$

b)   in the case the ground is liquefiable and $F_L > 1.0$, the soil in front of improved ground is assumed to partially liquefy

$$P_{DPH} = \min \begin{cases} K_p \cdot \frac{1}{2}\gamma'_c \cdot H_c^2 + P_{Rw} + r_u \cdot (1 - K_p) \cdot \frac{1}{2}\gamma'_c \cdot H_c^2 - P_{Dw} \\ K'_{EP} \cdot \left(\frac{1}{2}\gamma'_c \cdot H_c^2 + w \cdot H_c\right) + P_{Rw} \end{cases} \qquad (6.57b)$$

c) in the case the ground is assumed not to liquefy in front of improved ground

$$P_{DPH} = K_{EP} \cdot \frac{1}{2} \gamma_c' \cdot H_c^2 + P_{Rw}$$

(6.57c)

where

K_A : coefficient of static active earth pressure

$$K_A = \frac{\cos^2 \phi'}{\cos \delta \cdot \left\{ 1 - \sqrt{\dfrac{\sin(\phi' + \delta) \cdot \sin \phi'}{\cos \delta}} \right\}^2}$$

K_P : coefficient of static passive earth pressure

$$K_P = \frac{\cos^2 \phi'}{\cos \delta \cdot \left\{ 1 + \sqrt{\dfrac{\sin(\phi' + \delta) \cdot \sin \phi'}{\cos \delta}} \right\}^2}$$

K_{EA} : coefficient of dynamic active earth pressure

$$K_{EA} = \frac{\cos^2 (\phi' - \theta)}{\cos \theta \cdot \cos(\delta + \theta) \cdot \left\{ 1 + \sqrt{\dfrac{\sin(\phi' + \delta) \cdot \sin(\phi' - \theta)}{\cos(\delta + \theta)}} \right\}^2}$$

K_{EP} : coefficient of dynamic passive earth pressure

$$K_{EP} = \frac{\cos^2 (\phi' - \theta)}{\cos \theta \cdot \cos (\delta - \theta) \cdot \left\{ 1 + \sqrt{\dfrac{\sin(\phi' + \delta) \cdot \sin(\phi' - \theta)}{\cos(\delta + \theta)}} \right\}^2}$$

K_{EA}' : coefficient of dynamic active earth pressure incorporating pore water pressure generation

$$K_{EA}' = \frac{\cos^2 (\phi_{r_u}' - \theta)}{\cos \theta \cdot \cos(\delta_{r_u} + \theta) \cdot \left\{ 1 + \sqrt{\dfrac{\sin(\phi_{r_u}' + \delta_{r_u}) \cdot \sin(\phi_{r_u}' - \theta)}{\cos(\delta_{r_u} + \theta)}} \right\}^2}$$

K_{EP}' : coefficient of dynamic passive earth pressure incorporating pore water pressure generation

$$K_{EP}' = \frac{\cos^2(\phi_{r_u}' - \theta)}{\cos \theta \cdot \cos(\delta_{r_u} + \theta) \cdot \left\{ 1 + \sqrt{\dfrac{\sin(\phi_{r_u}' + \delta_{r_u}) \cdot \sin(\phi_{r_u}' - \theta)}{\cos(\delta_{r_u} + \theta)}} \right\}^2}$$

r_u : excess pore water pressure ratio

$$r_u = \frac{\Delta u}{\sigma'_v}$$

$$= F_L^{-7}$$

P_{Dw} : total dynamic water force per unit length (kN/m)

$$P_{Dw} = \frac{7}{8} \cdot k_h \cdot (\gamma_w + r_u \cdot \gamma) \cdot H_w^2$$

P_{Rw} : total residual water force per unit length (kN/m)
H_c : thickness of ground (m)
w : embankment pressure (kN/m^2)
Δu : excess pore water pressure (kN/m^2)
σ'_v : effective overburden pressure (kN/m^2)
γ'_c : unit weight of soil (kN/m^3)
ϕ' : internal friction angle
ϕ'_{r_u} : internal friction angle incorporating excess pore water pressure

$$\phi'_{r_u} = \tan^{-1}\{(1 - r_u) \cdot \tan \phi'\}$$

k_h : design seismic coefficient at bottom of improved ground
k_{h0} : design seismic coefficient at ground surface (at top of improved ground)
δ : friction angle of boundary of improved ground and unstabilized soil (°)

$$\delta = \phi'/2$$

δ_{r_u} : friction angle of boundary of improved ground and unstabilized soil
      incorporating excess pore water pressure (°)

$$\delta_{r_u} = \delta$$

θ : resultant angle of seismic coefficient (°)
      for soil upper than water level

$$\theta = \tan^{-1} k_h$$

for soil lower than water level

$$\theta = \tan^{-1}\left(\frac{\gamma}{\gamma'} \cdot k_{h0}\right)$$

7.3.5.2   Bearing capacity

In the bearing capacity calculation, it is assumed that the self-weight of unstabilized soil between the grid walls is supported by itself but the horizontal load induced by the seismic inertia forces is supported by the grid walls. The subgrade reactions at the bottom of the improved ground, t_1 and t_2 calculated by the force equilibrium as shown in Figure 6.68 and are assured lower than the allowable bearing capacity, as Equation (6.58). The subgrade reactions, t_1 and t_2, on the bottom of the improved ground are calculated by Equation (6.59).

$$t_1 \leq q_a$$
$$t_2 \leq q_a \tag{6.58}$$

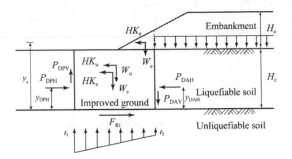

Figure 6.68 Bearing capacity calculation.

$$e = \frac{B_i}{2} - \frac{\begin{array}{c}(P_{DPH} \cdot y_{DPH} + P_{DAV} \cdot x_{DAV} + W_e \cdot x_e + W_s \cdot x_s + W_u \cdot x_u) \\ - (P_{DAH} \cdot y_{DAH} + HK_s \cdot y_s + HK_u \cdot y_u + HK_e \cdot y_e)\end{array}}{W_e + W_s + W_u + P_{DAV} - P_{DPV}} \tag{6.59a}$$

In the case of $e <= B_i/6$

$$\left.\begin{array}{l} t_1 = \dfrac{W_e + W_s + W_u + P_{DAV} - P_{DPV}}{B_i} \cdot \left(1 + \dfrac{6 \cdot e}{B_i}\right) \\[4mm] t_2 = \dfrac{W_e + W_s + W_u + P_{DAV} - P_{DPV}}{B_i} \cdot \left(1 - \dfrac{6 \cdot e}{B_i}\right) \end{array}\right\} \tag{6.59b}$$

In the case of $e >= B_i/6$

$$t_1 = \frac{2 \cdot (W_e + W_s + W_u + P_{DAV} - P_{DPV})}{3B_i} \tag{6.59c}$$

where
 t_1 : subgrade reaction at front edge of improved ground (kN/m²)
 t_2 : subgrade reaction at rear edge of improved ground (kN/m²)
 e : eccentricity (m).
 The allowable bearing capacity can be calculated by Equation (6.60).

$$q_a = \frac{1}{Fs} \cdot \left(c_{ub} \cdot N_c + \frac{1}{2} \cdot \gamma_b \cdot B_i \cdot N_\gamma\right) + \gamma_c \cdot H_i \cdot N_q \tag{6.60}$$

where
 B_i : width of improved ground (m)
 H_i : height of improved ground (m)
 Fs : safety factor
 N_c : bearing capacity factor of soil beneath improved ground

N_γ : bearing capacity factor of soil beneath improved ground
N_q : bearing capacity factor of soil beneath improved ground
q_a : allowable bearing capacity (kN/m^2)
c_{ub} : undrained shear strength of soil beneath improved ground (kN/m^2)
γ_b : unit weight of soil beneath improved ground (kN/m^3)
γ_c : unit weight of soft ground (kN/m^3).

7.3.6 Examination of the internal stability of improved ground

In the "internal stability analysis," the induced stresses in the improved ground are calculated based on the elastic theory. The width of improved ground and the thickness of grid wall are determined so that the induced stresses become lower than the allowable strengths of the stabilized soil. In the calculation, the stabilized soil is generally assumed to have a uniform property for the sake of simplicity even it contains possibly weaker zones due to construction process such as overlap joints.

According to the accumulated experiences in design, the internal stability evaluation at the four critical parts is considered: (a) subgrade reaction at the front edge of the improved ground, (b) average shear stress along a horizontal shear plane, (c) average shear stress along the horizontal plane of the rear most grid wall, and (d) average shear stress along a vertical shear plane.

7.3.6.1 Subgrade reaction at the front edge of improved ground

The subgrade reaction at the toe of the improved ground is calculated by Equation (6.22) and assured to be lower than the allowable unconfined compressive strength, q_{ua} as Equation (6.61). The q_{ua} is calculated by Equation (6.62).

$$t_1 \leq q_{ua} \tag{6.61}$$

$$q_{ua} = \frac{1}{Fs} q_{uf} \tag{6.62}$$

where
  Fs : safety factor
  q_{ua} : allowable unconfined compressive strength (kN/m^2)
  q_{uf} : unconfined compressive strength of in-situ stabilized soil (kN/m^2)
  t_1 : subgrade reaction at front edge of improved ground (kN/m^2).

It is widely known that the field strength is equal or lower than the laboratory strength in the case of clay. In the case of sand, however, it has been found that the field strength is in some cases higher than the laboratory strength. Therefore, selecting q_{ul} in Equation (6.63) instead of q_{uf} can be considered as safe side design.

7.3.6.2 Average shear stress along a horizontal shear plane

The horizontal shear failure of the improved ground is then assumed as shown in Figure 6.69, where the average shear stress induced along the horizontal plane should

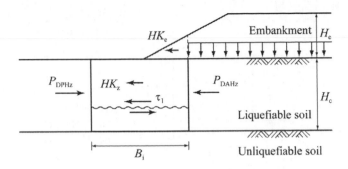

Figure 6.69 Examination of horizontal shear failure.

be lower than the allowable shear strength. The average shear stress, τ_1 is calculated for an assumed depth of horizontal shear plane by Equation (6.63).

$$\tau_1 = \frac{P_{DAHz} + HK_{sz} + HK_{uz} + HK_e - P_{DPHz}}{a_s \cdot B_i} \qquad (6.63)$$

where

τ_1 : average shear stress along horizontal shear plane (kN/m^2)

P_{DAHz} : horizontal component of total dynamic active force per unit length of soft ground and embankment above assumed horizontal shear plane (kN/m)

P_{DPHz} : horizontal component of total dynamic passive force per unit length of soft ground above assumed horizontal shear plane (kN/m)

HK_e : total seismic inertia force per unit length of embankment on improved ground (kN/m)

HK_{sz} : total seismic inertia force per unit length of stabilized soil above assumed horizontal shear plane (kN/m)

HK_{uz} : total seismic inertia force per unit length of unstabilized soil between stabilized grid walls above assumed horizontal shear plane (kN/m).

7.3.6.3 Average shear stress along the horizontal plane of the rear most grid wall

The punching shear failure of the grid wall is examined where the rear most grid wall is sheared by the horizontal forces of the soft ground and embankment (see Figure 6.70). The average shear stress induced in the rear most grid wall is calculated for an assumed depth of horizontal shear plane by Equation (6.64) and assured to be lower than the allowable strength.

$$\tau_2 = \frac{H_z \cdot L_{gr} + P_{DAHz} \cdot L_{gr} + HK_{sz} \cdot L_{gr} + HK_{uz} \cdot L_{gr} + HK_e \cdot L_{gr} - P'_{DPHz} \cdot L_{gr}}{B_{gr} \cdot L_{gr} + 2 \cdot B_{gr} \cdot z}$$

$$P'_{DPHz} = \frac{1}{2} \cdot K_0 \cdot \gamma_c \cdot z^2 + P_{Rw} \qquad (6.64)$$

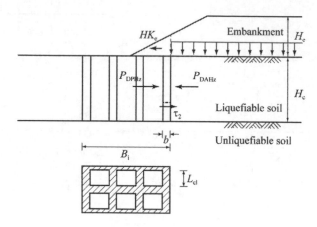

Figure 6.70 Examination of punching shear failure.

where

τ_2 : average shear stress along horizontal plane of the rear most grid wall (kN/m^2)

z : depth of assumed horizontal shear failure plane (m)

B_{gr} : thickness of grid wall (m)

L_{gr} : length of grid wall (m)

P_{DAHz} : horizontal component of total dynamic active force per unit length of soft ground and embankment above assumed horizontal shear plane (kN/m)

P'_{DPHz} : horizontal component of total dynamic passive force of unstabilized soil between stabilized grid walls above assumed horizontal shear plane (kN/m).

7.3.6.4 Average shear stress along a vertical shear plane

The induced average shear stress along the vertical shear plane, τ_v as shown in Figure 6.71 is calculated by Equation (6.65) and should be lower than the design shear strength of the stabilized soil.

$$\tau_v = \frac{\{(P_1 - W_1) \cdot a_s - W_{E_l} + P_{PV'}\} \cdot L_u}{H_i \cdot L_T} \tag{6.65}$$

where

H_i : height of improved ground (m)

L_u : unit length of improved ground (m)

L_T : thickness of grid wall of improved ground (m)

P_l : total subgrade reaction force at the part of B_l (kN/m)

P'_{PV} : total subgrade reaction force at the part of B_l (kN/m)

W_l : weight per unit length of improved ground at the part of B_l (kN/m)

W_{El} : weight per unit length of embankment at the part of B_l (kN/m)

τ_v : average shear stress along vertical shear plane (kN/m^2).

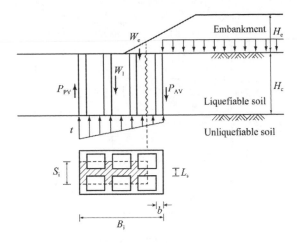

Figure 6.71 Examination of vertical shear failure.

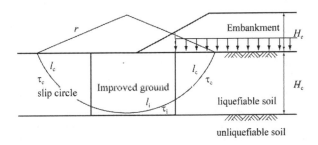

Figure 6.72 Slip circle analysis.

7.3.7 Slip circle analysis

A slip circle analysis is carried out to evaluate the overall stability of the improved ground, embankment and soft ground. As the strength of stabilized soil is usually a very high value, a slip circle analysis passing through the improved ground is not necessary in many cases. In the case where sufficient bearing capacity is assured in the bearing capacity analysis, a slip circle analysis is not necessary. The safety factor against slip circle failure is calculated by the modified Fellenius analysis (see Figure 6.72) with Equation (6.66). The design safety factor of 1.3 is adopted for the static condition in many cases.

$$Fs_{sp} = \frac{r \cdot (\tau_e \cdot l_e + \overline{\tau}_i \cdot l_i + \tau_c \cdot l_c)}{W_e \cdot x_e} \qquad (6.66)$$

where
Fs_{sp} : safety factor against slip circle failure
l_c : length of circular arc in soft ground (m)

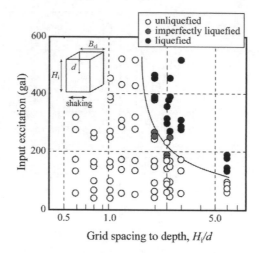

Figure 6.73 Relationship between input excitation and ratio of grid spacing to depth (Takahashi *et al.*, 2006).

l_e : length of circular arc in embankment (m)
l_i : length of circular arc in improved ground (m)
r : radius of slip circle (m)
W_e : weight per unit length of embankment (kN/m)
x_e : horizontal distance of weight of embankment from center of slip circle (m)
τ_c : shear strength of soft ground (kN/m²)
τ_e : shear strength of embankment (kN/m²)
$\bar{\tau}_i$: average shear strength of improved ground (kN/m²).

7.3.8 *Important issues on design procedure*

7.3.8.1 Effect of grid wall spacing on liquefaction prevention

As already shown in Figure 6.66, a quite simple guideline for the method is proposed in the design procedure in which the ratio of grid, B_{cl}/H_i should be less than 0.8 to prevent liquefaction against earthquake attack of amplitude 200 gal. However, this method does not take into account the different seismic behavior at different depths but evaluates the possibility of liquefaction only at the mid-depth. Furthermore, Kodaka *et al.* (2002) pointed out a limitation of the guideline, in which the grid wall spacing should be quite small when the thickness of sandy layer becomes small. Takahashi *et al.* (2006) carried out a series of centrifuge model tests to investigate the effect of grid wall spacing on the generation of pore pressure and seismic response in a sand layer. They revealed that (1) the improvement effect of grid type improvement on liquefaction prevention is influenced not only by the grid wall spacing but also by the magnitude of excitation and the concerned depth, (2) the current guideline should be modified by incorporating the effects of magnitude of excitation and the depth to be concerned. They proposed a new rational design guideline as shown in Figure 6.73.

REFERENCES

Adams, T., Filz, G. & Navin, M. (2009) Stability of embankments and levees on deep-mixed foundations. *Proc. of the International Symposium on Deep Mixing and Admixture Stabilization.*

Åhnberg, H. (2003) Measured permeabilities in stabilised Swedish soils. *Proc. of the 3rd International Conference on Grouting and Ground Treatment.* pp. 622–633.

Akamoto, H. & Miyake, M. (1989) Deformation characteristics of the improved ground by a group of treated soil columns, *Proc. of the 24th Annual Conference of the Japanese Society of Soil Mechanics and Foundation Engineering* (in Japanese).

Amano, H., Morita, T., Tukada, Y. & Takahashi, Y. (1986) Design for deep mixing method for high road embankment – Hayashima I.C. – , *Proc. of the 21st Annual Conference of the Japanese Society of Soil Mechanics and Foundation Engineering* (in Japanese).

Architectural Institute of Japan (2000) *Recommendations for Loads on Building.* Architectural Institute of Japan. 512p. (in Japanese).

Architectural Institute of Japan (2006) *Recommendations for Design of Ground Improvement for Building Foundations.* Architectural Institute of Japan (in Japanese).

Broms, B.B. (1999) Design of lime, lime/cement and cement columns. *Proc. of the International Conference on Dry Mix Methods for Deep Soil Stabilization.* pp. 125–153.

Cement Deep Mixing Method Association (1999) *Cement Deep Mixing Method (CDM), Design and Construction Manual* (in Japanese).

Coastal Development Institute of Technology (2002) *The Deep Mixing Method – Principle, Design and Construction.* A.A. Balkema Publishers. 123p.

EuroSoilStab (2002) Development of design and construction methods to stabilise soft organic soils. *Design Guide Soft Soil Stabilization. EC project BE96-3177.* 94p.

Filz, G.M. & Navin, M.P. (2006) Stability of column-supported embankments. *Final contract report, VTRC 06-CR13,* Virginia Transportation Research Council.

Fire and Disaster Management Agency (1995) *Criteria for a Specific Operation of Outdoor Storage Tank using Deep Mixing Method.* Fire and Disaster Management Agency. No. 150, 2p. (in Japanese).

Han, J., Parsons, R.L., Huang, J. & Sheth, A.R.Stockholm (2005) Factors of safety against deep-seated failure of embankments over deep mixed columns. *Proc. of the International Conference on Deep Mixing – Best Practice and Recent Advances,* Stockholm. pp. 231–236.

Kitazume, M. (1994) Model and analytical studies on stability of improved ground by deep mixing method. *Doctoral thesis, Tokyo Institute of Technology.* 73p. (in Japanese).

Kitazume, M. (2008) Stability of group column type DM improved ground under embankment loading. *Report of the Port and Airport Research Institute.* Vol. 47. No. 1. pp. 1–53.

Kitazume, M. & Maruyama, K. (2006) External stability of group column type deep mixing improved ground under embankment loading. *Soils and Foundations.* Vol. 46. No. 3. pp. 323–340.

Kitazume, M. & Maruyama, K. (2007) Internal stability of group column type deep mixing improved ground under embankment loading. *Soils and Foundations.* Vol. 47. No. 3. pp.437–455.

Kitazume, M. & Nagao, T. (2007) Studies of reliability based design on deep mixing improved ground. *Report of the Port and Airport Research Institute.* Vol. 46. No. 1. pp. 3–44 (in Japanese).

Kitazume, M., Miyajima, S., Ikeda, T. & Karastanev, D. (1996) Bearing capacity of improved ground with column type DMM. *Proc. of the Second International Conference on Ground Improvement Geosystems.* pp. 503–508.

Kitazume, M., Nakamura, T. & Terashi, M. (1991) Reliability of clay ground improved by the group column type DMM with high replacement. *Report of the Port and Harbour Research Institute.* Vol. 30. No. 2. pp. 305–326 (in Japanese).

Kitazume, M., Okano, K. & Miyajima, S. (2000) Centrifuge model tests on failure envelope of column type DMM improved ground. *Soils and Foundations*. Vol. 40. No. 4. pp. 43–55.

Kivelo, M. (1998) Stabilization of embankments on soft soil with lime/cement columns, *Doctoral thesis 1023, Royal Institute of Technology*, Sweden.

Kurisaki, K., Sugiyama, K., Izutsu, H., Yamamoto, M., Takeuchi, G., Ohishi, K., Katagiri, M., Terashi, M. & Ishii, T. (2005) Physical and numerical simulation of deep mixed foundation Part 1: Bearing capacity of treated soil resting on a rigid layer. *Proc. of the International Conference on Deep Mixing Best Practice and Recent Advances*. pp. 255–262.

Ministry of Construction (1999) Design and construction manual of liquefaction prevention techniques (draft). *Joint Research Report*. 186p. (in Japanese).

Ministry of Land, Infrastructure, Transport & Tourism (2007) *Technical Standards for Port and Harbour Facilities* (in Japanese).

Ministry of Transport (1989) *Technical Standards for Port and Harbour Facilities* (in Japanese).

Ministry of Transport (1999) *Technical Standards for Port and Harbour Facilities*. pp. 525–536 (in Japanese).

Miyake, M., Akamoto, H. & Aboshi, H. (1988) Sliding failure of the improved ground by a group of treated soil columns. *Proc. of the 23rd Annual Conference of the Japanese Society of Soil Mechanics and Foundation Engineering* (in Japanese).

Ogawa, M., Sakai, S. & Tanaka, M. (1996) Simple method for prediction on deformation of improved ground by deep mixing method as solidified columns. *Proc. of the symposium on Cement Stabilized Soil*. pp. 217–222 (in Japanese).

Ogawa, S., Yamamoto, Y. & Bessho, M. (1996) Deformation prediction method for improved ground by deep mixing method based on centrifuge model tests. *Proc. of the symposium on Cement Stabilized Soil*. pp. 211–216 (in Japanese).

Ohishi, K., Katagiri, M., Terashi, M., Ishii, T. & Miyakoshi, Y. (2005) Physical and numerical simulation of deep mixed foundation Part 2: Revetment on treated soil block underlain by a sandy layer. *Proc. of the International Conference on Deep Mixing Best Practice and Recent Advances*. pp. 281–288.

Ohno, M. & Terashi, M. (2005) Behavior of deep mixed foundation by numerical simulation. (in Japanese).

Public Works Research Center (1999) *Technical Manual on Deep Mixing Method for On Land Works*. 326p. (in Japanese).

Public Works Research Center (2004) *Technical Manual on Deep Mixing Method for On Land Works*. 334p. (in Japanese).

Public Works Research Institute (2007) *Manual of ALicc Method for Ground Improvement*. 89p. (in Japanese).

Swedish Geotechnical Institute (1997) Lime and lime cement columns – guide for project planning, construction and inspection, *Swedish Geotechnical Society SGF Report 4:95E*, 111 p. 4:95E.

Terashi M. & Kitazume, M. (1987) Interference effect on bearing capacity of foundations on sand. *Report of the Port and Harbour Research Institute*. Vol. 26. No. 2. pp. 413–436 (in Japanese).

Terashi M., Kitazume, M. & Nakamura, T. (1989) External forces acting on a stiff soil mass improved by DMM. *Report of the Port and Harbour Research Institute*. Vol. 27. No. 2. pp. 147–184 (in Japanese).

Terashi M., Kitazume, M. & Yajima, M. (1985) Interaction of soil and buried rigid structures. *Proc. of the 11th International Conference on Soil Mechanics and Foundation Engineering*. pp. 1757–1760.

Terashi, M. (2003) The state of practice in deep mixing method. Grouting and ground treatment, *Proc. of the 3rd International Conference, ASCE Geotechnical Special Publication*. No. 120. No. 1. pp. 25–49.

Terashi, M. (2005) Keynote Lecture: Design of deep mixing in infrastructure application. *Proc. of the International Conference on Deep Mixing Best Practice and Recent Advances.* Vol. 1. pp. 25–45.

Terashi, M. & Tanaka, H. (1981a) Ground improved by deep mixing method. *Proc. of the 10th International Conference on Soil Mechanics and Foundation Engineering.* Vol. 3. pp. 777–780.

Terashi, M. & Tanaka, H. (1981b) On the permeability of cement and lime treated soils. *Proc. of the 10th International Conference on Soil Mechanics and Foundation Engineering.* Vol. 4. pp. 947–948.

Terashi, M. & Tanaka, H. (1983) Settlement analysis for deep mixing method. *Proc. of the 8th European Regional Conference on Soil Mechanics and Foundation Engineering.* Vol. 2. pp. 955–960.

Terashi, M., Ooya, T., Fujita, T., Okami, T., Yokoi, K. & Shinkawa, N. (2009) Specifications of Japanese dry method of deep mixing deduced from 4300 projects, *Proc. of the International Symposium on Deep Mixing and Admixture Stabilization.* pp. 647–652.

Terashi, M., Tanaka, H. & Kitazume, M. (1983) Extrusion failure of ground improved by the deep mixing method. *Proc. of the 7th Asian Regional Conference on Soil Mechanics and Foundation Engineering.* Vol. 1. pp. 313–318.

The Building Center of Japan (1997) *Design and Quality Control Guideline of Improved Ground for Building.* 473p. (in Japanese).

The Overseas Coastal Area Development Institute of Japan (1991) *Technical Standards and Commentaries for Port and Harbour Facilities in Japan (English Version).*

The Overseas Coastal Area Development Institute of Japan (2002) *Technical Standards and Commentaries for Port and Harbour Facilities in Japan (English Version).*

The Overseas Coastal Area Development Institute of Japan (2009) *Technical Standards and Commentaries for Port and Harbour Facilities in Japan (English Version).* 1028p.

The Ports and Harbours Association of Japan (1989) *Technical Standards and Commentaries for Port and Harbour Facilities in Japan (Japanese Version).* (in Japanese).

The Ports and Harbours Association of Japan (1999) *Technical Standards and Commentaries for Port and Harbour Facilities in Japan (Japanese Version).* (in Japanese).

The Ports and Harbours Association of Japan (2007) *Technical Standards and Commentaries for Port and Harbour Facilities in Japan (Japanese Version).* (in Japanese).

Tsukada, Y., Kawamura, K., Murai, I., & Goto, T. (1988) Behavior of improved ground by deep mixing method under horizontal load (1) and (2), *Proc. of the 23rd Annual Conference of the Japanese Society of Soil Mechanics and Foundation Engineering.* pp. 2267–2270 (in Japanese).

Zen, K., Yamazaki, H., Watanabe, A., Yoshizawa, H. & Tamai, A. (1987) Study on a reclamation method with cement-mixed sandy soils – Fundamental characteristics of treated soils and model tests on the mixing and reclamation. *Technical Note of the Port and Harbour Research Institute.* No. 579. 41p. (in Japanese).

Chapter 7

QC/QA for improved ground – Current practice and future research needs

I INTRODUCTION

The quality of stabilized soil depends upon a number of factors including the type and condition of original soil, the type and amount of binder, and the execution process as described in Chapter 2. The quality control and quality assurance (QC/QA) practice which focuses upon the quality of stabilized soil was originally established in Japan and the Nordic countries and has been accepted worldwide for more than three decades. It comprises a laboratory mix test, field trial test, monitoring and control of construction parameters during execution and verification by measuring the engineering characteristics of stabilized soil either by unconfined compression tests on core samples or by sounding. Diversification of application, soil type, and execution system, together with the improved understanding on the behavior of improved ground necessitate our profession to review the current QC/QA practice.

Section 2 will discuss the importance of various QC/QA related activities along the work flow of a deep mixing project. Section 3 summarizes the current QC/QA procedures for stabilized soil and Section 4 discusses the technical issues to be considered.

The purpose of deep mixing is not only to manufacture a good quality stabilized soil but to create an improved ground which guarantees the performance of a superstructure. The improved ground by the deep mixing method is a composite system comprising stabilized soil columns and original soils. Section 5 discusses the QC/QA procedures for improved ground by the deep mixing method.

2 FLOW OF A DEEP MIXING PROJECT AND QC/QA

Geotechnical design procedure in deep mixing project differs for different application and the construction control items and values differ for different execution system. However the overall work flow exemplified in Figure 7.1 (Terashi, 2003) is common to all the in-situ stabilization projects. Parties involved in a deep mixing project are the project owner, design engineers, prime contractor, deep mixing contractor and soil investigation and testing firm. In the figure, the project owner's functions are shown in white frame (plain line frame), activities related to the geotechnical design are in slight gray frame (double line frame), activities related to the process design and actual execution with QC are in light gray frame (triple line frame), and accumulated

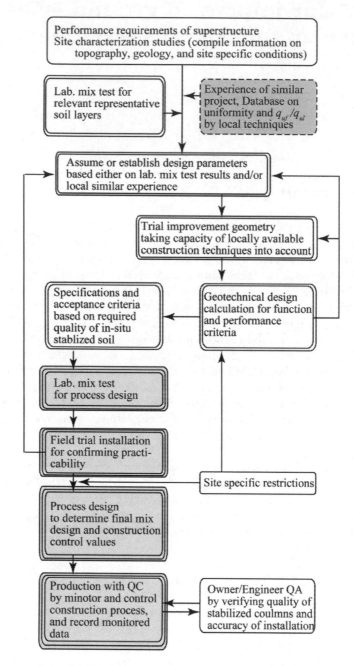

Figure 7.1 Workflow of deep mixing project (Terashi, 2003).

experience and database on locally available execution processes is shown in gray frame (dotted line frame).

The project owner defines the functional and performance requirements of the structure, carries out site characterization studies, provides information regarding the site-specific restrictions, and sometimes clarifies the purpose and requirements for ground improvement based on the conceptual design.

The geotechnical design calculation is to determine the size of improved area, installation depth, end-bearing condition, installation pattern and necessity of over-lapping so that the improved ground can satisfy the performance criteria of the superstructure. Before the geotechnical design calculation, the designer should assume/establish the required quality of in-situ stabilized soil and required level of accuracy of installation by taking into account the capability of deep mixing equipment available locally based on similar experience or a local database. It should be noted that the validity of assumptions is only confirmed after the field trial test or the actual production. The preparation of a contract document including acceptance criteria and verification procedures is, therefore, an important role of the geotechnical design. In a region (or for a soil type) without sufficient previous experience or database, laboratory mix test is the start point for the geotechnical design.

The process design is to determine the binder type, binder content, construction procedure, construction control items and construction control values in order to realize the required quality of in-situ stabilized soil (such as strength and uniformity) and to determine the construction procedure to realize the location, depth, contact with bearing layer, and reliable overlap of columns to the level of accuracy that the geotechnical design requires. A laboratory mix test and field trial test are often carried out for the process design. The deep mixing contractor is also expected to co-operate the owner's quality assurance and verification. The results of verification testing together with the laboratory test results will be accumulated to improve the local database. Quality assurance of the deep mixing method to fulfill the requirements of the geotechnical design cannot be achieved only by process control (QC) during construction conducted by the deep mixing contractor, but it should involve relevant activities that are carried out prior to, during and after construction by all the parties involved in the deep mixing project. Table 7.1 provides brief descriptions of QC/QA related activities along with the work flow. Usually the site investigation of the original ground, for example, is not considered as a part of QA but it is classified as one of the important relevant activities in Table 7.1. If the site investigation failed to identify the existence of a problematic layer, the laboratory mix test would not be undertaken for the layer, which might result in insufficient process design (including QC/QA methods/ procedures) and would cause difficulty in interpretation of the field trial stabilized soil columns and/or verification test of production columns.

3 QC/QA FOR STABILIZED SOIL – CURRENT PRACTICE

3.1 Relation of laboratory strength, field strength and design strength

Whatever the type of application and the function of stabilized soil columns, it is important to discuss the QC/QA procedures for the stabilized soil. As described in Chapter 2, the strength of stabilized soil is affected by many factors such as soil properties (natural water content, liquid limit, plastic limit, pH, organic matter content, grain size distribution and clay minerals), type and quantity of binder, mixing degree, and curing conditions. The effects of these factors are quite complex, making it difficult to directly determine field strength only by a laboratory mix test.

Table 7.1 Work flow of a deep mixing project and QC/QA related activities.

Activity	Description	Impact on QA/QC of in-situ mixing
Definition of a Project	Provide functional and performance criteria of the expected structure.	
Site characterization study by preliminary site investigation	Examine original ground to determine soil profile and soil characteristics of the site. Compile information on topography, geology, and site specific constraints.	
Conceptual or preliminary design	Conceptual design with and without ground improvement to identify the necessity and requirement of ground improvement.	
Additional or detailed Site Investigation	Detailed site investigation for geotechnical design including ground improvement should be planned and conducted based on conceptual design and preliminary site investigation.	Identify soft soil layer(s) to be improved by in-situ mixing
Laboratory mix test in the geotechnical design phase	Laboratory mix test should be planned for relevant representative soil layers identified by the site investigation. The lab test provides information for evaluating the feasibility of deep mixing and estimating the in-situ strength and/or binder mix design.	Pre-production QA
Assume Design Parameters for In-situ Treated Soil	Determine design parameters of original ground and assume design parameters of in-situ treated soil based either on laboratory mix test results and/or on local similar experience. Reasonable assumption is made possible only when the reliable local database for the relation between laboratory and field condition is available.	Contractual scheme may dictate when and who carries out Lab mix test(s) Database on the quality of in-situ treated soil by locally available execution system
Trial Improvement Geometry	Assume the improvement geometry for design calculation taking the capability of locally available construction techniques.	Define the target of of QA/QC quality of deep mixed soil and necessary installation accuracy
Geotechnical Design Calculation	Determine the column installation pattern, end-bearing condition, depth and extent of improvement so that the structure can satisfy its function and performance criteria.	
Specification and Acceptance Criteria	Define the practicable acceptance criteria based on required quality of in-situ treated soil and required accuracy of column installation. Specify the practicable QA processes and verification technique.	Pre-production QA
Laboratory mix test for Process design	Laboratory mix test should be planned in the same way as Lab mix tests conducted in the geotechnical design phase. The lab test provides information for planning field trial and/or for determining the production binder mix design.	Contractual scheme may dictate when and who carries out Lab mix test(s)
Preliminary Process Design	Examine construction process, construction control items and construction control values so that the improved ground may satisfy the specifications given by the geotechnical design. Preliminary process design is based on the information on the laboratory mix test, site investigation, deep mixing database and contractor's experience.	Depending on the contractual scheme, VE proposal for improving the quality and efficiency of ground improvement work
Field trial installation & Final Process Design	Confirm the practicability of assumptions made by geotechnical design and Determine the final mix design and construction process of production columns including the determination of construction control values so as to assure the quality of in-situ treated soil and accuracy of installation. Well programmed field trial installation (or initial production columns) will provide the basis of process design.	Pre-production QA
Production with QC/QA	Monitor & control the construction process during production of treated soil columns, and record the monitored data of control items for submission to the owner for QA purpose.	QA/QC during construction
Owner QA Verification and Acceptance	Investigate a required number of production columns to verify the quality of treated soil and accuracy of installation.	QA during and/or after the production
Monitoring during & after the construction of superstructure	Monitoring of the behavior of the improved ground provides the information on the quality of ground improvement including the validity of geotechnical & process design.	
Independent Review	In a difficult project, the expert consultant and/or academia may be involved in the above process to help the project owner evaluate the works conducted by different parties involved and coordinate them.	

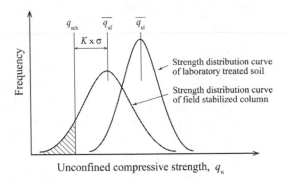

Figure 7.2 Field and laboratory strength of stabilized soil.

The deep mixing machines must be simple and tough enough to endure severe working conditions. Mixing time in practice must be as short as possible for economic reasons. Hence, in-situ mixing conditions and curing conditions are quite different from the standard laboratory testing, and the strength of the in-situ stabilized soil column is usually different from that in the laboratory. The in-situ stabilized soil columns have a relatively large strength variability even if the execution is done with the established mixing machine and with the best care. Average compressive strength, q_{ul} and the deviation of the laboratory specimen and the average strength, q_{uf} and the deviation of the in-situ column are schematically shown in Figure 7.2. Usually the in-situ stabilized soil column has smaller average strength and larger strength deviation than those of the laboratory specimen. The design strength, q_{uck}, is derived from q_{uf} by incorporating the strength deviation as Equation (7.1). The target strength of the laboratory specimen should be determined by incorporating the strength difference and the strength deviation. When using statistical measures for quality control, the following relationship between field strength and the design standard strength must be formulated if the field strength of the improved soil is assumed to have a normal distribution curve.

$$q_{uck} \leq \overline{q_{uf}} - K \cdot \sigma$$
$$\overline{q_{uf}} = \lambda \cdot \overline{q_{ul}}$$

(7.1)

where
  K : coefficient
  q_{uck} : design standard strength (kN/m^2)
  $\overline{q_{uf}}$: average unconfined compressive strength of in-situ stabilized column (kN/m^2)
  $\overline{q_{ul}}$: average unconfined compressive strength of laboratory stabilized soil (kN/m^2)
  σ : standard deviation of the field strength (kN/m^2)
  λ : ratio of q_{uf}/q_{ul} (see empirical value in Chapter 3).

3.2 Flow of quality control and quality assurance

To ensure sufficient quality of the stabilized column, quality control and quality assurance is required before, during and after construction. For this purpose, quality control

Table 7.2 Flow chart for quality control and quality assurance.

Time sequence	QC/QA activities	Objectives
Prior to production	Laboratory mix test	Selection of Binder type
		Determine Binder content
	Field Trial Test	Final Mix design
		Determine Process control value
During production	Store & Prepare binder	Maintain Quality of binder
	Control, monitor & record	Ensure Geometric layout
	construction parameters	(plan, verticality and depth)
		Ensure quality of stabilized soil
After production	Continuous Coring	Observe continuity & uniformity
	Test on selected sample	Verify the quality of stabilized soil

for the deep mixing method mainly consists of i) laboratory mix tests, ii) field trial test, iii) quality control during construction and iv) quality assurance after construction through laboratory test on core samples and pile head inspection, as shown in Table 7.2.

3.2.1 Laboratory mix test

A laboratory mix test is an important pre-production QA which may be carried out in a different phase or phases of a project either for the geotechnical design or for the process design. The laboratory mix test is the responsibility of the owner/engineer if the deep mixing work is awarded with detailed specifications, but is the responsibility of the deep mixing contractor if the contract is awarded by performance basis.

A laboratory mix test should be conducted on soil samples retrieved from all soil layers to be stabilized, in order to determine a suitable type of binder and a suitable quantity of the binder to ensure the design strength. Ordinary Portland cement and blast furnace slag cement type B (including 30 to 60% slag) are usually used as a binder both in the wet and dry method in Japan. Dozens of special binders are also available on the Japanese market for organic soils and extremely soft soil with high water content (Japan Cement Association, 2007) and they are used for the laboratory mix test when required.

Laboratory strength is influenced by many factors, such as mixing and molding conditions, curing condition, and testing conditions. To avoid the influence of these factors, the Japanese Geotechnical Society issued a draft standard laboratory mix test procedure in 1981 and later officially standardized the procedure in 1990 (Japanese Society of Soil Mechanics and Foundation Engineering, 1990), and made minor revisions in 2000 and 2009 (Japanese Geotechnical Society, 2000 and 2009). Almost all laboratory tests for practical and research purposes follow this standard in Japan, which makes Japanese engineers rely upon test results obtained by different parties. The procedure is described in Appendix A.

3.2.2 Field trial test

A field trial test is an important pre-production QA for deep mixing project especially when no comparable experience is available. It is recommended to conduct a field trial test in advance in or adjacent to the construction site, in order to confirm the actual strength and uniformity in the real construction condition and determine the operational parameters and final mix design for production.

In Japanese projects, however, the final mix design is often based on the laboratory mix test and the field trial test is rarely undertaken to confirm the field strength. The investigation of production columns in the beginning of deep mixing work at an early age, 7 days for example may be conducted as an alternate for a field trial test. This is due to the amazingly large number of projects done in Japan in the last three to four decades and the contractors' confidence in the correlation between field and laboratory test results.

The trial penetration of the deep mixing machine at the construction site without injecting the binder is a common practice in Japan to determine the process control value to confirm the end-bearing of columns to the stiff stratum if it is required. The change in the electric or hydraulic power consumption, change in torque and/or the change of penetration speed are measured during the trial installation to establish the construction control criteria for end bearing. A field trial installation for this purpose should be conducted in the vicinity of the existing boring to compare with the known soil stratification.

3.2.3 Quality control during production

During production, stabilized soil columns must be installed to satisfy both the geometric layout and the quality of stabilized soil specified by the geotechnical design. The rig operator should locates, control, monitor and record the geometric layout of each column (plan location, verticality and depth). When the termination depth is designated to ensure reliable contact to the underlying stiff layer, the rig operator should carefully identify the depth according to the construction control criteria established in the field trial test.

Quality control of the stabilized soil includes binder storage, binder or binder slurry preparation, and control of the mixing process. Storage and proportioning of binder, additives and mixing water are normally controlled, monitored and recorded at the plant placed in the construction site. Construction control parameters during column installation in the in-situ mechanical mixing systems by rotary mixing with vertical shaft(s) (see Table 1.4 in Chapter 1) include the continuous monitoring of penetration and withdrawal speed, rotation speed, quantity of binder, water/binder ratio (for the wet method). The construction control values are predetermined by the process design considering the results of the laboratory mix test, field trial test, and contractors' experience. Depending on the contractual scheme the construction control values may be modified during production based on the examination of the early installed production columns. During column installation, construction control values are controlled, monitored and displayed in the control room at the plant and/or cab of the mixing machine for the plant operator and rig operator to adjust the execution procedure when necessary. The mixing shaft and mixing tools are frequently observed for any possible defects during construction.

Reporting the recorded construction control parameters is an important QA during production. This is because the quality of a stabilized soil column may be consistent if the construction process in the same project site is consistent.

The mixing degree mostly depends on the rotation speed of the mixing blade and penetration and withdrawal speeds of the shaft. In Japan, an index named "blade rotation number", T has been introduced to evaluate the mixing degree. This number means the total number of mixing blade passes during 1 m of shaft movement and is defined by the following equation for the penetration injection method and withdrawal injection method respectively. According to the accumulated researches and investigations, "blade rotation number" should be around 270 or larger to assure a sufficient mixing degree for Japanese wet and dry methods, CDM and DJM (Cement Deep Mixing Method Association, 1999; Coastal Development Institute of Technology, 2002, 2008; Public Works Research Center, 2004). It is obvious that the required "blade rotation number" is influenced by many factors, such as the shape and arrangement of mixing blades, the rotation and moving speed of the blades, soil properties, and so on. It should be reminded that the blade rotation number should be determined for each mixing machine and soil conditions by accumulating the test results on production columns.

For penetration injection

$$T = \sum M \cdot \left(\frac{N_d}{V_d} + \frac{N_u}{V_u} \right) \tag{7.2a}$$

for withdrawal injection

$$T = \sum M \cdot \left(\frac{N_u}{V_u} \right) \tag{7.2b}$$

where
  N_d : number of rotation of mixing blades during penetration (N/min)
  N_u : number of rotation of mixing blades during withdrawal (N/min)
  T : blade rotation number (N/m)
  V_d : penetration speed of mixing blades (m/min)
  V_u : withdrawal speed of mixing blades (m/min)
  ΣM : total number of mixing blades.

At the bottom of the column, the blade rotation number is not automatically guaranteed. Careful bottom mixing process by repeating penetration and withdrawal while injecting the binder may be conducted to attain a sufficient level of mixing. When the quality of bottom end is critical, the quality should be confirmed during the field trial test.

3.2.4 Quality verification

After the improvement, the quality of the in-situ stabilized soil columns should be verified in advance of the construction of the superstructure in order to confirm the design quality, such as continuity, uniformity, strength, permeability or dimension. In Japan, the verification is usually carried out by means of observation and testing of the core samples of production columns. The frequency of core borings is dependent upon the total volume of the stabilized soil. In the case of on-land works, three core borings are generally conducted in the case where the total number of columns is less than 500. When the total number exceeds 500, one additional core boring is conducted for every further 250 columns.

(a) Working platform for core boring.   (b) Core boring machine.

Figure 7.3 Core boring in marine construction.

Figure 7.4 Example of core sample of cement stabilized soil.

In each core boring, core samples are taken throughout the depth in order to verify the uniformity and continuity of the stabilized soil by visual inspection of the continuous core. Determination of the engineering properties of the stabilized soil is based on unconfined compressive strength on samples at 28 day curing. In general, three core barrels are selected from three different levels and three specimens are taken from each core barrel and subjected to unconfined compression test for each core boring. Engineering properties other than unconfined compressive strength can be correlated with unconfined compressive strength as discussed in Chapter 3. Figures 7.3 and 7.4 show a platform and machine of core boring, and an example of core sample of cement stabilize soil respectively.

The reliability and accuracy of the unconfined compressive strength determined on a core sample depends upon the quality of the core sample, and the quality of

Table 7.3 RQD index and rock quality.

RQD	Description of rock quality
0–25%	very poor
25–50%	poor
50–75%	fair
75–90%	good
90–100%	excellent

sample depends upon the drilling and coring method and the drillers' skill. A Denison type sampler, double tube core sampler, or triple tube core sampler can be used for stabilized soil columns whose unconfined compressive strength ranges from 100 to 6,000 kN/m^2. It is advisable to use samplers of a relatively large diameter such as 86 mm or 116 mm in order to take high quality samples. The evaluation of the quality of the retrieved core in Japan varies from subjective judgment such as good or bad by visual observation to the strict requirement of core recovery ratio of 100% and the RQD value larger than 90% . The RQD (*Rock Quality Designation*) index is defined by the Equation (7.3). The RQD index measures the percentage of "good rock" within a borehole and provides the rock quality as shown in Table 7.3.

$$RQD = \frac{\sum \text{length of core pieces} > 10\,\text{cm}}{\text{Total core run length}} \cdot 100\ (\%) \qquad (7.3)$$

3.3 Technical issues on the QC/QA of stabilized soil

An international collaborative study was carried out by the participation of 45 organizations from 7 countries to identify the similarities and differences in the current QC/QA procedures employed in different parts of the world and to discuss the future research needs (Terashi and Kitazume, 2009, 2011). A part of the collaborative study is briefly introduced in this section.

3.3.1 Technical issues with the laboratory mix test

In the initial phase of a project, the applicability of deep mixing may be judged by laboratory mix test on the soils at the project site. In regions where deep mixing experience is rich enough, a laboratory mix test is not undertaken at this initial phase. The laboratory mix test is normally conducted once in a project either by the owner or contractor depending on the contractual scheme. The design engineer uses the laboratory mix test results for assuming/establishing design parameters, and the contractor uses the same test results for planning the field trial test or for the process design. Only when the laboratory test is conducted by the standardized procedure, a certain party involved in a project can rely on the test results obtained by a different party. However, nationwide official standard is scarce except for the one by the Japanese Geotechnical Society (Japanese Geotechnical Society, 2009).

Table 7.4 compares the test procedures, whether it is standardized or not, documented or undocumented, or adopted regionally or individually (Kitazume *et al.*,

Table 7.4 Summary of some existing laboratory mix testing procedures (Kitazume et al., 2009a).

	Standards/guidelines		
Protocol	JGS0821 (Japan), 2000	SGI protocol[1][2] (Sweden), 2006	EuroSoilStab (Europe), 2002
Application	DMM, etc.	Deep and mass stabilisations	Column Stabilisation	Mass stabilisation
Binder state	Dry/Wet	Dry	Dry/Wet	Dry
Binder type	Cement, lime, etc.	cement, lime, etc.	cement, lime, etc.	
Applicable soils	$D < 9.5$ mm for $\phi 50$ mm mould	Clay, silt, gyttja, sulphide soil and peat	Soft cohesive soils with organic matters	
Mould size	$\phi 50$ mm × $h 100$ mm keeping $h/\phi = 2-2.5$.	$\phi 50$ mm × $h 170$ mm for soil, $\phi 70$ mm × $h > 200$ mm for peat	$\phi 50$ mm × $H 100$ mm. Larger mould may be used.	$\phi 68$ mm × $H 200-300$ mm. Larger mould may be used.
Mixer	Machine mixer	Dough mixer or kitchen machine	Dough mixer or kitchen mixer. No specification. To be decided based on local experience.	
Soil sample preparation	Kept at natural water content. Remove particles larger than 9.5 mm. Homogenise well before adding binder.	Lost water should not be added. Remove coarse objects. Homogenise well. Store peat sample at 7°C.	Remove isolated roots and large fibres. For peats, limit homogenising time to avoid destroying fibres.	
Mixing	Until sufficiently homogenious (typically 10 min). Manual intervention during machine mixing is recommended.	5 min in mixer with manual intervention. For peat, mix within 15 min after taking out of cold room.	5 min recommended, possibly with manual intervention	5 min recommended
Moulding method	Place into mould in three layers. Air bubbles should be removed from each layer by tapping against floor, hammering, shaking, etc. Care should be taken for sands and volcanic loams.	Statically compress each layer with 100 kPa for 5 sec. For peat, no compaction if it is liquid, otherwise place into mould in 5–6 layers and knead/tamp lightly with tamping rod[1] or in layers of 0.5 dl, each compacted with fall-weight[2].	Static compaction in layers of 25 mm with 100 kPa for 2 sec (6 compactions per layer) in a special manner*. For liquid soil, just pour.	Pour into mould. If solid, compact in 5–6 layers. Scratch the layer interfaces.
Moulding time	As soon as possible	Within 30 min since binder is added	Within 30 min since binder is added	
Curing	Seal with, for example, polymer film. Cure in high-humidity environment. 20 ± 3°C. Period can be selected from 1, 3, 7, 10, 14, 28 days, etc.	Sealed and cured at 7°C. With lime, specimens are sometimes first kept at room temperature for a certain time. Cure for 14, 28, 91, 180 days, etc. Avoid moisture loss. For peats, a typical surcharge of 18 kPa is applied at 45 min after start of mixing and kept during curing. Cure at 20 ± 2°C in water bath.	Sealed, 18–22°C.	Sealed, 18–22°C in water bath, with surcharge of 18 kPa, etc.

(Continued)

Table 7.4 Continued.

Protocol	Finland	USA (Virginia Tech)		USA (Raito)	Cambridge (UK)
			Common practices		
Application	–	Dry column stabilisation	Wet deep mixing		–
Binder state	Dry	Wet	Wet	Wet	Wet
Binder type	cement, lime, etc.	cement. lime	cement		–
Applicable soils	–	Inorganic soils		–	
Mould size	φ42–50 mm, h125–170 mm for standard tests, φ 20 mm × h60 mm for index tests	φ 50 mm × h 100 mm		φ50 mm × h100 mm or φ75 mm × h150 mm	φ50 mm or φ 100 mm × H100 mm
Mixer	Dough mixer	Dough hook (for cohesive soils) and flat beater (for non-plastic soils)		typically Hobart mixer	Food mixer
Soil sample preparation	Homogenise for 6 min.	Homogenise for 3–4 min.	Remove particles >#4 sieve. Store in 100% in RH at 20°C. Homogenise for 3 min.	Keep or adjust to natural water content. Remove particles > 1/4 of mould dia. Homogenise.	Water content similar to that of natural soils to be treated. Homogenise.
Mixing	2–6 min, possibly with manual intervention	3–5 min	10 min	7–10 min	Not specified
Moulding method	Static compaction: 25 mm layers, 100 kPa for 6 sec, or tapping. Fall-weight compaction may also be used.	Pour and tap if liquid, compact if solid for each layer of 25 mm, 25 blow with 5 mm-dia. Rod and statically compact with up to 100 kPa for 2–5 sec.	3 lifts with rodding and tapping	Tap in 3 layers for φ50 mm and 5 layers for φ75 mm. If too solid, squash with thumb or wooden rod.	Tamping with rod, in various layers
Moulding time	Within 30 min since binder is added	Within 20 min since completion of mixing	Within 30 min since completion of mixing	Within 45 min since binder is added	Within 30 min since binder is added
Curing	In water basin, insulated box or plastic bag (RH > 95%). With lime, 20°C for first 0, 2, 7 or 14 days and then 6–8°C. With cement, 20°C for first 0, 2 or 7 days and then 6–8°C.	Seal and cure underwater or in 100%-humidity room at 20°C. Periods of 7, 14, 28 and 56 days are recommended.		23 ±2°C (ASTM C 192/C), RH > 95%. Cure for 7 and 28 days, and also 3 and 14 days if required.	20–21°C, RH = 98%

2009a). Among these, the Japanese standard can be singled out for domestic engineers' strong adherence to it. The prescriptions of the key elements, such as specimen size, mixing procedures, curing conditions, and mechanical tests that follow laboratory mix tests, were found fairly well regarded and accepted by many engineers. At the same time, however, shortcomings or insufficiency in the scope of the Japanese Geotechnical Standard were also raised, particularly concerning mixing and molding methods when dealing with more problematic soils such as peats, low water-content clays and volcanic loams.

The test procedure, especially the method of preparation and curing of specimens, differs from one region to another or from one organization to the other even in the same region. An international collaborative study was carried out to know differences in laboratory test procedures and the influence of these differences on the test result to avoid misinterpretation.

3.3.1.1 Effect of rest time

The chemical reaction between binder and water starts immediately after the preparation of binder slurry in the wet method. Excessive rest time before the addition of binder slurry to soil is anticipated to invite deterioration of the binder. The chemical reactions between soil and binder start when they are mixed together. Excessive rest time before molding is anticipated to cause difficulty in the molding due to the change of consistency of the soil binder mixture, and also on the breakage of chemical reaction products formed in the stabilized soil in the early stage. As shown in Figures 2.31 to 2.33, the range of rest time from the preparation to mixing with soil was up to 120 minutes and that from mixing to molding was up to 60 minutes in the concerted test program. Within these ranges no meaningful influence was observed on the 28 days strength of stabilized soil.

3.3.1.2 Effect of molding

The molding procedure of stabilized soil in a laboratory mix test influences its strength. According to the survey, several molding procedures have been adopted as shown in Table 7.4:

–   Dynamic compaction: Compact each soil binder mixture layer by a falling weight. The weight and fall height, number of blows are chosen based on local experience.
–   Static compaction: Compress each soil binder mixture layer by static pressure. The magnitude of pressure and time for press are chosen based on local experience.
–   Tapping: Tap the mold for each soil binder mixture layer. The number of tapping is chosen based on local experience.
–   Rodding: Tamp the soil binder mixture with a rod for each layer. The number of poking and the rod diameter are chosen based on local experience.
–   No compaction: Simply pour the mixture into a mold.

Figure 7.5 shows the influence of different molding procedure on the unconfined compressive strength, q_u of stabilized soil, in which four organizations, Port and Airport Research Institute (PARI), Dry Jet Mixing Association (DJM), Cambridge University and the Swedish Geotechnical Institute (SGI) carried out the tests on different soils and different types of binder (Kitazume et al., 2009b). In the figure, Bc is

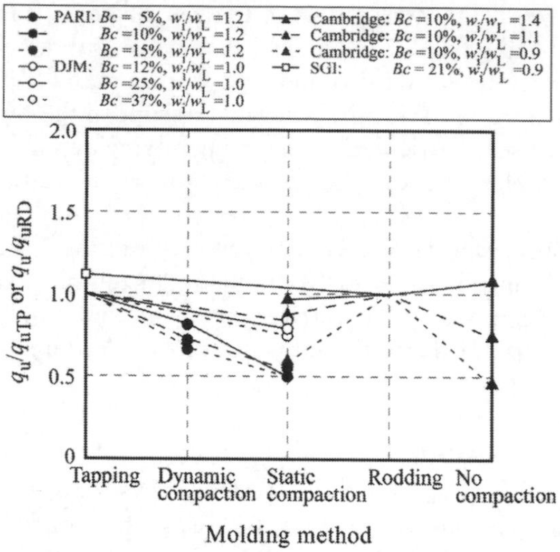

Figure 7.5 Influence of different molding condition on the strength of stabilized soil (Kitazume *et al.*, 2009b).

the binder factor, w_i is initial water content and w_l is the liquid limit of original soil respectively. For the comparison, strength data of PARI and DJM are normalized by the strength obtained on the specimen prepared by the tapping, q_{uTP}, and those of SGI and Cambridge University are normalized by the strength obtained on the specimen prepared by the rodding, q_{uRD}. Although the influence of molding procedure differs for different soil type and binder content, the dynamic compaction and the static compaction resulted in lower strength up to 50% than that obtained with the tapping. When comparing the static compaction and the rodding, the static compaction resulted in lower strength up to 40% than that with the rodding.

3.3.1.3 Effect of curing temperature

Important elements of curing conditions are temperature, humidity and application of surcharge or not. The rate of chemical reaction (strength development with time) depends on the combined influence of temperature and curing time often referred to as maturity. The relationship between the maturity and strength may be further influenced by soil type, binder type and binder content. For each soil type, the strengths of stabilized soils are normalized by the 28 days strength cured at 20°C and compared with time and temperature. In Chapter 2, Equations (2.5) shows four definitions of Maturity proposed by the previous studies. Figure 7.6 shows the relationship between normalized strength and one of the maturity equations, M_4 (Equation (2.5d)) (Kitazume *et al.*, 2009b). As far as the clayey soils tested are concerned, the combined influence of time and temperature on the normalized strength seems to be explained by a unique relationship with M_4 in logarithmic scale as shown in Figure 7.6(a). For peat, while M_4 is less than 562 which corresponds to 20°C-14 days, the results from both of Civil Engineering Research Institute for Cold Region (CERI) and SGI (except 40°C curing condition in SGI) show a linear relationship between the M_4 and normalized q_u as

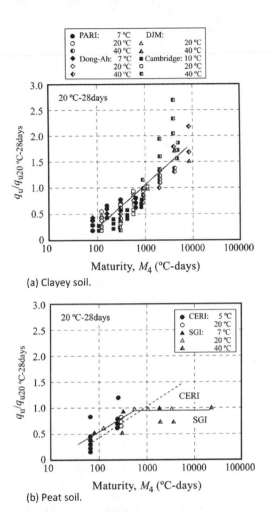

Figure 7.6 Relationship between normalized q_u and Maturity, M_4 (Kitazume *et al.*, 2009b).

shown in Figure 7.6(b). When M_4 is more than 562, normalized q_u has little or no correlation with the Maturity in the data obtained by SGI.

Regarding the influence of soil type on the relationship between the q_u and Maturity, it was reported that the proportional constants of q_u and Maturity differ greatly for soil type as shown in Figure 7.7 (Babasaki *et al.*, 1996). It must be noted that there is room for discussion and further investigation of the influence of soil type on the $q_u - M$ relationship is recommended.

3.3.2 *Impact of diversified execution system on QC/QA*

Table 7.5 shows a variety of techniques available for deep mixing projects worldwide (Terashi and Kitazume, 2009). The first column shows the method of introducing binder either by Wet (binder-water slurry) or Dry (dry powder). The second column

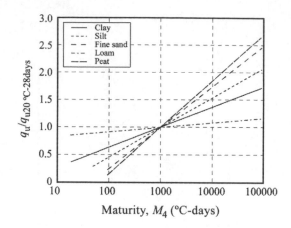

Figure 7.7 Relationship between normalized q_u and Maturity, M_4 (Babasaki *et al.*, 1996).

Table 7.5 A variety of techniques for deep mixing projects.

Binder Type	Type of shaft	Position of mixing	Representative system	Origin
Dry	Vertical rotary shaft	Blades at bottom end of shaft	DJM (Japan), Nordic dry method (Sweden)	Deep mixing
Wet A	Vertical rotary shaft	Blades at bottom end of shaft	CDM (Standard, MEGA, Land 4, LODIC, Column21, Lemni2/3) (Japan), SCC (Japan), Double mixing (Japan), SSM (USA), Keller (Central Europe), MECTOOL (USA)	Deep mixing
Wet B	Vertical rotary shaft assisted by Jet	Blades and high pressure injection at bottom of shaft	JACSMAN (Japan), SWING (Japan), WHJ (Japan), GeoJet (USA), HydraMech (USA), TURBOJET (Italy)	Deep mixing
Wet C	High pressure injection at bottom of shaft		Jet grouting – single fluid, double fluid, triple fluid (Japan), X-jet (Japan)	Deep mixing
Wet D	Vertical rotary shaft	Auger along shaft	SMW (Japan), Bauer Triple Auger (Germany), COLMIX (France), DSM (USA), MULTIMIX (Italy)	Diaphragm wall
	Horizontal rotary shaft	Vertical mixing by Cutter mixer	CSM (Germany, France)	or Trench cutter
	Chainsaw, Trencher	Continuous vertical mixing	Power Blender (Japan, shallow to mid-depth, down to 10 m), FMI (Germany, shallow to mid-depth), TRD (Japan, down to 35 m)	

shows the driving mechanism of mixing tools. The third column shows the type of mixing tool and its location. The fourth column shows the name of techniques followed by the country or region which was originally developed. The fifth column shows the roots of techniques either originally developed for deep mixing, modified from diaphragm wall or trench cutter, or those for shallow improvement.

Actual execution of the deep mixing is achieved either one of the locally available execution systems from those variations found in Table 7.5. There is a tremendous difference in the level of sophistication in the mixing tools. Some systems use one or two mixing blades attached to a single vertical rotating shaft, some uses mixing blades attached at several different levels of multiple vertical shafts, some employs mixing tools that rotate in the counter directions, and some employ a mixing tool that stem from the trench cutter. The quality of stabilized soil such as in-situ strength, uniformity and continuity highly depends upon the execution process. Further, some systems are superior to others in the ease of overlapping operation or ensuring end bearing. It should be noted that the quality control items and control values are different for different execution system.

The process design starts with the laboratory mix test and is achieved through field trial test (or by the examination of several stabilized soil columns in the early stage of construction) together with the contractors' proprietary information. In the planning of the laboratory mix test it should be noted that the stabilized soil produced by the rotation of vertical shafts remembers the soil stratification of original ground, and hence, the laboratory mix test should be programmed for each and all representative soil layers within the improvement depth. However for continuous mixing techniques such as the TRD method (Aoi *et al.*, 1996) and the Power Blender method, the soils in stratified layers are moved vertically and mixed together. For these methods, the laboratory mix test should be programmed for the mixture of representative soil layers by taking into account the thickness of each layer. Furthermore, the optimal viscosity of soil binder mixture should be carefully examined in a series of laboratory mix tests to guarantee the vertical movement of soil binder mixture along with the continuous mixing tools.

The laboratory mix test provides a good insight into the mix design of production columns but can never be an exact simulation of field execution because of the large variations in the execution process.

In order to assume/establish design parameters of stabilized soil, database, which compiles the relationship between laboratory mix test result and field strength, and the uniformity of stabilized soil in terms of coefficient of variation, is necessary. The ratio of q_{uf}/q_{ul} and coefficient of variation discussed in Chapter 3 are only applicable for the execution system by vertical rotary shafts with mixing blade. Contractors are encouraged to produce their own database for their own proprietary execution system. Until such a database become available, field trial installation should be the routine in deep mixing projects, especially when the construction is awarded by performance specification.

3.3.3 Verification techniques

The primarily used verification technique for the field strength is an unconfined compression test on drilled core samples both for the wet and dry methods in Japan and the US. That for the Nordic dry method is the column penetration test (Larsson, 2005). This difference in the preferred verification technique corresponds to the preferred field strength. Continuity of the stabilized soil column is verified by visual observation and the core recovery ratio of core run in Japan and the USA, and by column penetration or by reverse column penetration in Nordic countries.

A variety of verification test procedures to examine the engineering characteristics of stabilized soil have been proposed as shown in Table 7.6 (Hosoya *et al.*, 1996; Halkola, 1999; Larsson, 2005). However, actual practices rely upon traditional verification techniques such as the unconfined compression test on drilled core

Table 7.6 Verification techniques proposed for determining the quality of stabilized soil (Hosoya et al., 1996; Halkola, 1999; Larsson, 2005).

No	Verification Test	Method	Method Description, characteristics, corelation with strength, limitation, etc.
1	Laboratory test on drilled core sample	Unconfined compression test and/or Other lab tests	Retrieval of intact core of treated soil columns and store the sample under predetermined condition until laboratory testing, commonly unconfined compression test. The verification test results can be directly compared with the design assumption. Most of the alternate in-situ test procedures are calibrated against q_u test results on core samples.
2	Laboratory test on wet grab sample	Unconfined compression test	Retrieve "fresh" soil-binder mixture immediately after mixing by a special probe, molding it at site and store the specimen until laboratory testing. Sampling cylinder may tend to collect unmixed cement slurry rather than soil-binder mixture.
3	Sounding	Ordinary column penetration test	A probe equipped with two opposite vanes is statically pressed down into the center of treated soil column and continuous record of resistance is taken. Commonly used for Nordic Dry Method. Applicable for $q_u < 300\,kPa$ down to 8 m, for $q_u < 600$–$700\,kPa$ down to 20 or 25 m if pre-bored at the center. Swedish guideline for the test is available.
4		Reverse column, penetration test	A probe attached to the wire is placed at the bottom of the treated soil column during production and left there until testing. The probe is withdrawn from the column and the continuous re cord of resistance can be made. Applicable for $q_u < 1200\,kPa$ down to 20 m. Bearing capacity formula is used to evaluate undrained shear strength of treated soil, where $N_c = 10$ (Sweden), 10–15 (Finland)
5		Standard penetration test, SPT	Driving a split sampler into soil dynamically by hammering, and measure number of blows to penetrate 30 cm. Empirical corelation between SPT N value and q_u has been reported, $q_u = N$ where $N = 25$ to 33 for soils with $q_u < 1000\,kPa$.
6		Portable dynamic cone penetration test (Japan)	Driving a cone into soil by hammering, and measure blow count to penetra te 10 cm. Applicable for $q_u = 200$ to $500\,kPa$. Blow count N_d is correlated to unconfined compressive strength, $q_u = 29\,N_d - 258\,kPa$ for soils with $q_u < 1000\,kPa$.
7		Dynamic cone penetration test (Finland)	Driving a cone into soil by hammering and measure penetration depth for each blow. DCP index is correlated to CBR, $CBR = 292/DCP^{1.12}$ for the 60 degree cone angle.
8		Combined static-dynamic penetration test (Finland)	Combination of static penetration and hammering test. During penetration, the rod is rotated continuously by 12 rpm and torque is measured to calculate shaft friction. Applicable for $q_u < 4\,MPa$.
9		Cone penetration test, CPT	Cone is statically penetrated into ground and measures the penetration resistance, skin friction and pore water pressure. The undrained shear strength is corrected by $c_u = (q_t - s_{v0})/N_{kt}$, where, c_u is undrained shear strength, s_{v0} and N_{kt} are total overburden pressure and cone factor, respectively.

No.	Method	Description
10	Rotary penetration sounding Test, RPT (Japan)	A sensing rod equipped with a special drilling bit attached at the bottom end of drilling shaft is drilled into the treated soil column, and measure drilling speed R, rotation n, thrust W, torque T and water pressure at the drilling bit. Unconfined compressive strength q_u is correlated to measured data by $q_u = K R^a n^b W^c T^d$, where, K, a, b, c, d are constants.
11	Automatic Swedish weight sounding test, A-SST (Japan)	A screw point connected to a series of rods is driven statically into the ground to measure the number of half-rotations for every 25-cm penetration. Applicable for q_u <500 kPa. The equivalent number of rotations for 1-meter penetra tion, N_{SW} is converted to shear strength of column.
12	Column vane test (Finland)	Diameter of the vane is 130 or 160 mm and the height is one half of the diameter. Applicable for q_u < 400 kPa.
13	Geophysical method from bore hole	P- and S-wave velocities are measured either by down hole test or suspension method. Their distributions with depth reflect the uniformity of treated soil columns. Elastic modulus of the treated soil column at small strain can also be calculated from these velocities.
14	Electro-magnetic logging	Measuring electrical and magnetic properties of the ground to identify the soil layering, cavities and underground utilities. Application of these imaging techniques to the deep mixed ground seems still be in the research stage.
15	Loading test from borehole	A cylindrical probe is expanded radially onto the borehole wall and measure the pressure and a radial displacement. Elastic modulus and the strength of the soil are evaluated by the measurements.
16	Non-destructive tests at top of a column	Hammering top surface of column and measuring the reflected waves at the top surface to assess the continuity of treated soil columns or the shape of as-built columns. Applicable for more than 4 m long and with q_u > 1 MPa.
17	Impact acceleration test	A rammer is free fallen onto the treated soil ground surface and "impact acceleration" is measured, and to converted to unconfined compressive strength.
18	Plate loading test on top of a column	Rigid plate is statically loaded by step-wise to measure bearing capacity and deformation characteristics.
19	Stabilized pile loading test	Pile load test or compression test is carried out in-situ column or extracted column to determine the load bearing capacity of single treated soil column. These tests have been conducted so far for the research purpose and not for daily QA/QC undertakings.
20	Visual observation of whole column and testing	Retrieve of the full scale treated soil column by huge sampler and test by pocket vane or by phenolsulfonphthalein to determine the uniformity.

samples and/or the column penetration test. The unconfined compression test on a drilled core is admitted as the best technique. Most of the other procedures in Table 7.6 seem to be used only for the research purpose or for settling the non-compliance. This may be due to the unfamiliarity of these techniques both to owner/designer and contractor. Another reason may be the lack of direct correlation between the measured data from most of the verification test procedures and the design parameters. It is highly recommended for public organizations to carry out the comparative test program for promising techniques preferably under the international experts' involvement.

4 QC/QA OF IMPROVED GROUND – RESEARCH NEEDS

Current QC/QA procedures for deep mixing described in Section 3 place special emphasis on the quality of stabilized soil. It is because the designed properties of the stabilized soil, such as strength and uniformity, should be achieved in the field regardless the type of application and the function of stabilized soil columns. However, the purpose of deep mixing is not only to manufacture a good quality stabilized soil but to create an improved ground which guarantees the performance of the superstructure. The improved ground by the deep mixing method is a composite system comprising stabilized soil columns and original soils. When the type of application is different, different functions are expected to improved ground whose performance is not only governed by the quality of stabilized soil but also the characteristics of original soils, improvement geometry, installation accuracy of columns, end-bearing condition of columns and overlap of adjacent columns, *etc*. These effects should be carefully considered in the geotechnical design and should lead to different requirements on stabilized soil columns and original soils affected by the column installation in some cases, different acceptance criteria and different verification procedure for the deep mixing project. Quality assurance is an art to assure the quality of the product which was envisaged in the design. Therefore the required quality of stabilized soil column in a composite system should be discussed taking into account the reliability of design concept, design procedure, and selected design parameters. The quality of the improved ground (composite system) will be only achieved when the intents of the geotechnical design and process design are consistent each other. The flow of QC/QA which was shown earlier in Figure 7.1 and Table 7.1 will not change even if the QC/QA is addressed to the improved ground. What should be done is to reconsider each activity in Table 7.1 from the viewpoint of the performance of the improved ground which is a complicated composite system.

The correlation of geotechnical design and the QC/QA is discussed in this section for the group column type improvement for an embankment support and the block type improvement for a heavy structure.

4.1 Embankment support by group of individual columns

4.1.1 QC/QA associated with current design practice

The current design procedure employed in Japan (Public Works Research Center, 2004) involves two major modes of failure; one is the slip circle analysis to examine the internal stability of stabilized soil columns and the other is the external stability to examine the sliding of a stabilized soil zone as already shown in Figure 6.34 in Chapter 6.

The simple slip circle failure mode is associated with two assumptions. One is that the stabilized soil columns and original soil behave as a composite material which exhibits the weighted average shear strength. The other assumption is that the composite material always fails by shear irrespective of the location along the slip surface.

Slip circle analysis is useful in determining the width of the improved zone when slip circles pass entirely outside the improved zone. In determining the required strength and replacement area ratio of improved ground, slip circle analysis is effective only when the two assumptions mentioned above are satisfied. In order to satisfy the average shear strength concept, the commentary to the Japanese design guide emphasizes the importance of learning the previous successful case records in determining the size and location of improved zone and addresses the following notes: 1) The width to height ratio of the improved zone should be larger than 0.5 at least and preferably larger than 1.0, 2) The range of design strength in terms of unconfined compressive strength has been between 100 to 600 kN/m^2, and 3) Replacement area ratio is larger than 30% and often exceeds 50% .

For the majority of embankment support project, the size of project in terms of stabilization volume and budget is small. The current simple design procedure together with the simple QC/QA focusing on stabilized soil might be preferred in the future as well, because they have enjoyed happy harmony each other with the aid of commentaries based on the practitioners' experience. It should be noted that the current QC/QA procedure focusing upon the strength and uniformity of the stabilized soil column is effective when the assumptions on the design procedure are empirically satisfied.

From the viewpoint of geotechnical design, the commentaries to the design guide should mention that the tangent column are preferable column installation pattern underneath the slope for improving stability. It is highly recommended to compile the practitioners' experience into the database of case records by correlating the ground improvement geometry, the original soil conditions and the design conditions of superstructure such as the height of embankment, slope angle, allowable settlement and adopted safety factor, which will help the engineers with less experience reach the appropriate design.

4.1.2 QC/QA for sophisticated design procedure considering the actual failure modes of group column type improved ground

Recent physical and numerical investigations have shown that there exist several failure modes both in the external and internal instabilities for the group column type improved ground as explained in Section 2 of chapter 6. The dominating mode of instability for a particular situation will be strongly influenced by the replacement area ratio, location of improved zone relative to the superstructure, end bearing conditions, and as-built quality of stabilized soil columns.

Numerical analysis may be used as a design tool which can take the different failure modes into account (Han et al., 2005; Filz and Navin, 2006; Adams et al., 2009). In numerical analysis such by FEM, ordinary programs for use in geotechnical problems are designed to analyze compression, pure tension and shear but are often not appropriate to take bending failure into account, and tend to provide larger resistance against bending failure (Ishii, 2008). For the tilting failure mode, it may be better to introduce a joint element at the bottom end and periphery of the stabilized soil columns for allowing discontinuity (Kitazume and Maruyama, 2006, 2007). Kitazume (2008) proposed a design procedure by a set of simple calculations for different failure modes. The advantage of examining each mode of failure independently is that the design engineer can identify the most critical mode of failure on the trial geometry and easily arrive at the optimum solution. When employing these design procedures, it is important to examine whether they can correctly model the end-bearing condition and the moment capacity of stabilized soil columns that influence the tilting and bending failure modes.

When these design procedures are adopted, QC/QA indices can be determined appropriately for each mode of failure. For shear failure mode, the current QC/QA procedure may apply. When the bending failure of a deep mixed stabilized soil column is critical, the moment capacity of the deep mixed stabilized soil column becomes the major characteristics to be quality assured. The moment capacity is a function of the diameter, stiffness, and tensile strength of the deep mixed stabilized soil column under bending mode. The bending strength of the small deep-mixed soil specimen can be correlated with the unconfined compressive strength. However the bending strength of the full scale column may be heavily influenced by the variability of strength exist in the cross section of the full scale column. It is desirable to carry out the full scale tests for the bending situation in a similar way to that done by Futaki *et al.* (1996) for compression and correlate the design tensile strength and unconfined compressive strength of core samples taking the coefficient of variation into account. The coefficient of variation may become one of the important QA indices and invite the increased number of tests.

The tilting and bending failure modes are also influenced by the strength profile of unstabilized soil such as the existence of a dry crust at the ground surface and the end-bearing condition of stabilized soil columns (Kitazume, 2008). The accurate characterization of the unstabilized soil between columns and evaluation of possible disturbance due to column installation become important. For the end-bearing condition of deep-mixed column, construction process at the bottom end should be carefully planned and verified through a field trial test.

When the sliding of the improvement zone is found the dominating failure mode, the accurate characterization of the unstabilized ground outside the deep mixed stabilized soil columns increases the importance in addition to the current QC/QA of deep mixed stabilized soil. The possible disturbance of the original soil due to deep mixing execution may invite a reduction of the safety factor. For example, in an execution system whose binder outlets locate above the mixing tool, 100 to 200 mm thick soft layer at the bottom end of the mixing tool is not stabilized but simply disturbed by the mixing tool. The careful treatment at the tip should be planned and be verified through field trial test as one of the QA measures.

4.1.3 Practitioners' approach

When the problem involves a complicated design and QC/QA, one of the practitioners' approaches may be bringing the problem simpler, easier to handle with the available technology. When stability is the major issue in the project, it is better to employ the panel or grid improvement pattern rather than the group column type improvement pattern. This is the approach what the EuroSoilStab (2002) recommends and the Japanese practitioners already have done by employing the larger area replacement and align the columns perpendicular to the slope by the tangent column layout or overlapped column layout (see Chapter 4 or Terashi *et al.*, 2009).

4.2   Block type and wall type improvements for heavy structures

The engineering behavior of the block type and wall type improvements was discussed earlier in Section 2 in Chapter 6. Simple design calculations examining the external and internal stabilities of the block type and wall type improved grounds are explained in Sections 5 and 6 in Chapter 6 and it is also easier for numerical simulation due to its simple geometry. For this approach, the evaluation of the overlapped portions becomes important in both the design and QC/QA procedure in addition to the QC/QA of stabilized soil. The efficiency of the overlapped portion may be easily evaluated by numerical simulation by specifying arbitrary shear characteristics to the overlapped portion. The reliabilities of the overlapped portion and end-bearing are currently guaranteed only through the monitoring and recording of depth, location and verticality, although inclined coring to capture the overlap face or coring down into the bearing layer are undertaken in some cases. It seems necessary to develop the verification technology to quantify the efficiency of overlap or end-bearing. For the overlap operation it is important to select the appropriate execution system. Obviously a multiple shafts machine is superior to the single shaft machine due to its rigidity and ability of reducing the number of overlapped portions. Further the use of slow-hardening binder may improve the overlap operation. When contact to the stiff underlying layer is required, the record of torque or energy consumption change with depth of driving mixing blades can supplement the verification test by continuous coring. In any case, it is recommended to select an experienced contractor and appropriate execution system and confirm the reliability through a field trial test.

Economy concerns often tempt clever design engineers to adopt sophisticated column installation patterns such as honeycomb or arch to reduce the volume of stabilized soil. It should be reminded that even the local and limited noncompliance may lead to unacceptable performance. Whereas the local noncompliance may be easily amended in case of the traditional block type improvement.

5   SUMMARY

The recent worldwide acceptance of deep mixing technology sometimes necessitates international bidding of geotechnical design and/or construction of deep mixing. The QC/QA procedures should be discussed in the same terminology and same concept in order to avoid any misunderstandings among international players. The activities related to deep mixing projects especially of the QC/QA procedure should preferably be standardized in order to avoid any misunderstandings between the owner, engineer and contractors.

Deep mixing improved ground is a composite system, comprising stabilized soil columns, original soil, and often bearing stratum stiffer and reliable. The function expected to stabilize soil columns differs for different applications (and associated difference in column installation patterns such as group column, panel, grid, and block types improvement). The behavior of the composite system is influenced by many factors such as the diameter and end-bearing condition of stabilized soil columns, disturbance of original soil due to construction of the columns and the reliability of overlapped portion between adjacent columns in addition to the engineering property of intact stabilized soil columns. These factors are dependent on the construction parameters and execution system to be employed in the project even when the same type and same amount of binder are used. It is necessary for the design engineer to understand the capability of the execution system locally available when he/she places the assumption in the design and in writing specifications and acceptance criteria. There is strong need of a database for the performance of a variety of execution systems.

In order to maintain the quality of deep mixing improved ground, it is necessary for all the parties involved in the project to share the responsibility according to the adopted contractual scheme. The contract document should correctly specify the quality indices and acceptance criteria together with the practicable verification procedure. The contractor should propose the best construction sequence and QC procedures to assure the requirement and demonstrate their capability and limitation. The owner/engineer should coordinate all the parties' appropriate involvement and responsibilities.

REFERENCES

Adams, T., Filz, G. & Navin, M. (2009) Stability of embankments and levees on deep-mixed foundations. *Proc. of the International Symposium on Deep Mixing and Admixture Stabilization.* pp. 305–310.

Aoi, M., Komoto, T. & Ashida, S. (1996) Application of TRD method to waste treatment on the ground. *Proc. of the 2nd International Congress on Environmental Geotechnics*, IS-Osaka '96. pp. 437–440.

Babasaki, R, Terashi, M., Suzuki, T., Maekawa, A., Kawamura, M. & Fukazawa, E. (1996) Japanese Geotechnical Society Technical Committee Reports: Factors influencing the strength of improved soil. *Proc. of the 2nd International Conference on Ground Improvement Geosystems.* Vol. 2. pp. 913–918.

Cement Deep Mixing Method Association (1999) *Cement Deep Mixing Method (CDM), Design and Construction Manual* (in Japanese).

Coastal Development Institute of Technology (2002) *The Deep Mixing Method – Principle, Design and Construction.* A.A. Balkema Publishers.

Coastal Development Institute of Technology (2008) *Technical Manual of Deep Mixing Method for Marine Works.* 289p. (in Japanese).

EuroSoilStab (2002) *Development of Design and Construction Methods to Stabilise Soft Organic Soils.* Design Guide Soft Soil Stabilization. EC project BE96-3177. 94p.

Filz, G.M. & Navin, M.P. (2006) Stability of column-supported embankments. *Final contract report, VTRC 06-CR13, Virginia Transportation Research Council.*

Futaki, M., Nakano, K. & Hagino, Y. (1996) Design strength of soil-cement columns as foundation ground for structures. *Proc. of the 2nd International Conference on Ground Improvement Geosystems.* Vol. 1. pp. 481–484.

Han, J., Parsons, R.L., Huang, J. & Sheth, A.R. (2005) Factors of safety against deep-seated failure of embankments over deep mixed columns. *Proc. of the International Conference on Deep Mixing – Best Practice and Recent Advances*, Stockholm. pp. 231–236.

Hosoya, Y., Nasu, T., Hibi, Y., Ogino, T., Kohata, Y. & Makihara, Y. (1996) Japanese Geotechnical Society Technical Committee Reports: An evaluation of the strength of soils improved by DMM. *Proc. of the Second International Conference on Ground Improvement Geosystems.* Vol. 2. pp. 919–924.

Ishii, T. (2008) Personal communication on the use of numerical simulation.

Japan Cement Association (2007) *Soil Improvement Manual using Cement Stabilizer (3rd edition).* Japan Cement Association. 387p. (in Japanese).

Japanese Geotechnical Society (2000) *Practice for Making and Curing Stabilized Soil Specimens without Compaction. JGS 0821-2000.* Japanese Geotechnical Society. (in Japanese).

Japanese Geotechnical Society (2009) *Practice for Making and Curing Stabilized Soil Specimens without Compaction. JGS 0821-2009.* Japanese Geotechnical Society. Vol. 1. pp. 426–434 (in Japanese).

Japanese Society of Soil Mechanics and Foundation Engineering (1990) *Practice for Making and Curing Stabilized Soil Specimens without Compaction. JGS T 821-1990.* Japanese Society of Soil Mechanics and Foundation Engineering (in Japanese).

Kitazume, M. (2008) Stability of group column type DM improved ground under embankment loading. *Report of the Port and Airport Research Institute.* Vol. 47. No. 1. pp. 1–53.

Kitazume, M. & Maruyama, K. (2006) External stability of group column type deep mixing improved ground under embankment loading. *Soils and Foundations.* Vol. 46. No. 3. pp. 323–340.

Kitazume, M. & Maruyama, K. (2007) Internal stability of group column type deep mixing improved ground under embankment loading. *Soils and Foundations.* Vol. 47. No. 3. pp. 437–455.

Kitazume, M., Nishimura, S., Terashi, M. & Ohishi, K. (2009a) International collaborative study Task 1: Investigation into practice of laboratory mix tests as means of QC/QA for deep mixing method. *Proc. of the International Symposium on Deep Mixing and Admixture Stabilization.* pp. 107–126.

Kitazume, M., Ohishi, K., Nishimura, S. & Terashi, M. (2009b) International collaborative study Task 2 Report: Interpretation of comparative test program. *Proc. of the International Symposium on Deep Mixing and Admixture Stabilization.* pp. 127–139.

Larsson, S. (2005) State of Practice Report – Execution, monitoring and quality control. *Proc. of the International Conference on Deep Mixing – Best Practice and Recent Advances.* Vol. 2. pp. 732–785.

Public Works Research Center (2004) *Technical Manual on Deep Mixing Method for On Land Works.* 334p. (in Japanese).

Terashi, M. (2003) The state of practice in deep mixing method. Grouting and ground treatment, *Proc. of the 3rd International Conference,* ASCE Geotechnical Special Publication No. 120. Vol. 1. pp. 25–49.

Terashi, M. & Kitazume, M. (2009) Keynote lecture: Current practice and future perspective of QA/QC for deep-mixed ground. *Proc. of the International Symposium on Deep Mixing and Admixture Stabilization.* pp. 61–99.

Terashi, M. & Kitazume, M. (2011) QA/QC for deep-mixed ground: current practice and future research needs. *Journal of Ground Improvement.* Issue 164. No. GI3. pp. 161–177.

Japanese laboratory mix test procedure

1 INTRODUCTION

The shear strength of stabilized soil is considered the most important geotechnical characteristic which leads to the improved stiffness, homogeneity and long term stability of stabilized soil. In general, the shear strength of stabilized soil is influenced by many factors, including characteristics of soil (water content, organic matter content, *etc.*), non-uniformity of soil (due to complex natural soil structure), type and amount of binder, curing period and temperature, and the degree of mixedness (Babasaki *et al.*, 1996). Hence, it is difficult to predict the strength of field stabilized soil precisely solely by soil investigations prior to mixing. In order to determine the mix design for actual production it is very important to perform a laboratory mix test which examines the unconfined compressive strength of stabilized soils prepared in the laboratory, q_{ul} by changing the type and amount of binder, curing time, and water-cement ratio. This mix design process also contributes to quality control at the construction site. It is important to recognize that the strength of laboratory mixed stabilized soil, q_{ul} is not always same as the strength of field mixed stabilized soil, q_{uf}. This knowledge may prevent troubles encountered at the construction site. The strength of laboratory mixed stabilized soil is influenced by the procedure of making and curing stabilized soil. According to the recent questionnaire survey regarding protocols for laboratory mix test procedures, molding methods and curing conditions exhibit notable international differences (Kitazume *et al.*, 2009).

In this Appendix, a procedure of making and curing stabilized soil specimen is introduced which is frequently applied in Japan to obtain the mixing condition to assure the target strength, and to develop new binder. This procedure conforms to the Japanese Geotechnical Society Standard (Japanese Geotechnical Society, 2009).

2 TESTING EQUIPMENT

2.1 Equipment for making specimen

2.1.1 Mold

The standard mold size is 50 mm in diameter and 100 mm in height. However, depending on the soil characteristic, the specimen diameter may be varied. In the case of clayey

(a) Disposable plastic mold.   (b) Disposable metal mold.

Figure A.1   Standard – sized lightweight mold.

or sandy soil without gravels and when the amount of soil is limited, the diameter less than 50 mm has been used. Conversely, if the soil contains a large amount of gravels or decayed plants, a diameter larger than 50 mm can be accepted. In the both cases, the height of specimen is set to be 2.0 to 2.5 times the diameter.

The material for the mold is usually either cast-iron, plastic, or tin. The latter two types of mold are referred to as lightweight molds and are popular choices today. The merits of lightweight molds are that they are easy to tap against the surface of a table or floor to remove air bubbles and easy to remove the specimen out of the mold. Also, the specimen can be cured in the mold without the risk of the mold rusting. Figure A.1 shows photos of a standard-sized lightweight mold with 100 mm in height and 50 mm in diameter. Splittable cast-iron molds are also available in various sizes based on JIS A 1132 (Japanese Industrial Standard, 2006).

2.1.2 Mixer

A mixer should be capable of mixing soil and binder uniformly. An electric mixer consisting of three basic parts: motor, stirring blades, and mixing bowl is specified in the JGS standard, because the electric mixer is suitable for most types of soil: clayey, organic, and sandy soils in most cases. Figure A.2 shows an example of electric mixer which has been often used in Japan. The capacity of bowl ranges from 5,000 to 30,000 cm^3. Different types of mixing paddles are available as shown in Figure A.3, but for most of the case hook type is preferred for uniform mixing. In this particular soil mixer, the paddle revolves at 120 to 300 rpm with planetary motions of 30 to 125 rpm. The stand of the mixer enables the raising and lowering of the bowl during mixing.

2.1.3 Binder mixing tool

When binder in slurry form is used, use a mixing bowl (typically a metal bowl) and rubber spatula or spoon to mix the binder and water.

Figure A.2 Electric mixer.

(a) Hook type.   (b) Flat type.   (c) Whipper type

Figure A.3 Examples of mixing paddle.

2.2 Soil and binder

2.2.1 Soil

For laboratory mix test for actual construction purpose, it is a basic principle to collect soil samples from all soil layers to be stabilized. In order to collect soil samples from deeper layers, a thin-walled sampler is typically used. Sampled soil should be stored at its natural water content. The soil samples are classified based on their observation records and soil testing results. Natural water content, consistency limits, organic matter content, pH and grain size distribution are good indices for the classification (see Chapter 2). The soil samples are separated into the identified layers. However, the soil sample in a thick layer is sometimes further divided into sub-layers to take variation in water content into consideration. Conversely, in the case where a layer is thin and its soil characteristics are similar to those of its neighboring layer, these layers are combined to reduce testing complexity. Each grouped soil sample is sieved through a 9.5-mm sieve. In the case the diameter of mold used is less than 50 mm, the soil sample is sieved through an appropriate size sieve so that the maximum grain size of the sieved sample should be less than 1/5th of the inner diameter of mold. While sieving, large obstacles such as shells and plants should be removed. If it is clearly found that the grain size is less than 1/5th of the inner diameter of the mold and the sample does not contain any obstacles, this procedure can be skipped. Then, each grouped soil sample is stirred by a mixer and its water content is measured. If it is considered that the water content of the soil sample has been changed during the process of sampling, transportation, and storage, the water content of the soil sample should be adjusted to its natural water content.

The required amount of soil sample is about 500 g for a standard-sized specimen. The total number of specimens to be tested is determined by the variations in binder types, binder factor (or binder content), curing period (curing time), and other construction control values (such as the influence of water/binder ratio), or a combination thereof. Three or more specimens should be prepared for each mixing conditions and curing period. It is desirable to have an extra amount of soil samples for the case of follow-up tests or repeated tests (due to procedural errors).

Note: The sampling strategy mentioned above is applicable for mechanical mixing with vertical rotary shafts and blades. For the shallow mixing techniques or chainsaw type deep mixing system which involve the vertical movement of soil-binder mixture in the actual production, soil samples may be prepared to simulate the in-situ mixing condition such by combining the soils taken from different layers according to the weighted average.

2.2.2 Binder

The quality of binder should be stringently assured. In general, it is desirable to use fresh binder for the test. However, if aged binder is unavoidably used, it should be inspected thoroughly for any quality degradation. For instance, degraded cement becomes grainy. The binder form in the mixing test is roughly divided between the slurry form or powder or granular form. Chemical additives are sometimes used together with the binder, which provide a specific effect, such as accelerating or decelerating the rate of hardening. For instance, retarding chemical additives may be used for the ease of overlapping process of stabilized soil columns.

The required amount of binder is determined by binder factor (or binder content) and number of specimens. Similar to the required amount of soil sample, it is desirable to have an extra amount of binder. Tap water is generally used to make binder slurry. However, seawater may be used for marine construction.

3 MAKING AND CURING OF SPECIMENS

3.1 Mixing materials

An optimal duration to mix soil and binder varies due to many factors such as the type and amount of soil, the type and amount of binder and the consistency of soil-binder mixture. The JGS standard specifies that the binder should be mixed with the soil thoroughly to achieve uniform mixture and notes that about 10 minutes is the ordinary practice. However 10 minutes is accepted as de facto standard. When the mixing duration is too long, it becomes difficult to remove air bubbles from stabilized soil in a mold since the stabilized soil may begin to harden.

Notes: It is desirable to suspend the mixing after about 5 minutes, to detach the mixing bowl from the mixer, and to pour the stabilized soil in the mixing bowl and

Figure A.4 Tapping technique in molding procedure.

that adhered to the stirring blades to another container using a rubber spatula, to mix it briefly by hand, then to return it to the mixing bowl, and to restart to mix it by the mixer for another 5 minutes. Another option is to suspend the mixing every two minutes and to mix the soil in the mixing bowl by hand. These procedures can provide uniform mixing of the soil including the soil stuck on the mixing bowl and blades.

In the case of slurry form binder is used, splashing of the slurry may occur when starting the mixer right after pouring the binder slurry to the soil in the mixing bowl. It is desirable to mix the soil and the slurry by hand briefly before starting the mixer.

3.2 Making specimen

A thin layer of grease may be applied on an inner surface of mold to allow easy removal of the specimen out of the mold after curing. Then the stabilized soil is filled in a mold in three separate layers. After filling each stabilized soil layer, air bubbles should be removed. Typical methods for removing air bubbles are (1) lightly tapping the mold against a table or a concrete floor (Figure A.4), (2) hitting the mold with a mallet, and (3) subjecting the mold to vibration. The air removing procedure is terminated once air bubbles are no longer found on the soil surface.

In general, it is hard to remove air bubbles from stabilized soil with low consistency. Also, some stabilized soils decrease in volume over time, resulting in insufficient specimen height. To assure the proper specimen height, a sheet of hard polymer film of 10 to 15 mm taller than the mold height is placed around the inner perimeter of the mold so that stabilized soil can be filled above the top edge of the mold and be sealed by sealant as shown in Figure A.5. The hard polymer film also functions to protect a specimen when it is removed from the mold.

The water content of stabilized soil is measured for each mixing bowl. By comparing the water contents before and after mixing, any mistakes in material amounts can be spotted in the early experimental stage.

Some stabilized soils become hard quickly to cause removing air bubbles difficult. In such a case, the stabilized soil mixture should be filled in molds as quickly as possible by increasing the number of personnel and/or dividing the making into several times by reducing the quantity of one batch.

Figure A.5 Sealing by plastic film.

The sandy soil and binder sometimes separate easily during mixing and filling into molds. Especially it happens in the case of slurry form binder. This causes a strength decrease of laboratory mix stabilized soil, which is thought one of the reasons for the high strength ratio of the field strength q_{uf} to the laboratory strength q_{ul} (Sasaki *et al.*, 1996; Ishibashi *et al.*, 1997). In order to prevent the separation, mix the stabilized soil by hand in a mixing bowl and scoop and fill it in molds quickly.

In the case of uniform sandy soil with less fines content being mixed with slurry form binder, excessive tapping of the mold for air removal may cause the density and strength decrease. In the case of loam or clayey soil with sand being mixed with powder form binder, the mixture can form lumps during mixing by an electric mixer. If it happens, the lumps should be broken and filled in molds.

3.3 Curing

The specimen in the mold is covered by sealant to prevent the change of water content and cured at $20 \pm 3°C$ over a prescribed curing period. The curing period may be selected from 1, 3, 7, 14, 28, and 91 days, etc. depending on the purpose of test and type of binder, while it is common and desirable to include 7 and 28 days.

The following are desirable curing procedures: (i) Sealed mold and/or specimen should be placed in a temperature and humidity-controlled container (Figure A.6(a)), (ii) Sealed mold and/or specimen should be placed in a humid box (relative humidity above 95%) and the box should be placed in a temperature controlled room (Figure A.6(b)). The best care should be paid to prevent tears in the sealant material to assure tight sealing. The reason for not curing the specimen directly under water is that the actual stabilized soil is mostly cured underground with negligible migration of water.

3.4 Specimen removal

Once the strength of a stabilized soil specimen is found to reach sufficient level, the specimen may be taken out of the mold during the curing process. Figure A.7 shows an

(a) Temperature and humidity-controlled container.   (b) Humid controlled box.

Figure A.6 Examples of curing container and box.

Figure A.7 Removal of specimen by splitting along pre-processed slits of lightweight plastic mold.

Figure A.8 Unconfined compression test on stabilized soil.

Table A.1 Example format for test report.

Specimen condition	Binder Type: Binder factor (%) Binder content (kg/m^3) Slurry mixing water type* Water/binder ratio* (%) Curing period (days)	Additive* Additive amount** (%) Number of specimens Mold dimension (volume) ϕ cm × cm (cm^3)

Raw soil water content	Container No. m_a (g) m_b (g) m_c (g) w (%) Average $\overline{w}$ (%)		

Required amount of materials per one batch	Soil, m_T (g) Water*, m_W (g)	Binder, m_H (g) Additive*, m_A (g)

Stabilized Soil water content	Container No. m_a (g) m_b (g) m_c (g) w (%) Average $\overline{w}_s$ (%)		

For saturated soil, determine the required material amounts by the equations below.

① Soil mass, m_T (g) :

$$\text{Wet density, } \rho_T \text{ (g/cm}^3) = \{1 + \boxed{}^{\overline{w}}/100\}/\{1/\boxed{}^{\rho_s} + \boxed{}^{\overline{w}}/100\}$$

$$=$$

$$m_T \text{ (g)} = \underset{\text{\# of specimens}}{\boxed{}} \times \underset{\text{volume of mold}}{\boxed{}} \times \underset{\rho_t}{\boxed{}} \times \underset{\text{extra***}}{\boxed{}}$$

② Binder mass, m_H (g):

$$\text{Soil dry mass, } m_D \text{ (g)} = \underset{m_T}{\boxed{}}/(1 + \underset{\overline{w}}{\boxed{}}/100) = \boxed{}$$

$$m_H \text{ (g)} = \underset{m_D}{\boxed{}} \times \underset{\text{binder factor}}{\boxed{}}/100 = \boxed{}$$

Binder content (kg/m^3) = $\{10\rho_T/(1 + \overline{w}/100)\}$ × Binder factor

③ Mass of slurry mixing water, m_W (g)

$$m_W \text{ (g)} = \underset{m_H}{\boxed{}} \times \underset{\text{water to binder ratio (\%)}}{\boxed{}}/100 = \boxed{}$$

④ Additive mass, m_A (g)

$$m_A \text{ (g)} = \underset{m_H}{\boxed{}} \times \underset{\text{Additive amount (\%)}}{\boxed{}}/100 = \boxed{}$$

notes:
* only if used
** ratio w.r.t. binder mass
*** normally 1.1~1.2

Table A.2 Unconfined compressive strength of various stabilized soils.

Soil		Grain Size Composition							Binder			Unconfined Compressive Strength q_u (kN/m²)	
Sample Location	Soil Type	Sand (%)	Silt (%)	Clay (%)	Water Content (%)	Liquid Limit (%)	Plastic Limit (%)	Organic Content (%)	Type	Powder/Slurry (W/C)	Binder/Soil Ratio (amount of binder)	7 days	28 days
Yokohama Bay	Marine Soil	6.4	37.5	56.1	97.9	95.4	32.3	3.6	NP	C slurry (60%)	13.5 (100)	2,140	2,870
									BF		13.5 (100)	1,180	1,990
									NP		27.0 (200)	4,050	5,490
									BF		27.0 (200)	3,690	5,960
Osaka Bay	Marine Soil	3.5	30.8	65.7	93.9	79.3	30.2	2.7	NP	C slurry (60%)	13.1 (100)	950	1,400
									BF		13.1 (100)	980	1,470
									NP		26.2 (200)	1,490	2,750
									BF		26.2 (200)	3,150	4,890
Imari Bay	Marine Soil	2.2	44.5	53.3	83.3	70.4	24.2	4.3	NP	C slurry (60%)	12.0 (100)	540	830
									BF		12.0 (100)	490	830
									NP		24.0 (200)	1,130	2,060
									BF		24.0 (200)	2,190	4,250
Tokyo Prefecture	Land Soil	5.0	53.0	42.0	54.0	44.7	23.9	3.8	NP	C slurry (80%)	4.6 (50)	530	730
									BF		4.6 (50)	160	350
									NP		6.8 (75)	1,260	1,760
									BF		6.8 (75)	580	1,090
									NP	CB slurry (200%)	22.8 (250)	700	1,510
									BF		22.8 (250)	1,110	2,410
Funabashi, Chiba	Land Soil	95.6	3.1	1.3	14.2	–	–	–	NP	CB slurry (80%)	15.3 (300)	460	910
									BF		15.3 (300)	560	1,800
									Slag		15.3 (300)	1,110	2,860

(Continued)

Table A.2 Continued.

Soil									Binder			Unconfined Compressive Strength q_u (kN/m²)	
Sample Location	Soil Type	Water Content (%)	Grain Size Composition			Liquid Limit (%)	Plastic Limit (%)	Organic Content (%)	Type	Powder/Slurry (W/C)	Binder/Soil Ratio (amount of binder)	7 days	28 days
			Sand (%)	Silt (%)	Clay (%)								
Fujishiro, Ibaragi	Land Soil	236	–	–	–	251	92.7	25.2	NP	C slurry (80%)	72.5 (250)	130	190
									BF		72.5 (250)	140	160
									For Organic Soil		72.5 (250)	490	780
Nangoku, Kouchi	Land Soil	295	–	–	–	272	69.1	17.6	NP	C slurry (80%)	85.0 (250)	140	250
									BF		85.0 (250)	98	200
									For Organic Soil		85.0 (250)	590	1,570
Haneda	Reclaimed Land Soil	160	1.0	33.0	66.0	99.1	39.7	4.8	Quicklime	Powder	10 (–)	540	740
											20 (–)	640	1,370
Yokohama	Reclaimed Land Soil	102.5	9.9	44.6	45.5	78.8	39.1	2.95	Quicklime	Powder	10 (–)	1,670	2,740
Naruo, Hyogo	Marine Soil	90.2	2.0	26.1	71.9	83.0	31.4	–	Quicklime	Powder	20 (–)	2,350	3,720
											10 (–)	250	690

Notes:
1) NP: ordinary Portland cement; BF: blast furnace cement type B
2) C slurry: cement slurry; CB slurry: cement-bentonite slurry
3) W/C: water/cement ratio
4) Binder/Soil ratio (%): Ratio of binder mass to dry soil mass; Amount of binder: binder mass (kg) per m³ of test soil
5) The unconfined compressive strengths of stabilized soil with quicklime is obtained from the figures (Terashi et al., 1997).
6) Organic contents of soil are determined according to JGS T 231 "Testing Procedure for organic content of soil" (chromic acid oxidation method)

example of removal of specimen by splitting along pre-processed slits of a lightweight plastic mold. The exposed end of the specimen must be trimmed properly before removing out of the mold. The removed specimen should be put in a polyethylene bag or wrapped with a sheet of high polymer film (such as food storage-type plastic wrap) and placed back in the curing container to complete the curing process. To avoid possible deformation due to excess load, the wrapped specimens should not be stacked.

4 REPORT

In the report, it is desirable to report both the binder factor as well as the binder content, as they are most commonly used. There are other expressions on the binder amount such as (1) the ratio of the dry weight of binder to the wet weight of soil and (2) the ratio of volume of binder-slurry to the volume of soil.

The type and amount of chemical additives should be reported if used. Also, it is desirable to report any data on the amounts of all materials such as soil sample and binder measured during the preparation procedure. Table A.1 shows an example format for the specimen preparation report (Japanese Geotechnical Society, 2009).

5 USE OF SPECIMENS

The stabilized soil specimens are mostly used for the unconfined compression test. However they can also be used for the triaxial test, simple tensile strength test, splitting tensile strength test, cyclic triaxial test, and fatigue strength test.

REFERENCES

Babasaki, R., Terashi, M., Suzuki, K., Maekawa, J., Kawamura, M. & Fukazawa, E. (1996) Factors influencing the Strength of improved soils. *Proc. of the Symposium on Cement Treated Soils*. pp. 20–41 (in Japanese).

Ishibashi, M., Yamada, K. & Saitoh, S. (1997) Fundamental study on laboratory mixing test for sandy ground improvement by deep mixing method. *Proc. of the 32nd Annual Conference of the Japanese Geotechnical Society*. pp. 2399–2400 (in Japanese).

Japanese Geotechnical Society (2009) *Practice for making and curing stabilized soil specimens without compaction. JGS 0821-2009*. Japanese Geotechnical Society. Vol. 1. pp. 426–434 (in Japanese).

Japanese Industrial Standard (2006) *Method of making and curing concrete specimens, JIS A 1132: 2006* (in Japanese).

Kitazume, M., Nishimura, S., Terashi, M. & Ohishi, K. (2009) International collaborative study Task 1: Investigation into practice of laboratory mix tests as means of QC/QA for deep mixing method. *Proc. of the International Symposium on Deep Mixing and Admixture Stabilization*. pp. 107–126.

Terashi, M., Okumura, T. & Mitsumoto, T. (1977) Fundamental properties of lime-treated soils. *Report of the Port and Harbour Research Institute*. Vol. 16. No. 1. pp. 3–28 (in Japanese).

Subject index

additive, xix
admixture stabilization, 4 (Table 1.2), 6, 13
  classification of, 14(Table 1.3)
angle of internal friction, 86, 135
axial strain at failure, 82, 135

bearing capacity, 150, 155, 165, 302, 318,
  343, 358
bearing capacity failure, *see* failure mode
bending failure, *see* failure mode
binder, xix, 13, 41
  characteristics of binder, 30(Table 2.1), 31,
    41
  binder content, xix, 44
  binder factor, xix, 34
  influence of amount of binder, 32, 38, 56
  minimum binder content, 56
  influence of type of binder, 44
binder slurry, xix, 18, 56, 201, 203, 205,
  207, 210, 375
binder water slurry, *see* binder slurry
blade
  blade rotation number, 198, 208, 376
  excavation blade, 203
  mixing blade, 17, 19, 203
block type improvement, 148–149, 165–168,
  171–174, 264–265, 391
  design procedure for, 309, 330
  engineering behavior of, 260–273

calcium
  calcium aluminate, 31
  calcium content distribution, 102–104,
    132
  calcium hydroxide, 31, 40, 74
  calcium oxide, 31, 74
  calcium silicate, 31
  leaching of calcium, 100, 104

CDM, 18(Table 1.4), 143, 154, 189
  Ordinary CDM, 201
  CDM Column 21, 18
  CDM Double-mixing, 18
  CDM-Land 4, 18
  CDM Lemni 2/3 method, 217
  CDM LODIC method, 210
  CDM MEGA, 18
cement, 40, 41(Table 2.4), *see also* binder
  blast furnace slag cement, 44, 98, 374
  ordinary Portland cement, 44, 98, 374
  cement based special binder, 42(Table 2.5),
    45, 114, 115
cement content, *see* binder content
cement factor, *see* binder factor
cement hydration, 13, 40, 75
cement stabilization
  mechanism of cement stabilization, 13, 41
chemical additives, 19, 237, 398,
    see also additive
coefficient of variation, 125–126, 138, 174,
  385, 390
column, *see* stabilized soil column
column installation patterns, 143–150,
  264–266, 294, 392
  see also block type improvement
  see also group column type improvement
  see also grid type improvement
  see also tangent columns
  see also wall type improvement
consolidation characteristics, 105, 137
  coefficient of consolidation, 107, 137
  coefficient of permeability, 110, 137
  coefficient of volume compressibility, 107,
    137, 304
  consolidation yield pressure, 106
consolidation/ dewatering, 4, 8
COV, *see* coefficient of variation

creep strength, 88
curing condition
 influence of curing temperature, 61
 influence of curing period, 39, 59,
  see also strength ratio
 influence of maturity, 63
cyclic strength, 90, 136

deep mixing method, 17, 18(Table 1.4)
 development of, 8
 statistics of, 12
densification, 3
density
 change of density, 76–78
design strength, 156, 281, 347, 371
deterioration, *see also* long term strength
 depth of deterioration, 104
 exposure test, 105
 strength decrease, 100, 136
dewatered stabilized soil, 22
DJM, 18(Table 1.4), 143, 154, 189,
 191(Table 5.1)
DMM *see* deep mixing method
DM machine, xix, 189, 201
double fluid technique, *see* high pressure
 injection mixing
Dry Jet Mixing, *see* DJM
dry method of deep mixing, 18(Table 1.4),
 154, *see also* DJM
dynamic property, 87
 damping ratio, 88, 136
 equivalent shear modulus, 87–88
 initial shear modulus, 88, 136

earthquake
 liquefaction prevention, 162–164,
  167–171, 350
 seismic design, 331
 seismic performance of improved ground,
  176–184
element, 144, *see* stabilized soil element
elution of contaminant, 113
embankment
 embankment support, 295, 388
 embankment stability, 154, 159, 284
end bearing capacity of columns, *see* failure
 mode
end bearing condition, 188, 266, 284, 291,
 294, 329, 375, 388, 391
entrained rotation, 192, 202, 213, 219, 225
exposure conditions, *see* deterioration

ex-situ mixing, 14(Table 1.3), 19
external stability, xix, 266–268, 271–277,
 314, 337, 353, *see also* failure mode
extrusion failure, *see* failure mode

failure mode, 263, 269, 278, 291, 389
 bearing capacity failure, 267(Figure 6.2),
  269–274, 318, 343, 358
 bending failure, 283, 291, 308, 390
 end bearing capacity of columns, 281, 292
 extrusion failure, 279, 325, 348
 irregular slip, 269, 291
 overturning failure, 267(Figure 6.2), 268,
  315, 335
 shear failure, 268, 282
 sliding failure, 267(Figure 6.2), 287, 291,
  299, 315, 338, 390
 slip circle failure, 269, 281, 287, 291, 300
 tilting failure, 276, 283, 288, 291, 390
 vertical shear along lap joint, 277, 278,
  290, 321, 322, 345
field strength, xix, 68, 128, 323, 373, 385
 variability of, 124, 138, *see also* coefficient
  of variation
 ratio of field strength and laboratory
  strength, q_{uf}/q_{ul}, 128, 373, 385
field trial test, 187, 196, 206, 228, 294, 374
fixed type, xix, 196, 208, 275, 297, 303,
 307, 328
 see also end bearing condition
floating type. xix, 275, 298, 305, 307, 328
free blade, 192, 202, *see also* entrained
 rotation
freezing, *see* thermal stabilization

geotechnical design, 25, 187, 263, 292–294,
 369–371
grain size distribution
 influence of grain size distribution, 35, 49,
 112
grid type improvement, 147, 149(Table 4.1),
 162, 168, 309, 351, 365
group column type improvement, 143, 145,
 149(Table 4.1), 158–162, 265, 295,
 388–390
 design of, 295
 engineering behavior of, 280–292
group of individual columns, *see* group
 column type improvement
grouting, 5

hazardous substance, 113, 119
heating, *see* thermal stabilization
high pressure injection mixing, 12, 153, 157, 174, 177, 188, 235
  single fluid technique, 236
  double fluid technique, 239, 244
  triple fluid technique, 247, 251
humic acid
  influence of humic acid, 36, 50
hydration, 13, 30, 40, 66, 73, 75, 76, 78

ignition loss, 51, 68
improved ground, xix, 187
improvement area ratio, 145, 156, 298, 304, 307
improvement pattern, *see* column installation patterns
improvement type
  see fixed type
  see floating type
influence of type of water, 45
initial shear modulus, *see* dynamic property
in-situ mixing, 14(Table 1.3)
  deep mixing, *see* deep mixing method
  shallow mixing, 16, 398
  surface mixing, 16
in-situ strength, *see* field strength
internal friction angle, *see* angle of internal friction
internal stability, xix, 269, 272, 277, 279, 320, 344, 360, *see also* failure mode
in-water works, 222, 232
ion exchange, 6, 13, 30, 78, 135

JACSMAN, 255
Jet grouting, 6, 12, 18(Table 1.4), *see also* high pressure injection

laboratory mix test, 39, 56, 187, 293, 374, 378, 395
laboratory strength, xix, 128, 138, 324, 371
light weight geo-material, 20
lime
  hydrated lime, 10, 30, 31, 32, 34, 45, *see also* calcium hydroxide
  slaked lime, 31, *see also* calcium hydroxide
  quicklime, 30, *see* calcium oxide
  quality of quicklime, 10, 32
  lime-based special binder, 31–32
lime columns, 6, 143

lime stabilization, 13
  mechanism of lime stabilization, 30
liquefaction mitigation, *see* earthquake
liquefaction prevention, *see* earthquake
liquefaction resistance, 6, 168
liquefaction resistance value, 356
long term strength, 96, 136
  strength increase, 29, 97–100
  strength decrease, *see* deterioration

marine construction, 19, 150, 171–174, *see also* in-water work
maturity, *see* curing condition
mechanical mixing, 14, 19–20, 150, 187
mixing conditions, 29, 38, 56
modulus of elasticity, 83

oil tank, 264
on-land work, 154–156, 189, 200, 280
organic matter content, *see* ignition loss
overlap, 42, 45, 133, 147, 148, 188, 265, 323–324
  overlapped portion, 131–134, 138, 324, 391
  overlap joint, 274, 277
overturning failure, *see* failure mode

panel type, *see* wall type improvement
penetration injection, 196–197, 207, 228, 376
permeability, *see* consolidation characteristics
pH, 120, 138
  influence of pH, 30(Table 2.1), 36, 51
plasticity index, 31, 78
Pneumatic flow mixing method, 23
Poisson's ratio, 84, 135
Power Blender, 18(Table 1.4)
pozzolanic reaction, 13, 31, 40
pozzolanic reactivity, 34, 45
Premixing method, 19
process control, 25, 371, 375
process design, 187, 188, 195, 369–371, 375, 385
pseudo pre-consolidation pressure, *see* consolidation yield pressure

quality assurance, 188, 200, 215, 369, 371, 372 (Table 7.1), 374(Table 7.2), 388
quality control, 187–188, 199, 209, 238, 374(Table 7.2), 375
quality verification, 376, 386

*q*uf/*q*ul, *see* field strength
quick lime, *see* lime
quicklime pile method, 13
*q*u28/*q*u7, *see* strength ratio

reinforcement, 7
replacement method, 3
residual strength, 80, 83
rest time, 54, 58, 381
Rock Quality Designation, 378

sand fraction, *see* grain size distribution
sea water, *see* influence of type of water
secant modulus of elasticity, 83
seismic inertia force, 309, 315, 338
settlement
    settlement reduction, 146, 154, 158
    settlement reduction factor, 304
    consolidation settlement, 303, 328
    rate of settlement, 307
shallow mixing, 15
shear failure, *see* failure mode
sheet pile, 8, 20, 22, 157, 177, 255
shield tunnel, 153, 157, 174
single fluid technique, *see* high pressure
    injection mixing
size effect, 128, 138
sliding failure, *see* failure mode
slip circle failure, *see* failure mode
stabilized soil column, xix, 144
stabilized soil element, 144
stabilizing agent, *see* binder
standard deviation, 128
strain at failure, *see* axial strain at failure
strength ratio, *q*u28/*q*u7, 61(Table 2.9)
stress–strain curve, 79, 135
stress concentration ratio, 304,
    see also settlement reduction factor
surface treatment, 15

tangent columns, 146, 282, 391
    tangent block, 146, 157, 159
    tangent wall, 146
tap water, *see* influence of type of water
tensile strength, 94–96, 136, 390
    allowable tensile strength, 323
    design tensile strength, 348, 390
tensile strength test
    split tension test, 94
    simple tension test, 94
    bending test, 94
thermal stabilization
    freezing, 4(Table 1.2), 7
    heating, 4(Table 1.2), 7
tilting failure, *see* failure mode
TRD, 18(Table 1.4), 385
triple fluid technique, *see* high pressure
    injection mixing

unconfined compressive strength, 7, 9, 25, 29
unit weight
    change of unit weight, 76

verification technique, 385
void ratio – consolidation pressure curve,
    105, 137

wall type improvement, 147, 309, 391
    design procedure for, 309, 330
water content
    influence of water content, 37, 54
    change of water content, 73, 123, 134
water to binder ratio, 205
water to cement ratio, W/C, 210,
    see also water to binder ratio
wet method of deep mixing, 11, 18(Table
    1.4), 154, 200, 222
withdrawal injection, 196–197, 203, 228,
    376